Aquaculture in the United States

Aquaculture in the United States

A Historical Survey

ROBERT R. STICKNEY
University of Washington

JOHN WILEY & SONS, INC.
NEW YORK / CHICHESTER / BRISBANE / TORONTO / SINGAPORE

Library of Congress Cataloging in Publication Data:

Printed in the United States of America

10 9 8 7 6 5 4 3 2 1

Preface

Aquaculture has been defined in a variety of ways. One among the better recent definitions describes it as "the planned or purposeful intervention in the production of aquatic animals."[1] I've defined it as the rearing of aquatic organisms under controlled or semicontrolled conditions.[2] The second definition is a bit more exhaustive in that it includes plants, which in some countries represent a significant proportion of the organisms harvested by aquaculturists. In this book, we'll be examining the history of aquaculture in the United States, which has been primarily, though not exclusively, associated with the production of aquatic animals, so we'll use the second definition. (Besides, that's the one I came up with, so I'm a bit partial to it.)

The subject matter discussed in this book is broad. Included is information on sport fish, minnows, ornamental fish, frogs, and even alligators in addition to a good deal on invertebrates and finfish grown directly for human food. I harbor an interest in all of those subjects, plus the role of aquaculture in the restoration and enhancement of native populations. Not everyone feels that way. In fact, one of my pet peeves as a scientist has involved the parochialism of some people with whom I've interacted professionally. While that parochialism isn't universal, it seems to be quite widespread. Let me provide you with a few examples.

A number of years ago I attended a meeting, the object of which was to establish some national guidelines for evaluating the toxicity of what we used to call "dredge spoils" but now refer to as "dredge material" (a less disgusting, perhaps, and more *politically correct* term that was invoked long before that overused expression was developed). It became obvious to me early in the deliberations that dredge material from one location could differ considerably from that obtained at another location. I considered the mud being dredged from the bays along the southeastern Atlantic to differ quite a bit from the sandy sediments that fill the channels in Florida. It was also apparent to me that the water chemistry associated with dredge material from the fresh waters of the Great Lakes might be different from that of the saline waters along the Texas Gulf coast. One attendee, who was particularly obnoxious, just couldn't discuss the subject without referring to one particular river that empties into the Pacific Ocean and forms part of the border between Washington and Oregon. It was his only frame of reference, and many of the situations that existed with respect to dredge material in other parts of the nation just didn't register with this guy.

Many of the students with whom I've come in contact have also taken the parochial view. "Well, if it doesn't involve _____ [you fill in the blank: *salmon, trout, shrimp, catfish,* and *oysters* are common responses], then I'm not interested." My view is that you darn well better be interested because when you graduate there may be no job involving _____. Instead, you might have to work with *something else!*

So what does this little tangent have to do with the history of aquaculture in the United States? Simply this. Because of a high degree of parochialism among individuals, even those within agencies of the state and federal governments, aquaculture may be seen as relating only to one of the following:

- animals (and perhaps plants) being raised for human food
- sport fish reared for stocking
- commercial fish reared for stocking
- the rearing of endangered or threatened species

The U.S. Department of Agriculture tends to think only in terms of food production—after all, they are agriculturists—while folks in state fish and game agencies and at least some in the U.S. Fish and Wildlife Service and the National Marine Fisheries Service look at the production of fish for enhancement stocking (either for recreation or commercial capture) as the primary focus of aquaculture. Some groups or individuals within those agencies are involved exclusively with endangered species.

None of the four categories listed include ornamental fish production or the rearing of bait. There are several hundred ornamental fish producers in Florida alone; bait production is an enormous industry in Arkansas and is significant in a number of other states. Both ornamental species and minnows are integral parts of aquaculture and shouldn't be ignored.

Neither the culture of ornamentals nor bait has been overlooked in this volume. Still, I must admit, that the bulk of the material presented in this book deals with aquatic organisms produced specifically for direct human consumption or for release to augment sport and commercial fisheries. In the historical context, the culture of aquatic organisms directly for human consumption is a relatively recent phenomenon in the United States. The roots of U.S. aquaculture lie in enhancement stocking programs.

That is not true of the rest of the world where aquaculture was developed— perhaps as long ago as 4,000 years in China—to put fish in the human diet. Throughout most of the world the goal of aquaculture has remained the same over the millennia. Much of the early interest in aquaculture development within the United States also focused on food production, though the emphasis was on stocking marine fish to enhance capture fisheries in the nation's freshwater and marine environments since many fisheries were already beginning to fail by the late nineteenth century and some much earlier. Recreational fish enhancement developed hand in hand with interest in commercial fisheries enhancement, and in many cases, there were species in common to both. The surge in commercial aquaculture did not become very apparent until the 1960s, by which time much of the technology required for controlled production and growout had been developed by scientists working in state and federal agencies.

The information related here was drawn primarily from the published literature and to a much lesser extent from personal knowledge and discussions with colleagues. I've used a numbering system that refers to the references that can be found at the end of the book to provide interested or incredulous readers with the sources of the information.* I've used that contrivance in order to keep from interrupting the flow of the narrative with names and dates, which would be the form used in the scientific literature and the one to which I am most accustomed. During my perusal of the historical literature, and particularly with respect to articles written during the last century and early in this one, I found that the material was often amusing, frequently insightful, and always delightful to read. The specific information presented in this book has been selected from a much wider array of examined material. Someone else might well have selected different topics, incidents, and individuals to discuss. I admit that my selections were based to a large extent on topics and incidents that were particularly entertaining or on individuals who, to my mind, made significant contributions. Also, I've included some information from my personal experiences. In many cases I've used the words of the authors to make or underscore particular points. Direct quotes become less frequent as one progresses down the time line toward the present. The colorful and often exhaustive narratives so common during the early years are no longer accepted in the scientific literature. While that makes reading the modern literature less entertaining, it allows the reader to cut to the chase. In many of the early papers it often takes a careful reader to dig out the important points.

Deciding on how to present the information was somewhat difficult. My first inclination was to segregate the material into chapters and subsections by entities (U.S. Fish and Fisheries Commission, U.S. Bureau of Fisheries, U.S. Fish and Wildlife Service, universities, professional societies, and so on), by species (for example, salmon culture, oyster culture, carp introductions, and bass and bluegill culture), or by personality (following the careers of selected individuals). As I became increasingly involved in developing material from which to draw, it became apparent that none of those approaches would be satisfactory. The tale is like a spider's web with the various pieces so intricately intertwined that taking any simplistic approach would not do justice to the art, science, and personalities involved. So, I decided to mix and match. Events are basically presented in chronological order, but in some cases it seemed appropriate to follow the development of different species indepen-

---

*There are also a few footnotes scattered about that are intended to explain technical terms, provide ancillary information, identify people from whom I obtained information in the form of personal communications, and to drop in a personal opinion or two.

dently, so there is some jumping back and forth. The people "jumped around," much as does this manuscript, so while I've tried to fill in some background information on several people, we may revisit them from time to time. For those of you who feel left out because I didn't put in details on your particular career, I can only indicate that it was not possible to include everyone.

It is revealing and very humbling to examine the work that was accomplished by some of the early aquaculturists. As we shall see, they were spawning animals a century ago that we are often unable to get to reproduce in captivity today. The old-timers didn't always reveal how they accomplished their feats, and in some cases those that followed didn't bother to look up the information when it was available, so there has been a lot of reinvention. One instance of that nature slapped me across the face like the tail of a dead fish in about 1971.

At the time, I was working at the Skidaway Institute of Oceanography in Savannah, Georgia, where I had a Sea Grant (U.S. Department of Commerce) research program that involved the rearing of flounders. Dave White and I were working with summer and southern flounder postlarvae that we captured in plankton nets from a river adjacent to the laboratory. We were looking at how to get them to convert from natural foods to prepared feed pellets, and how their salinity requirements changed with age, among other things.

After we had been working for a year or so, the Sea Grant Program sent a site review team to evaluate our program. Such reviews have a lot to do with whether or not a project receives continued funding, so we spent a lot of time preparing our presentation. Dave and I put on our dog and pony show along with our plans for the future should we be provided with additional funding.

The review team listened politely. Then one member said, "We knew all of that 10 years ago."

My response was, "I thought I had searched the literature thoroughly. I certainly haven't seen published information concerning what we've just presented."

"That's true," was the response from the review team member. "I was part of a group of people who were interested in flounder research several years ago. We were from various East Coast states and would get together every so often to compare notes. We came to many of the same conclusions you did, but we never wrote anything down because we thought we were the only people interested in the subject."

That little episode brings us back full circle to the subject of parochialism. In 1985, I came to the Pacific Northwest to find a cadre of salmon biologists, among the world's leaders in that field, whom I knew virtually nothing about. I was familiar with fisheries managers in both the inland and coastal states throughout much of the country. Many were people I had met and gotten to know through my activities in the American Fisheries Society and other professional organizations. But I didn't know these salmon guys. It soon became apparent why. When I looked at their publication records, I found that many of them rarely or never published in the referred literature. They limited their publications to what we call the "gray literature"—technical reports and annual reports to funding agencies and organizations.

When I asked some of these people why they didn't try to put their findings in the primary journals, they responded that no one outside of Alaska would be interested in their work. It was like beating my head against a stone wall trying to convince them that perhaps some of their ideas, approaches, innovations, and views on how the salmon world worked might influence the thinking of someone working with flounders in Georgia or with bass in Kansas.

The tunnel vision of parochialism can be extended far beyond the examples that have been presented. Many of the ideas that have influenced my research activities came from disciplines well outside of aquaculture or even fisheries science. Adapting and adopting a technique that has been used on terrestrial animals by agriculturists or wildlife biologists is sometimes the best way to address a fish-related problem. Scientific curiosity must extend beyond a person's primary field of interest.

Enough philosophy. It's about time to take a look at how aquaculture developed in the United States. But first, a few words about the dreaded topic, taxonomy. I have elected to stick with common names throughout this book. The taxonomy of both the fishes and invertebrates has changed so much in the last one hundred plus years that referring to the Latin names would be very confusing. Many fishes that were thought to be separate species have been lumped together, others have been split out, and the scientific names of many of the species discussed have been changed. I've tried to, when I could figure them out, use the currently accepted common names. This convention should make the text easier to read, and if I've erred in my assumptions that such-and-such a fish in 1880 is presently known as something else, I apologize in advance and am sure I'll hear about it.

One other comment: I've employed the English system of weights and measures throughout the text. While I and many readers are perfectly comfortable with, and may even prefer, the metric system, I've used the English system in keeping with the common-name convention.

Finally, I must admit to being unable to avoid mixing a few stories from my own experience into the manuscript. I've tried to keep them brief, to use them primarily to show how the more things change the more they stay the same, and to provide information on my personal relationships with some of the people who helped shape U.S. aquaculture in the past few decades.

Acknowledgments

I am indebted to Arden Trandahl, who was the moving force behind establishing the National Fish Culture Hall of Fame in Spearfish, South Dakota and who continues to be its steward, for copies of the information engraved on the plaques and for additional unpublished documents. Others who sent me biographical information on some of the heros of U.S. aquaculture were Harry Dupree (Fish Farming Experimental Station in Stuttgart, Arkansas), Roy Heidinger (Southern Illinois University at Carbondale), and Brian Duncan (Auburn University, Auburn, Alabama). I am also indebted to the taxpayers of the state of Washington for providing support for the marvelous collection of books in the University of Washington library. Those volumes, some of which made me sneeze every time I opened them because they were so musty, provided hours of interesting and enjoyable reading. For all of you who have been my colleagues over the past two decades, I thank you for many memorable, and usually enjoyable times. There were also some frustrating experiences, but some of those who were involved remain my colleagues and sometimes my friends. Finally, I owe much to the person who has shared the past 30 plus years with me, my wife, Carolan.

All citations of copyrighted material of fifty or more words are used with permission from *Transactions of the American Fisheries Society, Aquaculture Magazine, Proceedings of the World Mariculture Society,* and *Journal of Shellfisheries Research.*

Contents

Aquaculture in the United States

CHAPTER 1

The Beginnings of U.S. Aquaculture

ESTABLISHMENT OF THE FISH COMMISSION

As the twentieth century draws to a close, U.S. fishery resources appear to be in serious trouble. The cod fishery off New England has collapsed, the 2-million-ton-per-year pollock fishery centered in the North Pacific off Alaska is in decline, we are being told that sharks have been overexploited in the Gulf of Mexico, bans on commercial fishing and bans or severe restraints on sportfishing have been imposed in various areas, and the Great Lakes fisheries continue to elude the best intentions of managers to affect their recovery. It's not only a U.S. problem, of course. It is truly global in scope.

To many, the decline in U.S. fisheries is a recent phenomenon, though it has been predicted for some time that total world harvest from the sea (and by implication, the fresh waters of the world) has an upper limit. Among the most widely cited sources of what that limit might be is the one John Ryther published in 1969. He was among the first to predict that the world's fishery harvest would peak at about 100 million tons.[3] Statistics gathered over the past few years by the Food and Agriculture Organization of the United Nations indicate that the 100-million-ton figure has been reached through a combination of commercial fishing and aquaculture. Total world landing from capture fisheries has been in decline since about 1990, so the early predictions of Ryther and others are now upon us.

Fishing has always been important to the United States, and while we are not the world's leading fishing nation, we have enjoyed the bounty of broad and productive continental shelves teeming with fish. In 1873, a government official wrote, "The importance to the U.S. of the fisheries on its coasts can scarcely be exaggerated. . . ."[4]

Declines in U.S. fisheries may seem like a recent phenomenon, but striped bass and sturgeons apparently were eliminated from the Exeter River, New Hampshire, by as early as 1762 as a result of overfishing. The alewife spawning runs in the same river were destroyed by 1790 because of dam construction.[5] Other rivers and species were similarly affected, and while there may have been no general outcry about diminishing fishery resources, some naturalists were well aware of the problem. One of the more vocal and visible of them was Spencer F. Baird, who summarized the situation:[4]

> A few years ago, in view of the enormous abundance of fish originally existing in the sea, the suggestion of a possible failure would have been considered idle; and the fisheries themselves have been managed without reference to the possibility of a future exhaustion. The country has, however, been growing very rapidly. . . . The object of those engaged in the fisheries has been to obtain the largest supply in the shortest possible time. . . .

As a result of such observations, Baird approached Henry L. Dawes, representative from Massachusetts and chair of the House Appropriations Committee. Baird, who was assistant secretary at the Smithsonian Institution, wanted funding placed in the Smithsonian earmarked to study the fishery decline. Dawes indicated that the Smithsonian was not mandated to undertake such investigations, so on January 3, 1871, Baird approached Dawes with a different idea.[5] Dawes submitted a joint resolution to Congress, and after having to withdraw it once, obtained passage in the House on January 22 and in the Senate a few days later. The legislation that created the U.S. Fish and Fisheries Commission was signed by President Ulysses S. Grant.* Baird, who was appointed by Grant to be the first commissioner, described the event:[6]

> Responding to the appeals made from numerous quarters, and for the purpose of settling the question as to the facts, a resolution was passed by Congress on the 9th of February, 1871, directing the President to appoint some one of the civil officers of the Government competent to the task, to serve, without salary, as Commissioner. . . .

So who was this person who not only thought the nation's fishery resources were in decline but also was placed in a position to begin doing something about the problem? Like so many others of his era, he was an extremely competent scientist who had virtually no credentials that we would recognize today.

Spencer Fullerton Baird was born on February 3, 1823.[7] He attended Dickinson College in Carlisle, Pennsylvania, entering at the age of 13. After graduation he looked into a career in medicine but found it not to his liking. Instead, he elected to become a naturalist. At that time, formal scientific training was generally not available, so Baird learned through experience, becoming so proficient in his field of choice that he gained the confidence of John J. Audubon, who approached Baird when he needed a bird identification. In 1845, Baird was elected to the position of Honorary Professor of Natural History at Dickinson College. He became a full professor a year later. In 1850, when Baird was 27, he was appointed to the staff of the Smithsonian Institution in Washington, D.C. He remained at the Smithsonian for

---

*Throughout this book the U.S. Fish and Fisheries Commission is often called the Commission with a capital "C," while reference to the various state agencies of the same name use the lower case "c," for example, the California Fish Commission.

Spencer F. Baird (1823–1887) at age 17 (1840)

37 years and was Secretary for 10 years. He effectively maintained the positions of secretary and Commissioner of the U.S. Fish and Fisheries Commission simultaneously for several years. The work he initiated as Commissioner between the time of his appointment and his death on August 17, 1887, set the stage for much of the government-sponsored marine biological, oceanographic, and aquaculture research and development that has continued to the present. Baird died at Woods Hole, Massachusetts (often spelled *Wood's Holl* by Baird and others), where he had spent many summers conducting research, and where he had established a permanent Commission facility that was later to become the Woods Hole Oceanographic Institution. During his professional career Baird published 1,065 articles and books. The

lack of television and many other diversions we now take for granted undoubtedly played a role in his productivity, though if he had had access to a good word processor, his publication numbers might have been even more prodigious.

Though Baird was convinced that America's fisheries were in decline, part of his original charge as Commissioner was to, in fact, determine if there had been a diminution in the number of foodfishes in U.S. coastal waters and further, to suggest how the situation could be remedied. He wasted no time in meeting the obligations of this new position. In fact, a good deal of activity was initiated in 1872 as documented in the *Report of the Commissioner* for that and the following year. Baird had the vision to recognize that the mere study of fish stocks was insufficient if the

Spencer F. Baird at age 27 (1850)

Spencer F. Baird during his term (1871–1887) as Commissioner of U.S. Fish and Fisheries Commission, with his wife and daughter Lucy.

problem was to be properly understood and addressed. Therefore, he was a strong proponent of studies designed to explore various aspects of marine science, recognizing that to understand the fisheries it was also necessary to have[6]

> . . . thorough knowledge of their associates in the sea, especially of such as prey upon them or constitute their food. It is well known that the presence or absence of particular forms of animal life in certain localities determines the occurrence of many kinds of fish; and it was thought best to make an exhaustive inquiry in this direction.

Baird's annual reports to Congress (the first one actually covered two years), often approached or exceeded 1,000 pages and set the stage for later publications by the Commission and its descendents. Each volume contained a summary of activities written by the Commissioner, detailed accounts of activities by the various people involved, and importantly, reprinted articles of interest that had been published in other countries. Many of the latter were translations from Norwegian, German, and other languages. Baird elucidated the goals he had for the Commission in a report published in 1880:[8]

1. The preparation of a series of reports upon the various groups of aquatic animals and plants of North America, especially those that have a direct relation to the wants or luxuries of mankind. . . .
2. The utilization of the very extensive facilities at the command of the commission in the interest of educational and scientific establishments in the United States, by securing large numbers of specimens of aquatic animals and plants which, after reserving the first series for the National Museum, will be distributed, properly labeled, to colleges . . . and scientific societies.
3. A complete account of the physical character and conditions of the waters of the United States, as to chemical composition, temperature, &c., with special reference to their availability in nurturing the proper species of food-fishes.
4. A history and description of the various methods employed in North America, in the pursuit, capture, and utilization of fishes and other aquatic animals, with suggestions as to imperfections of existing methods and the presentation of devices and processes not hitherto adopted by the United States.
5. Statistics of the various branches of the American fisheries from the earliest procurable dates to the present time, so as to show the development of this important industry and its present condition.
6. The establishment . . . of a thoroughly reliable and exhaustive system of recording fishery statistics for the future. . . .
7. The bringing together in the National Museum not only of a complete collection of aquatic animals and plants . . . , but of illustrations of all apparatus or devices used in the prosecution of fisheries at home and abroad, together with specimens of the results.
8. An investigation of the movements and habits of the various kinds of fish, to serve as a basis of legislation, either by the general government or by the states.
9. . . . to determine what regulations shall be made by the general government or by the States in respect to close seasons or intermissions of capture, the size of fish to be caught, the enforced use of fishways . . . , &c.
10. The stocking the various waters of the United States with the fish most suited to them, either by artificial propagation or transfer, and the best methods and apparatus for accomplishing this object.

Baird's approach to reversing the decline in the nation's fisheries is couched in items 9 and 10, that is, regulation and enhancement stocking.

In the Commissioner's report for 1872 and 1873, Baird provided his view on declines in the coastal and river fisheries of New England. He acknowledged that states in the region had recognized the problem and were already involved in restoring fish stocks. Baird pointed out that lumbering in Maine, manufacturing in New Hampshire and Massachusetts, overfishing, and improper construction and maintenance of fishways associated with dams hampered restoration of river fisheries. Baird saw a relationship between declines in cod, which remain at sea throughout their lives, and species that spawn inshore or in freshwater. He concluded[6]

. . . that the reduction in the cod and other fisheries, so as to become practically a failure, is due, to the decrease off our coast in the quantity, primarily of alewives; and secondarily, of shad and salmon, more than to any other cause.

Whatever may be the importance of increasing the supply of salmon, it is trifling compared with the restoration of our exhausted cod-fisheries; and should these be brought back to their original condition, we shall find, within a short time, an increase of wealth on our shores, the amount of which would be difficult to calculate. Not only would the general prosperity of the adjacent States be enhanced, but in the increased number of vessels built, in the larger number of men induced to devote themselves to maritime pursuits, and in the general stimulus to everything connected with the business of the sea-faring profession, we should be recovering, in a great measure, from that loss which has been the source of so much lamentation to political economists and well-wishers of the country.

Cod restoration was clearly a major priority for Baird, and one that he felt would bolster the economy of New England. At the same time, Congress had directed that the investigations of the Commission should include not only the nation's coastal waters, but also the Great Lakes. Baird appointed James W. Milner as Assistant Commissioner and instructed Milner to collect data on the status of foodfish stocks in the Great Lakes. If Milner's studies showed that the fisheries were in decline, he was to ascertain the causes of that decline and make recommendations for a recovery program. As we shall see, Baird enlisted other individuals to undertake assignments in other regions of the country. Our concentration is on aquaculture, but as his 10-point list of objectives, portions of which you have already seen, clearly show, Baird was involved in a wide variety of pursuits.

Almost coincident with the establishment of the U.S. Fish and Fisheries Commission was the establishment of the American Fish-Culturists' Association (which later became the American Fisheries Society, one of the premier scientific societies for fisheries scientists in the world). As discussed by Baird, the 1872 meeting the Fish Culturists' Association put forth the suggestion that the U.S. should[6]

. . . take part in the great undertaking of introducing or multiplying shad, salmon, and other valuable food-fishes throughout the country, especially in waters over which its jurisdiction extended, or which were common to several States, none of which might feel willing to incur expenditures for the benefit of the others.

That "suggestion" certainly played right into the hands, or at least mirrored the philosophy that had been developed by Professor Baird. As an attendee of the meeting, he may (one might conclude) have had some influence on the deliberations of the membership. The association put some muscle behind their suggestion. Baird again:[6]

A committee, of which Mr. George Shepard Page was chairman, was accordingly appointed to present the subject to Congress, and to do whatever was in its power to secure the desired object. This gentleman visited Washington, and appeared before the Committee on Appropriations to urge the measure and secure its favorable action. A clause appropriating $10,000 was accordingly put into the appropriation bill for the purpose in question; but this was rejected by the House. Subsequently, however, the subject was considered by the Senate committee, who took an equal view of it with the House committee, and an amendment appropriating $15,000 was introduced and carried successfully through Congress; its disbursal being placed under my charge.

Congress could probably have saved some money if the House had approved the committee's recommendation the first time around. In that regard the political process hasn't changed very much. In any event, Baird consulted with the fish commissioners of the New England states and with members of the Fish Culturists' Association a few days after Congress approved the appropriation, and it was decided to use the $15,000 to secure the services of some individuals who would be charged with producing fish for distribution. He approached established fish culturists who were working in the private sector and managed to obtain the services of two of them, who were assigned to spawn shad. It was important to pursue that activity immediately since the money had come while the spawning season was already underway. Baird then turned his attention to other species and other individuals who could assist him in their culture.

Baird's primary charge was to restock the coastal waters and the Great Lakes, but when he put forward his initial list of species to be cultured, he was obviously expanding on that assignment. The list of fish that he thought worthy of consideration by the Commission included:[6]

Shad
Alewife
Atlantic salmon
Quinnat salmon*
Lake trout (also referred to as salmon trout)
Brook trout
Whitefish
Carp

Baird also mentioned a sea trout from the St. Lawrence River; the Danube salmon (possibly a race of brown trout); gourami (from China and the Indian Ocean); the nerfling, orfe, or golden tench (a European fish); sterlet (a Russian sturgeon); and salmon hybrids as possibilities for culture. Not all of the fishes Baird identified ultimately panned out, but the vast majority received a good deal of attention by the Commission.

Since the establishment of the U.S. Fish and Fisheries Commission did not precisely coincide with the beginnings of U.S. aquaculture, it seems appropriate to step back and take a look at the people who were in place when the national program was implemented. Most of them were operating in the private sector, and none had been active for more than about a decade.

PIONEERS OF U.S. AQUACULTURE

According to James W. Milner, one of the important aquaculturists of the late 1800s, the first published account of artificial fish spawning in the United States was associated with a paper read by the Reverend John Bachmann of Charleston, South Carolina, before the South Carolina State Agricultural Society in 1804.[9] Bachmann reported on fertilizing and hatching the corporal (identity unknown) and brook trout. Hatching and subsequent fry growth in a pond were reported for the trout.

---

*This is the original name of the California salmon currently called chinook or king salmon and known to range from California to Alaska on the Pacific coast of North America.

That was apparently the end of it for Rev. Bachmann, who may have turned back to the pulpit rather than continuing his activities in the hatchery. Apparently, little of note occurred for another few decades, with the exception of moving fish about from one place to another as transfers but not as the result of actual culture.

Publications about fish culture from Europe were available in the United States and in 1853, after reading some of them, Theodatus Garlick, M.D., and Professor H. A. Ackley successfully fertilized the eggs of brook trout and were able to incubate some of them successfully.[5,9] This early fish culture activity took place at Ackley's farm, which had three ponds totalling something less than two acres. The two pioneers placed the fish they had produced on display at the Ohio State Fair and received accolades for their remarkable accomplishment.[5] Brook trout were the focus of a number of other early fish culturists. Even Samuel Colt, whose name is still carried on firearms, got into the act by establishing a spawning facility that produced about 4,000 fry in 1860. Stephen H. Ainsworth, a resident of West Bloomfield, New York, produced sufficient numbers of brook trout that he could sell some to pay expenses, stock area streams and ponds, and have enough left over for his own needs.

In 1864, perhaps the most famous of the early U.S. fish culturists began his work. Seth Green (the same Mr. Green who was approached by Professor Baird to initiate shad spawning activities in 1872) was a private entrepreneur who established his first facility (Spring Brook) in 1864 with advice and help from Stephen Ainsworth. Green purchased a site for $2,000 near Caledonia, New York, where he constructed a series of ponds. The water supply was 150,000 gallons per minute.[5]

Green achieved only about 25% fertilization with his first several batches of eggs.[10] When Ainsworth visited, he told Green that 25% was as good as anyone else was achieving. Green wasn't satisfied, and began examining what he might be doing wrong. He decided to reduce the amount of water used during fertilization and increase the amount of milt. His percentage of fertilization quickly increased to 95%. His trout hatchery was financially successful from the start. While he had initially intended to produce fish of eating size,[5] the value of eggs was sufficiently high ($8–$10 per thousand), that producing fingerlings or foodfish did not make great economic sense.

Seth Green had been involved in commercial fishing prior to embarking on fish culture. He began working in the fishing industry in 1837 and continued in that pursuit until 1860.[11] He must have been successful since he employed as many as 100 people in both capture and sales. In any event, he had apparently developed an interest in fish culture many years before he actually became involved.

Seth Green was born on March 19, 1817, in Rochester, New York, the son of Adonijah Green, who was an early Rochester settler, farmer, and tavern keeper.* Seth did not receive any formal training at the college level, but he was an astute observer of nature. At the age of 21, he developed a belief, based on his observations of local

---

*Details of Seth Green's life not otherwise referenced come from a biography written by Don Longacre of Caledonia, New York, dated May 28, 1987. The biographical sketch was graciously provided by Mr. Arden Trandahl, who manages the National Fish Culture Hall of Fame located at the D. C. Booth Hatchery, Spearfish, South Dakota. Seth Green was inducted into the Fish Culture Hall of Fame in 1987.

Seth Green (1817–1886)

fish, that trout could be spawned and reared under captive conditions. At his hatchery at Spring Brook, established in 1864, he produced brook trout, rainbow trout, lake trout, Atlantic salmon, grayling, lake herring, whitefish, carp, and goldfish. In addition to producing fish for stocking U.S. waters, Seth Green shipped brook trout eggs to England and France and introduced black bass to Great Britain. Among the many people with whom Seth Green corresponded over the years was Brigham Young, who was interested in obtaining advice as to which fish might be suitable for stocking into the Great Salt Lake. Horace Greeley featured Green and his exploits in several articles published in the *New York Times*. Mr. Green died on August 20, 1886.

The first fish and game commissions were established in various New England states in the early 1860s. Those commissions set the examples for the establishment of similar bodies in other states,[12] and they undoubtedly served as models for the U.S. Fish and Fisheries Commission as well. The New York commission was involved in the culture of bass, walleye, and perch as early as 1874.[13]

In 1867, the commissioners of fisheries from four of the New England states approached Seth Green to discuss whether he would be willing to attempt hatching shad.[10] Green proceded to Holyoke on the Connecticut River, where he told the locals that he was going to spawn shad. The locals thought he was a nut. After some false starts, Green developed a hatching system that worked. The commissioners were delighted. Green managed to hatch about 15 million eggs in 15 days, and the locals probably still thought he was a nut, though that last part is speculative. The shad fry were released in the river, and the commissioners reported that in 1870 the return was 60% higher than it had been in 1802.* The first shad to be introduced to the West Coast were transported there and stocked into the Sacramento River by Green in 1871.[9]

Even prior to obtaining any supporting evidence, Green was obviously confident that the release of newly hatched shad would contribute greatly to the recovery of depleted fish stocks, for in 1868, two years before the reported success associated with the Connecticut River releases, Green told a General Spinner in Washington, D.C., that for a cost of $2,500 the Potomac River could be stocked with shad that would, within three years, return the fishery to historical levels. Spinner approached Congress for funding but was unable to obtain it. Green said that, had the appropriation been made, ". . . the Potomac River would to-day [1875] be the greatest shad river in the world."[11] Green was not modest about his accomplishments with shad:[11]

> At that time [1868] I was the only man who had ever hatched shad to know anything practically about it, and did it all with my own hands. Now it is different. I have eight practical men, and every one of them is able to run a shad hatching establishment. There is but one other in the country who can, and that is Charles Smith, who was with me at Holyoke in 1867. Shad hatching is a trade by itself. A man may know how to hatch other kinds of fish, and make a perfect failure at hatching shad. We have been these for the last seven years, and I known that our great lakes can be abundantly stocked in a few years.

Green didn't stop with brook trout and shad. In 1869 he began hatching whitefish. He also worked with salmon,[10] and in 1867 was granted a patent for spawning and hatching mackerel.[14] He has long been recognized as a true pioneer in American fish culture. Milner wrote about Green, "More than those of any other person in the United States, his labors have popularized the subject and extended the new indus-

---

*Reported successes were not often based on careful scientific study, so in many cases it can be argued that increases in catch rates during a particular year or period of years subsequent to the initiation of a stocking program represents part of a natural cycle or may, in fact, be related to some other activity (fishway construction, imposition of new regulations, and so forth).

try throughout America, at the same time greatly improving and perfecting methods of work."9

One of the more colorful early fish culturists was the Reverend Livingston Stone,[15] who began his fish culture work in 1866 in Charlestown, New Hampshire, where he initially focused on trout.[5,9] Stone was born in Cambridge, Massachusetts, on October 21, 1836, a descendant on his father's side of Plymouth colony settlers who settled 200 years earlier.[15] After graduating from Harvard in 1857, he attended the Meadville Theological School, becoming a Unitarian minister in 1860. Always in

Livingston Stone (1836–1912)

poor health since a child,* he was prevented from becoming a chaplain during the Civil War. In 1864, he decided to quit his ministry in Charlestown to find work that would allow him to spend time outdoors. He found that opportunity in fish culture, though how that decision was made and what happened between 1864 and when he established his hatchery two years later are not clear. He was one of the founding members of the American Fish-Culturists' Association and was its first secretary.

Stone's hatchery was established with the objective of producing food fish for the marketplace. Prices at the Fulton Fish Market in New York ranged from $0.75 to $1.25 per pound, so the opportunity for profit appeared good. Since there was also a growing market for the early life stages of fish, Stone expanded his marketing activity by selling Atlantic salmon eggs and bass fry.[9] In addition, Stone produced yellow perch. His success and experience prior to 1871 placed him in a position to be selected by Spencer F. Baird to head up Pacific salmon spawning operations when the U.S. Fish and Fisheries Commission was formed. Details of his activities in California and his travels between the East and West Coasts are subjects that are taken up later in this chapter.

In 1897, Stone requested transfer to the Cape Vincent hatchery in New York. He made the move not because he wanted to leave his work in California but to have an opportunity to be with his family on a permanent basis. He retired at age 70 in 1906 and died on Christmas Eve, 1912.

Among the first to spawn Atlantic salmon was Samuel Wilmot, a Canadian, who collected salmon eggs from Lake Ontario in 1866. By 1870, Wilmot was selling eggs in various states. Also in 1866, Dr. W. W. Fletcher was charged by the New Hampshire fisheries commissioners to obtain salmon eggs from New Brunswick, Canada, and return them to the United States, where he was to begin restoration efforts on the Merrimack River.[9,16] Dr. Fletcher's activities were augmented by those of Livingston Stone in 1868, who also collected eggs in Canada. In 1870 Maine applied for eggs from Wilmot, who obtained a price of $40 (in gold) per thousand. That price was considered excessive since, in 1868, Stone sold eggs to New Hampshire and Massachusetts for $16 per thousand and no producer of salmon eggs in New Brunswick had asked over $20 per thousand.[16]

In 1871, Mr. Charles Grandison Atkins was the first to spawn Atlantic salmon in U.S. waters.[9,16] Atkins was born in New Sharon, Maine, on January 19, 1841.[17] After graduation from Bowdoin College in Maine, he taught school in Green Bay, Wisconsin, in 1862 and 1863, after which he apparently returned to Maine. The Maine legislature, in its session 1866–1867, authorized the appointment of two fish commissioners to look at the possibility of restoring that state's fisheries, which had been heavily impacted by dam building, overfishing, and to a lesser extent pollution. Charles Atkins was given one of the two appointments and remained a commissioner for the state of Maine until his term ended in 1872, coincidental with the time that Baird appointed him to the U.S. Fish and Fisheries Commission.

Atkins had first tried hatching Atlantic salmon in 1867, but only one egg hatched out of several thousand that he had obtained. In 1870, largely because of the high

*His health must have improved, for as we shall see, the rigors faced by Stone in conjunction with his fish-spawning activities were prodigious.

Charles G. Atkins (1841–1921)

cost of Canadian salmon eggs, Atkins decided to attempt spawning native fish. He was able to purchase fish from commercial fishermen in June and July, but spawning did not occur until November. He placed the fish in a pen that had been constructed in a large holding pond, and while losses were fairly high, he was able to produce 70,500 eggs on his first attempt. His productivity increased with time, and we will hear more about him as we proceed.

While Atkins did not confine himself entirely to the production of Atlantic salmon—he worked on cod hatching at Woods Hole in 1886 and 1887 and devoted some of his time to bass distribution in 1869—the majority of his activity was devoted to salmon production and distribution. When the Craig Brook hatchery was established in 1889 (and it was the first U.S. Government Atlantic salmon hatchery), Atkins was placed in charge.

In 1914, Atkins was appointed Fish Culturist at Large, a position he held until his retirement in 1920. He died a year later at the age of 80.

Thus, by 1871, many of the species on Baird's interest list had been successfully spawned, the eggs incubated, and fry released. Hatchery methods had been worked out that would provide at least acceptable hatching rates, and at least rudimentary methods for shipping eggs and fry had been developed. In 1874, Milner compiled a list of American fish culturists and those who had indicated an interest in fish culture. The list of active fish culturists contained the names of 107 people.[9]

By the time the U.S. Fish and Fisheries Commission came into being, a number of states had already established their own commissions. The state commissions established between 1856 and the end of the century were as follows:[18]

1856	Massachusetts
1865	New Hampshire, Vermont
1866	Connecticut, Pennsylvania
1867	Maine
1868	New York
1870	New Jersey, Rhode Island
1871	California, Alabama
1873	Ohio, Wisconsin
1874	Iowa
1875	Minnesota, Virginia
1876	Kentucky
1877	Kansas, Colorado, Nevada, West Virginia
1878	Tennessee, Utah
1879	South Carolina, Nebraska, Texas, Wyoming

THE FISH COMMISSION GOES TO WORK

The two men that Spencer F. Baird approached with respect to rearing shad during 1872 were Seth Green and W. Clift.[6] Green was really under the gun, as the shad spawning season in New York was about over. Clift, working farther north in Connecticut, where the spawning season began later in the year, had a bit more time. Both men were successful in their endeavors. Green hatched shad eggs for the states of New York and Vermont and provided 55,000 fry to the Commission. The Commission's fish were stocked in the Allegheny River at Salamanca, New York, and in the Mississippi River near St. Paul, Minnesota. Fish produced by Clift also ended up in the Allegheny River, as well as in the White River at Indianapolis, Indiana, and in the Platte River in Colorado.

Charles Atkins was asked by Baird to expand his facilities at Bucksport on the Penobscot River so he could produce Atlantic salmon for the Commission. Baird

William Clift (1817–1890)

supplied the funds for expansion and obtained a share of the eggs that were subsequently produced, the others having been spoken for by various states. To hedge his bet in the event that Atkins experienced a failure that year, Baird attempted to obtain additional Atlantic salmon eggs from Germany. He was pleased to obtain a positive response to his request, and arrangements were made for a shipment. Exposure to warm weather and delays related to steamship malfunctions delayed delivery, but eventually the eggs arrived.[6]

> The boxes, sixty in number, occupying nearly 300 cubic feet of space, were transferred to the hatching-houses of Dr. Slack, near Bloomsbury, N.J., and the contents imme-

diately assorted, but of the 750,000 eggs only four or five thousand were sound. These were successfully hatched out, and ultimately introduced into the Musconetcong, a tributary of the Delaware, and on which Bloomsbury is situated.

Atkins actually did much better that year, producing over one and a half million eggs, so Baird's backup plan was not needed, not the mention that it didn't work out according to plan.

There was apparently a good deal of interest in stocking Pacific salmon in the waters of New England. Baird asked Livingston Stone if he would go to California and look into the matter. Stone was not only able to get a primitive facility established on the McCloud during 1872, in October; he actually shipped eggs to the East Coast.[6]

Whitefish was the last species to be targeted for production in 1872. The person solicited by Baird to produce those fish was Nelson W. Clark of Clarkston, Michigan. Spawning activities were initiated in the fall, and Clark was able to place over 500,000 eggs in his hatchery.[6] Poor results were obtained from a lot of 200,000 eggs shipped to California early in 1873, possibly because of exposure to excessive cold or because the eggs were improperly packed. A second shipment arrived in California in excellent condition. After hatching, the young fish were introduced into Clear Lake.

The fish culture activities of the U.S. government had developed with amazing rapidity after the establishment of the Fish and Fisheries Commission. By 1874, the Commission was concentrating on shad, alewife, striped bass, salmon (both Atlantic and Pacific), whitefish, and carp. There was also activity with other species. For example, Fred Mather had collected eggs of the black sea bass and fertilized them (a storm terminated that particular experiment). Baird felt that the production and distribution of fish was of extreme importance. He was dedicated to getting the highest possible yields from the nation's waters:[19]

> This has rightly been considered an object of the greatest importance in view of the rapidly-increasing population of the United States and the almost corresponding diminution in the average yield of vegetable food by the farming-lands, and it is not considered exaggeration to say that the water can be made to yield a larger percentage of nutriment, acre for acre, than the land.

In Baird's time there was no thought given to whether introductions of exotic species might be harmful to native flora and fauna—after all, modern ecological theory had yet to be envisioned. The life histories of many species had yet to be worked out, so there was also little thought given as to whether a particular species might be able to adapt to the habitat into which it was introduced (the fish culturists did, of course, recognize the temperature tolerance limits of cold-water as opposed to warm-water species). Those who carried fish from one coast and introduced them on the other, as well as at many points in between, were acting in what they thought was the best interest of the nation. They should not be criticized for their ignorance. Their intentions were certainly honorable, and their objectives were clear. In the words of Baird, whose comments could easily have been made today:[6]

> . . . the great object is to increase the supply of food to the nation at large, and every capture, whether in Ohio or Louisiana, will tend to accomplish the same general result.

THE BEGINNINGS OF U.S. AQUACULTURE

After any species of fish has become permanently established in a given body of water, their continuance therein will depend in great measure upon the enactment of suitable laws, securing their access to suitable spawning-grounds, and protecting them during the critical period of their existence, from capture or unnecessary destruction. Otherwise the methods of artificial propagation must be resorted to indefinitely.

The remainder of this chapter provides some insight into the activities and accomplishments of the aquaculturists who were active during the early years of the Commission. The feverish activities from 1871 to 1880 or a bit beyond are the focus of our attention.

Fish Distribution Begins

The U.S. Commission wasted no time in not only propagating but also distributing fish, and those fish did not always go into waters in the vicinity of the hatcheries. Thirty-three states and two territories had received fish by 1874.[19] Baird indicated that it might not be wise for some states to stock their own streams because the fish would end up leaving those states for others. He gave the example of fish stocked in Minnesota, Ohio, or Pennsylvania entering the Mississippi River drainage and ending up in the southern states. He also thought that young fish stocked in the South wouldn't perform well because they required the cooler northern waters at that stage. He argued that it would be prohibitively expensive for states to set up their own hatcheries when the federal government could collect eggs of a given species from a limited geographic area and distribute them around the country more economically.[19] The states didn't necessarily agree and, in fact, fish culture facilities had already been established by fish commissions in the states of Massachusetts (1867) and Maine (1872).[18]

International Fish Distribution

Shipments were not restricted to the various states and territories. There was also interest expressed by foreign countries in obtaining eggs from the U.S.[19] By 1880, a total of 41 hatching stations in operation where fish eggs were collected and incubated. They had distributed over 100 million shad and over 50 million salmon. Some 44 million shad were released at the place where they were hatched, and the remainder were distributed to other regions.[20]

Trout had been sent to Canada, England, France, Holland, Germany, New Zealand, Australia, and the Sandwich Islands by 1880.[20] Foreign shipments continued for many years. As an example, Fred Mather provided the following information in his report on the shipment of fish eggs to foreign countries during the winter of 1882–1883:[21]

Germany*	25,000 brook trout eggs
	100,000 lake trout eggs

*All but the lake trout arrived in good condition (the lake trout eggs largely hatched en route and did not survive).

	6,500,000 whitefish eggs
	25,000 Atlantic salmon eggs
	A small number of largemouth bass juveniles
France	20,000 brook trout eggs
	50,000 lake trout eggs
	200,000 whitefish eggs
	15,000 Atlantic salmon eggs
England	10,000 brook trout eggs
	10,000 Atlantic salmon eggs
South America	6,000 brook trout eggs to Colombia

In 1877 salmon eggs collected by Livingston Stone at the McCloud River Station in California were sent to New Zealand. Five lots of 80,000 each were sent.[22]

The trip to New Zealand began with 22 miles of rough road to the nearest railroad station in heat measured at 104°F in the shade and 125°F in the sun. The eggs rode the train 300 miles to San Francisco, where they waited two nights and a day for the ship. Stone continues:[22]

> . . . and then [they] are carried seventy-six [hundred] miles, most of which is through the tropics and across the equator, and at the end of that distance are taken out; and I think it is perfectly surprising that the eggs of any fish whatever can be carried so far . . . and come out alive.

Baird indicated that Stone had been

> . . . authorized to give a small number of the eggs to the New Zealand colonies, and that lot of eggs was divided into two, one part of which hatched out very satisfactorily, but the other failed. In 1876 that experiment was renewed in New Zealand. . . . not less than 75 to 90 per cent. of the eggs have been turned into healthy, vigorous fish.

The first attempt to ship shad to Europe was made in 1874 by Fred Mather. He left New York for Germany by steamboat on June 5 with 100,000 one-day-old shad fry, 10,000 of which had been placed in each of 10 milk cans. Mr. A. Anderson assisted Mather on the trip. Each man stood six-hour watches and provided fresh water to the fish every hour for the first six days and every half hour for the next four days of the journey. Aeration was provided by removing half the water and pouring it from one pail to another and then replacing it. Dead fish were removed each morning: ". . . this was accomplished by swirling the water with a dipper, which caused all dead fish to collect in the center of the can."[23] the dead fish were then siphoned off. All the fish died by the night of June 14:[23]

> The fish, in my opinion, died from starvation; hatched on the morning of the 4th, they were probably looking for food about the third or fourth day after, but appeared strong until the morning of the 12th, when we first noticed signs of weakness by a slow motion, and many alive resting on the bottom of the can.

A log maintained by the two culturists chronicled fish losses and their attempts to ameliorate the problem:[23]

Aug 5	500 dead
6	200 dead
7	1,000 dead
8	20 dead
9	100 dead
10	3,000 dead
11	500 dead (fish appeared to be weakening—water exchange increased)
12	1,200 dead
13	5,000 dead (tried to feed them with a piece of raw beef)
14	All dead but 1,000 by noon; remainder dead that night

Mather thought that loading eggs and hatching them at sea would be practical, particularly if the temperature could be kept in the low 60s (Fahrenheit) to slow development. He indicated that he had designed a hatching container that might work well for that purpose: ". . . I do not regard the transportation of shad-fry as at all practicable on a journey occupying over eight days, unless we can discover some method of feeding them."[23]

James Milner was interested in the same problem and had some ideas about how the feeding of shad might be effected:[24]

> To devise a method for feeding them will require the services of a microscopist familiar with the lower forms of invertebrates and the eggs and larvae of higher groups, which are the principal minute organic forms available as food in the waters where the fish breed naturally.
>
> If food can be found among these forms, experiments as to the feasibility of breeding them *en route* will be in order. Many of them have been developed in numbers by naturalists for purposes of study, and with some it is very easily accomplished.

Charles Bell, who was an assistant to Fred Mather, suggested how shad eggs might be incubated on shipboard. That suggestion, made in 1875, involved placing the eggs in a funnel with a screen bottom and the use of upflowing water to keep the eggs in suspension.[19] The system appeared to work well on an experimental basis and may be the model on which the upwelling incubators used in many hatcheries today are based.

The First Commission Fish Hatching Ship

Hatching stations on land could be established in locations where anadromous fishes could be expected to reappear each year prior to spawning, or they could be placed in a central location to which eggs were delivered. To collect spawning marine fishes often meant searching around for the proper collection areas, which might or might not be in handy proximity to a shore-based hatchery. One way around the problem would be to employ a mobile hatchery. Such a facility was constructed during 1880 by the U.S. Fish and Fisheries Commission. The vessel, which was also used extensively for natural history and oceanographic sampling, was christened the U.S.S. *Fish Hawk*. It was a twin-screw, coal-fired steamship of 156 feet, 6 inches in length. It was equipped with various types of dredges and fishing gear along with fish hatching equipment.[25] The hatching facilities were designed for use with shad, though modifications were made to allow for the hatching of cod as well. One of the procedures used was to purchase adult shad from commercial fishermen. In May 1881, the ship passed through Albemarle Sound and

Pamlico Sound, to Hatteras Inlet. On May 5 it was anchored at the mouth of the North East River. On May 9, the following incident occurred:[26]

> . . . a fisherman called on behalf of the gillers, and stated that, at an informal meeting, they had decided to furnish the Commission with eggs, whether paid for or not, but, as they were put to some inconvenience and extra labor thereby, they requested him to see if I could not procure them the usual compensation of twenty-five cents for each spawning fish.

Compensation was arranged the next day.

Typical activity aboard the vessel during its first year of operation was as follows:[26]

> On the 10th [of May] . . . 341,000 eggs were procured during the day. On the 11th, 913,000 were taken, and on the 12th 979,000; 664,000 young fish were deposited in the river near the ship on the latter date. On the 13th 265,000 eggs were taken and 1,660,000 young fish deposited in the North East River and at the mouth of the Susquehanna.

And so it went, as did attempts to do more than just dump hatched embryos:[26]

> A considerable number of young fry, hatched on the 11th [still May 1881] from eggs taken on the 7th, were retained in a cone for the purpose of ascertaining how long they could be kept alive after absorption of the yolk bag, which disappeared from the naked eye on the 15th instant. On the 16th the fish which had hitherto remained on the surface went down from four to six inches or more, where they appeared to be feeding upon minute particles collected on the surface of the cone. On the 23d they were still thriving, but few dead ones having been seen. It was an undoubted fact that they were feeding and developing normally.

Observations taken aboard ship even hinted at how shad might be fed: "One of the young shad, fifteen days old, was examined under a microscope to-day [May 26]. Minute crustacea were found it in its stomach."[26]

Enter the Railroads

Distribution of fish from hatching stations to nearby stocking locations could be accomplished relatively easily by messengers employed by the Commission who rode on trains. Freight haulers were also utilized. Properly packed, fish eggs could be shipped relatively easily by whatever mode of transportation was available. The transportation of fish fry, fingerlings, and marine invertebrates was often a different matter. In addition, with the desire of the Commission to ship fish from coast to coast, logistics became an issue.

By the time the Commission was established in 1871, development of the nation's railroad system was well underway. While you couldn't get everywhere by railroad— wagon trains were still plying their way from St. Louis to the West Coast, and the stage coach was a convenient means of long-distance transportation—it was becoming possible to get from the East Coast to the West within a few days. The railroads provided an obvious means of shipping fish throughout much of the nation efficiently and quickly.

Pioneers in development of the McCloud River hatcheries. From left, Myron Green, Livingston Stone, and Willard T. Perrin.

In 1873, Baird charged Livingston Stone with the job of delivering a so-called aquarium car to California from the East Coast. The car was furnished by the Central Pacific Railroad Company.[19] It was fitted with holding tanks, ice chests, and bunks for the use of the culturists who would make the trip. A water reservoir on the car could be filled from water-tank spouts along the route (the water tanks being there to provide water for the steam engines). The group of culturists who accompanied Stone in the aquarium car were W. T. Perrin, Myron Green, and Edward Osgood.[27] The shipment consisted of the following:[19]

Black bass	60
Walleye	11
Yellow perch	190
Bullheads	12
Catfish	110
Tautogs	20
Eels	41,500
Trout	1,000 (probably brook trout)
Lobsters	162 (American lobsters)
Oysters	1 barrel (American oysters)

The car left Charlestown, New Hampshire, on June 3, 1873, and progressed in its westward journey without significant incident for the first few days. Livingston Stone described what happened thereafter:[27]

After leaving Omaha [on June 8], we stowed away as well as we could the immense amount of ice we had in the car [one and a half tons]; and, having regulated the temperature of all the tanks, and aerated the water all round, we made our tea and were sitting down to dinner, when suddenly there came a terrible crash, and tanks, ice, and everything in the car seemed to strike us in every direction. We were, every one of us, at once wedged in by the heavy weights upon us, so that we could not move or stir. A moment after the car began to fill rapidly with water, the heavy weights upon us began to loosen, and, in some unaccountable way, we were washed out into the river. Swimming around our car, we climbed up on one end of it, which was still out of the water, and looked around to see where we were. We found our car detached from the train, and nearly all under water, both couplings having parted. The tender was out of sight, and the upper end of our car resting on it. The engine was three-fourths under water, and one man in the engine-cab crushed to death. Two men were floating down the swift current in a drowning condition, and the balance of the train still stood on the track. . . .

A trestle over the Elkhorn River in Nebraska had collapsed under the weight of the train. Stone was obviously disappointed:[27]

One look was sufficient to show that the contents of the aquarium car were a total loss. No care or labor had been spared in bringing the fish to this point, and now, almost on the verge of success, everything was lost.

The lids had come off the tanks holding the various aquatic animals being carried in the aquarium car. Thus, the fish and invertebrates were inadvertently stocked in the waters of the Elkhorn River. Baird speculated that many could have survived, but the tautogs, lobsters, and oysters were probably goners.[19] That assumption seems accurate since none of those marine species seems to have become established in Nebraska.

A train wreck and the associated death of at least one trainman notwithstanding, Stone made his way with all possible haste to the nearest telegraph office, from which he contacted Baird in Washington, D.C. His instructions were ". . . to return east immediately, with my assistants, and take on a shipment of young shad to California. . . ."[19]

Lacking an aquarium car, the second trip was made using eight cans containing 40,000 newly hatched shad. The trip began on June 25, 1873. Changing water required development of an apparatus that would allow removal of water while retaining the fry. This was basically a siphon with one enlarged end that was screened (similar to what can be purchased in aquarium stores today). Turbulence had to be avoided, or the fish might be killed. No aeration was used, and the cans were tall with narrow necks so diffusion of atmospheric oxygen into them would have been severely restricted. Also, the water was filled to the neck to keep it from sloshing around. Oxygen was provided by frequent water exchange:[27]

Water was obtained at watering stations along the railroad. Passengers and railroad workers were consulted about the condition of the water at each stop. . . . if all accounts agree that the water is lime or alkaline water, or otherwise unsuitable, it is given up; but if nothing is learned against it, it is then tasted, and, if this first tasting is favorable, a supply is taken on board. It is then more carefully and deliberately tasted, and, if traces of lime or alkali are discovered, it is thrown away; if not, a few fish are placed in a tumbler full of it, and their movements watched. . . . if, at the end of [an hour or two], the fish appear to be doing well, it is considered safe to use the water.

The trip was not nearly as eventful as the earlier one, though one member of the group traveling with Stone did leave the train at Omaha (the last stop before the wreck of the aquarium car). In any case, on July 2, 1873,[27]

. . . 35,000 shad from the Hudson River, New York, were deposited safely and in good order in the Sacramento River, at Tehama, Cal.; and we turned away from the river toward our hotel, feeling as if a load of incalculable weight had been lifted from us.

How many railroads can you name?* There aren't many of them around today, but in Baird's time, the number was large and growing. Baird approached the nation's railroads and asked their cooperation in the transport of fish for stocking. The response was highly rewarding:[28]

All the railroads of the country with scarcely an exception, when applied to, gave instructions to allow the transportation, in baggage-cars, free of extra charge, of the cans containing the young fish, and granted access to the same on the part of the messengers; instructions being given, in many cases, to stop the car at stations near rivers or streams to allow the introduction of fish therein.

Baird was not exaggerating when he said the railroads were cooperative. A total of over 100 of the "iron horse" companies agreed to assist with fish deliveries.

While the railroads would agree to allow several cans of fish to be carried in a baggage car, what was needed was complete carloads of fish. The first dedicated fish hauling car (No. 1), adapted from a baggage car, hauled shad and was outfitted with assistance from President Hinckley of the Philadelphia, Wilmington, and Baltimore Railroad.[28] By the end of 1882, two fish cars had been put into operation on the railroads. Car No. 2 was built for the Commission by the Baltimore and Ohio Railroad Company. It carried the label "Baltimore and Ohio, No. 2, United States Fish Commission."[29] Construction costs for each car were $8,000. The vital statistics of the two cars were as follows:[29]

Length	car 1 = 51'2" (without platform); 57'6" (with platform)
	car 2 = 59'9"
Width	car 1 = 9'10"
	car 2 = 10'
Height	car 1 = 14'1 1/2"
	car 2 = 14'7/8"

Car No. 2 could carry up to 20,000 pounds of cargo. Four ice boxes held up to 3,000 lbs of ice. Fold-up sleeping berths were located in the middle section of the car. There was also an office area fitted with a sleeping berth, washroom, and heater. A kitchen and pantry were also included for the comfort of the staff.[30] In December 1881, car no. 2 delivered carp to Texas, Arkansas, Louisiana, and Missouri.[29]

*Amtrack doesn't count!

Salmon Spawning

The Commission concentrated its activities on both Atlantic and Pacific salmon, as well as on trout. Five hatcheries were established during the 1870s, where eggs were taken and/or hatched:[28]

Buckport, Maine	1871
McCloud River Salmon Station	1872
Northville, Michigan	1874
Grand Lake Stream near Calais, Maine	1875
McCloud River Trout Station	1879

Atlantic Salmon. As one of the first pair of fish commissioners appointed in the state of Maine,[17] Charles G. Atkins had, as we have seen, established himself as the American leader in Atlantic salmon spawning. Thus, when Baird appointed Atkins to head up salmon spawning efforts on the East Coast,[6] facilities and protocols had already been established at Bucksport, Maine, on the Penobscot River. Two distinct types of Atlantic salmon were recognized by the Commission—those that migrated from the sea into East Coast rivers to spawn and a landlocked strain. Atkins was initially responsible only for spawning the migratory Penobscot salmon, but interest also developed in a landlocked group called the Schoodic salmon. That fish, which Atkins was later charged with producing, was named for the Schoodic lakes region of Maine and New Brunswick, Canada. The fish was also known as the Sebago salmon, Glover's salmon, Win-ni-nish, and other names, depending upon the location in which it was found.[8] Scientists were reaching the conclusion that the landlocked and migratory fish were all of the same species, and that the Schoodic was merely a smaller form of the Atlantic salmon.

Salmon were collected by hatchery workers employed by Atkins and purchased from commercial fishermen who operated weirs from which they collected migrating salmon.[31] During the first year of operation (1872), nearly 1.25 million eggs were distributed and 876,000 fry were released. By 1874, a total of 6,376,000 eggs had been distributed and 4,667,000 fry released.

Unlike their Pacific counterparts, Atlantic salmon do not all die following spawning, though the mortality rate can be quite high. Baird suggested that Atkins should tag some of the fish after spawning. By having a numbered tag in place, fish that were recaptured at a later date could be evaluated for growth and migration, as well as post-spawning survival. Atkins described his first attempts at tagging spawning as follows:[31]

> The first mode adopted was the use of an aluminum tag about half an inch long and a quarter of an inch wide, stamped with a number which corresponded with a record showing the sex, length, and weight of the fish, and the date of liberation. This tag was at first attached to a rubber band that slipped on over the tail of the fish.

The first method was a failure. Loose rubber bands came off, and tight ones killed the fish. The technique was modified whereby the tag was attached to the rear of the first dorsal fin with a platinum wire. Tags appeared to stay on pretty well for several months, but after a year and a half, no tagged fish were recovered. Atkins tried platinum tags attached to the rear margin of the dorsal fin with platinum wire, and retention was improved;[32] however, returns were disappointing. In the fall of 1880, Atkins tagged 274 fish. He offered a two-dollar reward and got back tags from four females and a male between June 20 and 23, 1882. The fish had increased in weight from 45 to 127% between the time of release and recapture.

Following early successes with migratory Atlantic salmon, problems in obtaining eggs occurred, prompting Baird to suspend spawning and hatchery operations in 1877 and 1878 so the Commission could evaluate whether its activities were having an impact. In his report for 1878, Baird wrote:[8]

> I am happy to say that during the present year the indications of success have been so unquestionable as to warrant the re-establishment of the Bucksport station, and it is hoped that the result for 1879 will show a good progress in this connection. It may be stated in general terms that nearly every stream on the Atlantic coast as far south as the Susquehanna in which young salmon were introduced as far back as 1874 and 1875 has proved to contain adult spawning fish in 1878.
>
> . . . in the Merrimack . . . salmon of late years have only been seen at very rare intervals. As the result of the action of the commissioners of Maine and New Hampshire, large numbers of salmon were observed while ascending the fishway in the dam at Lowell for the purpose of performing the function of spawning in the headwaters of the rivers, . . . where many young were afterward seen.

The Schoodic salmon was much sought after by commissioners in the various states, as well as by fishing clubs and fish culturists.[8] The Commission preference for Schoodic salmon was based on ease of collection and general hardiness of the fish. The Schoodic was said to be adaptable to a number of habitat types. Baird decided to establish a hatchery for Schoodic salmon on Grand Lake Stream, near Calais, Maine. Charles Atkins was superintendent of the facility, that is, in addition to his activities at Bucksport. Those duties apparently weren't enough to keep Atkins sufficiently occupied, because he was asked in 1878 by Baird to look into establishing a spawning facility for the Sebago Pond variety of the Schoodic salmon. The impetus for that activity was a request from Mr. E. M. Stilwell, a Maine fish commissioner who lauded the Sebago's large size as making it a fish of interest. Baird described what transpired:[8]

> Unsuccessful efforts were made some years ago to obtain spawning fish from Sebago Pond, for the purpose of securing their eggs. It was imagined that, owing to the protection afforded by recent legislation and the removal of certain obstructions in the water, a new effort might be more satisfactory. Acting on this impression, Mr. Atkins was directed to establish a station, for the purpose of an experiment, which he accordingly did, leaving Mr. Buck, one of his assistants of long experience, in charge. After giving the matter a fair trial, the enterprise was abandoned, as, with all the devices in the way of nets, &c., only ten males and six females were captured, and the entire number of fish entering the river for the purpose of spawning was estimated at scarcely more than 50.

The take of Penobscot salmon in 1879 was a disappointment to Atkins.[33] Only 200,500 fish were distributed. Unusually high mortality of adults was attributed to damage sustained during transit to the holding pond and to subsequent bad weather. The report on Schoodic salmon production for 1879 was a good bit better. Over 1,100,000 eggs were taken, 75% of which hatched into fry that were released.[34]

Atlantic salmon operations were well established by the end of the 1880s. Atkins's facilities operated smoothly, and large numbers of fish were being distributed to a variety of recipients. After the Schoodic salmon eggs obtained in the fall of 1881 hatched, deliveries were made in California, Connecticut, Indiana, Iowa, Maine, Maryland, Massachusetts, Michigan, Minnesota, Missouri, New Hampshire, New Jersey, New York, Ohio, Pennsylvania, Tennessee, and Vermont.[35] All the spawners were obtained by the hatchery workers. Over two and a half million Penobscot eggs were taken in 1881.[36] Federal facilities received about a third of the eggs that were shipped, along with the states of Connecticut, Maine, Massachusetts, Minnesota, New Hampshire, New York, and Pennsylvania.

There were some mortalities in the holding enclosure for Penobscots. Fish that ultimately died were observed to have opaque eyes that subsequently swelled and burst. Yet Atkins felt things were running in normal fashion: "The routine work of the season went on with so little novelty that there is not much to report beyond the summaries of work accomplished."[36] That statement is in stark contrast to what was happening during the same period on the West Coast. The work of Livingston Stone and his assistants was anything but routine during the first years of operation. Even the wreck of the aquarium car was an event that was not all that unusual by comparison.

Pacific Salmon. On August 1, 1872, soon after being recruited to head up Pacific salmon spawning operations, Livingston Stone headed for California. His challenge was somewhat greater than that of Charles Atkins, because, as Baird admitted, little was known about Pacific salmon:[6]

> The experiment was of course uncertain, in the entire absence of any reliable information bearing upon the natural history of the species. It was not even known at what period they spawned, although Mr. Stone was assured by professed experts, on his arrival in California, that this occurs late in the month of September.

Scientists recognized that there were more than one species of salmon on the Pacific coast, but felt that there was only one species in California.

State fisheries officials and the president of the California Fish-Culturists' Association met with Stone and instructed him to select a location on the Sacramento or Columbia River. Stone learned from Robert B. Redding, the California fish commissioner, that salmon were being speared by Indians on the McCloud River, a stream in the Sierra Nevada range that emptied into the Pitt River some 320 miles north of San Francisco. It was to the McCloud that Stone proceeded, accompanied by Mr. John G. Woodbury. Upon arrival they immediately began constructing a hatching facility. The fact that the McCloud Station was 4 miles from the nearest road and was in an area where summer daytime temperatures often exceeded 100°F presented unique challenges.[6] Those were not, however, the only problems that Stone faced during the early years on the McCloud River:[37]

> He not only had to be a fish culturist, a biologist, a linguist, and a prophet; he had to be a pioneer—pioneer in the fish field and pioneer of the West. The West had not yet been

won and as soon as one set foot outside the newly populated centers he came in direct contact with the wild and woolly.

Stone tried to communicate with the local Indians with the idea of hiring some of them to assist in his activities. He had been assured that obtaining the services of these Native Americans would be easy, but was frustrated during 1872.

By literally working day and night, a spawning station was established, and while the spawning run was almost over by the time the hatching troughs were ready, and many of the eggs collected were destroyed by the extreme heat, some 30,000 were ultimately packed in moss and sent east.[6] By the time the eggs reached New Jersey, many had already hatched. The 7,000 that had not were placed in a hatchery, and the fry were maintained until March 1873. Some 5,000–6,000 fingerlings were ultimately stocked in the Susquehanna River.

Stone's first year taught him that he would have to open the McCloud River facility by the 20th of July, catch and confine the salmon he would need for spawning prior to August 20, and arrange to supply the hatchery with river water that was of suitable temperature.

You will recall that Stone was involved in a train wreck during 1873 and that he returned to the East Coast to pick up a load of shad that were delivered to California and stocked in the Sacramento River on July 2, without incident.[27] He continued by train to Redding, California, and from there proceeded by stagecoach to the McCloud River Station a few miles from the hatching station.

The camp was in good order, which was a bit of a surprise to Stone since it was in Indian country. Some lumber was missing, but that was used in an "emergency" by an agent of the California and Oregon Stage Company who quickly settled the bill with Stone. Stone went back to San Francisco, where he hired two fishermen to assist in capturing fish. The McCloud River Station was fully complemented by early August with Stone accompanied by[27]

> . . . John G. Woodbury, foreman; Myron Green, head-fisherman; Oliver Anderson, man of all work; George Allen, carpenter; Benjamin Eaton, steward; A. Leschinsky, fisherman; J. Leschinsky, fisherman; . . . [and] Indians, Lame Ben, Uncle John, One-eyed Jim, and others.

Stone had obviously found a means to communicate with the locals, and was able to solicit help from some of them, though as he reported, harmony did not prevail:[27]

> Our attempt to locate a camp on the river-bank was received by the Indians with furious and threatening demonstrations. They had until this time succeeded in keeping white men from their river, with the exception of one settler, a Mr. Crooks, whom they murdered a few weeks after I arrived. Their success thus far in keeping white men off had given them a good deal of assurance, and they evidently entertained the belief that they should continue, like their ancestors before them, to keep the McCloud River from being desecrated by the presence of the white man.

Confrontation with the Indians led to some anxious moments:[27]

> They assembled in force, with their bows and arrows, on the opposite bank of the river, and spent the whole day in resentful demonstrations, or, as Mr. Woodbury expressed it,

in trying to drive us off. Had they thought they could succeed in driving us off with impunity to themselves, they undoubtedly would have done so, and have hesitated at nothing to accomplish their object; but the terrible punishments which they have suffered from the hands of the whites for past misdeeds are too vivid in their memories to allow them to attempt any open or punishable violence.

Once Stone had established a relationship with the Indians, individuals from the tribe told him that he was stealing their salmon and occupying their land.

Stone related a conversation that he had with one of the Indians, during which he was told that ". . . white men had lands and fish in other places, that the Indians did not go there and steal their lands and salmon, and that white men ought not to come here and take what belonged to the Indians."[27] Stone admitted that the Indian's arguments seemed sound. He told the Indian that he was a friend and did not mean to take the land or the salmon and only wanted to spawn the fish. He indicated his intention to freely give the Indians the fish after taking the milt and roe. That apparently didn't satisfy the Indian, but when the hatchery men actually began giving the spawned-out fish to the tribe, relationships improved.

Temperature problems with the water supply used in 1872 led Stone and his coworkers to construct a paddle wheel equipped with buckets that lifted water into the hatchery building. The original design was subsequently modified with respect to both the channel in which the wheel was placed (trees were felled and rocks blasted) and the wheel itself (every other bucket had to be removed because the amount of water to be raised weighed too much for the wheel to turn in the available current). The wheel, when it finally began to function, raised 1,080 pounds of water 10 feet every minute.[27]

To remove turbidity, the water was passed through filter boxes equipped with screens and flannel. The hatching troughs were 16 feet long and were placed in two parallel series (10 troughs per series with a several-inch fall between the outfall of one and the inlet of the next).

The men were ready for fish by the August 19. The fish were kept in confinement while the 30 × 60 foot hatchery tent and paddle-wheel construction projects proceeded. Egg taking began on August 26, and by September 22, an estimated 2 million eggs had been put in hatching troughs. In all, about 1,000 salmon were caught, varying from 1/2 pound to 29 pounds. Stone described the hatching troughs:[27]

> Most of the eggs rested on the charcoal bottom of the troughs; but I used trays to a considerable extent formed of iron-wire netting; coated with asphaltum, and found them satisfactory for maturing eggs in for shipment, though I do not think fish hatched in the asphaltum troughs are as healthy as those hatched in the charcoal troughs.

With Seth Green's permission, Stone also tried shad-hatching boxes as a means of incubating salmon eggs. They worked well, but egg picking was not as simple as when the eggs were in troughs.* Stone attributed egg losses to

---

*Dead salmon eggs become opaque and are white in color. If not removed, they become covered with a fungus that will then attack live eggs. Egg picking is the process by which the dead eggs are removed and discarded.

- suffocation
- exposure to direct sunlight
- exposure to diffused sunlight
- inherent causes (infertile eggs, eggs from dead fish)
- excessive agitation (a problem early in the incubation process)
- lack of fertilization ("want of impregnation")

On September 20, Stone shipped 300,000 eggs to the east coast. Another 600,000 along with Stone, left the McCloud River Station on September 30. A third lot of 250,000 and a fourth lot of about the same number were sent in October. The eggs were packed for shipment in moss and crated in 2 × 2 foot boxes 1 foot deep. Each box contained a wooden partition and 75,000 eggs. The cost of obtaining the 1,500,000 eggs and preparing them for shipment was about $4,000.[27]

Stone was back on the McCloud River in 1874. His party consisted of nine white men, including a secretary and photographer. A Chinese cook, Ah Sing, was hired in San Francisco, and there were various numbers of Indian workers employed at different times during the spawning and hatching season.[38]

A total of 5,752,500 eggs were taken. They were packed in boxes and carried by wagon to Redding, California, where they were put on a train. At a stop in Sacramento, Stone bought some necessary supplies for the trip east. Those supplies included a pail, a dipper, and a thermometer. Procuring ice seemed to be a recurring problem. When the train arrived in Omaha on a Sunday night, Stone couldn't get much help in obtaining ice since the only agent at the station was intoxicated.

A few additional people had moved into the vicinity of the McCloud Station between the 1873 and 1874 seasons. The new neighbors included George Allen and his wife, who maintained the stage station a mile from the camp, Mr. O'Conner, who was the ferryman at a river crossing 4 miles downstream, and two others: Dr. Silverthorne and Mr. Campbell, each living a few miles away and both with Indian wives. Two Indian chiefs, named Concholooloo and Jim Mitchell, also lived nearby.

Some of Stone's more interesting journal entries included the following:[38]

July 9	Visit from Conchoolooloo, the Indian chief. Mercury in thermometer, in the sun, 159°—in the sand near the house. Chinaman very sick.
July 10	Mr. Woodbury killed a rattlesnake, making seven that have been killed in the neighborhood this summer.
July 11	The Chinaman went out in the boat and was carried over the rapids but not injured.
July 12	Unpleasantness between the Chinaman and Indians.
July 21	Rattlesnake was killed opposite the house. Twenty minutes exposure to the rays of the sun this afternoon cooked an egg.
July 22	Blew up rocks in the river-channel, below the wheel, with giant powder.
July 24	Two rattlesnakes were encountered.
July 27	A little gold-digging was done to-day, and some gold found.
August 10	The dam across the McCloud River, obstructing the ascent of the salmon, was completed to-day.
August 18	An Indian woman came to the camp for protection, being pursued by an Indian, whose brother she had killed.

| August 19 | The Indian in pursuit arrived in camp this morning, armed with a six-shooter. Danger of another murder. The Indian, after some flourishing of his revolver, was peremptorily ordered to leave the camp, which he did. |
| August 30 | Another rattlesnake killed. |

In a presentation to the American Fish Culturists' Association in 1874, Stone provided additional insight about day-to-day life at the McCloud Station. With respect to the Indians,[39]

We found them very serviceable in assisting about our work, although they were provokingly freakish. When they worked they worked well, but when they did not want to work, they were as obstinate as mules or as alevins—those who are accustomed to hatching fish will appreciate this last allusion, I know—and then they would not lift a hand to help us, however urgent the circumstances might be. I employed them to help run the seine, to chop wood, to cook, to build dams, to work in the water, to pick out dead eggs, and to do various odd jobs.

Stone began to develop a dictionary of Indian words once he began to communicate with local tribal members. The following is a sample:[40,41]

Chaark	Male salmon	*Poo-oop*	Salmon eggs
Chil-chilch	Bird	*Poo-tar*	Grandmother
Cow	Cow	*See-ee*	Teeth
Déek-et	Fish	*See-okoos*	To brush
Horse	Horse	*Shoohoo*	Dog
Kaáy-ell	Spear	*Syee-oolott*	Trout
Kelly-kelly	Knife	*Too-too*	Mother
Khlark	Rattlesnake	*Tu-lich*	To swim
Ko-lool	Bow, gun	*Weh!*	Come here!
Ko-raisch	Female salmon	*Win!*	Look!
Noo-oohl	Salmon	*Wintoon*	Indian
Nott	Arrow	*Woor-ous*	Fish-spawn
Pahn-ee-tus	Handkerchief	*Yi-patoo*	White man
Péss-sûs	Money	*Yorkos*	Gold

The same word was used for both "bow" and "gun," but the word could not have been a general term for weapon since "arrow," "spear," and "knife" were quite different. It is likely that the word *ko-lool* referred to the implement that launched a projectile. Stone recognized that several words in the Indian language had come from the Spanish,[40] but seemed to miss *péss-sûs* (looks pretty much like *pesos*), which he did not attribute to Spanish. It is reasonable that the Native Americans adopted words like *horse* and *cow* since both were introduced species.

Stone commented about the local scene:[39]

We frequently saw emigrant wagons dragging wearily along, some going from California to Oregon, and some the reverse, both hoping to make a change for the better. Twice every twenty-four hours the Oregon stage with its six galloping horses made its fast time over the stage road on the hills above us, carrying the mail from San Francisco, California, to Portland, Oregon, and back.

The food in camp was also discussed by Stone:[39]

Our table was usually supplied with venison, trout and salmon grilse; the small grilse of the fall run generally being good eating. We also had occasionally quails, squirrels, rabbits and fresh vegetables. Our staples to fall back upon when in want of something better were bacon, potatoes and baked beans. We had no domestic meat whatever.

In only a few years the McCloud station developed into a major egg-taking enterprise. Some 30,000 salmon eggs were collected and shipped East in 1872, but by 1876, nearly 9,000,000 eggs were taken. Stone reported to the American Fish Culturists' Association meeting of 1877 that since the McCloud station had been established, egg collection had reached 25 million. While most of the eggs were sent East, a portion were restocked in the headwaters of the Sacramento River drainage.[22]

Problems faced by Stone and his colleagues during 1878 included flooding that carried off some of the station's buildings.[8] Baird arranged for an allocation of $2,500 to allow Stone to restore the station. The loss of buildings apparently didn't inconvenience Stone too significantly, since the number of eggs taken that year reached a prodigious 14 million.

The Baird Post Office was established on the McCloud River on May 3, 1878. Authorized by the Postmaster General, and undoubtedly pushed for by Stone to honor Commissioner Baird, the Baird Post Office provided workers at the McCloud River Station and their neighbors with U.S. mail service. Previously, the nearest post office was located 22 miles away in Redding, California.[8]

Salmon canners on the Columbia River had expressed concern that the supply of fish was diminishing. As a result, egg-taking stations were established on the Clackamas and Rogue Rivers, Oregon Territory, in 1877.[18,42,43] Egg take on the Clackamas River that year was disappointing since the effort did not begin until the spawning season was nearly over,[8] but the next year a considerable number of fish were spawned.[42] The Rogue River hatchery was established by Mr. R. D. Hume, who had established a cannery at Ellensburgh. About 100,000 fry were stocked during the first year of hatchery operation.[43]

Eggs that were shipped East often ended up in Maryland and Michigan hatcheries from which fry were distributed. Because fry distribution often involved shipping the fish long distances, Baird began searching for a location in the southern states where a salmon hatchery could be established. On the surface it would appear that establishing a salmon hatchery in the South would be nonsensical. One might say, "if salmon were supposed to live in the southern United States, there would be salmon existing there." Baird wasn't a fool. He indicated that a suitable site would have to provide ". . . an ample supply of pure spring water of a temperature as much under 60° as possible; . . . and [be] convenient [in] relation to a railroad center from which the fish can be distributed. . . ."[8] In 1878, potential sites were located in Alabama, Mississippi, and Tennessee. A yellow fever outbreak in the South prohibited action being taken with respect to hatchery development that year. It doesn't appear, in fact, as though the plan was ever implemented.

The hatchery crew that worked at the McCloud River Station during the 1877–1878 season was faced with a significant flooding problem as a result of January and February rains, which caused the river to rise nearly 15 feet above its normal summer level. Heroic efforts by the men helped save most of the buildings:[44]

> During all the time of the high water, the men in charge, viz, Myron Green, Patrick Riley, and J. A. Richardson, together with four or five Indians who helped them, worked with great resolution and courage. During the whole of two days and one night they were in the water, sometimes up to their necks, and often in danger of their lives, guiding drift-wood so that it would pass through the fishery premises with the least danger. They worked so persistently and skillfully that the houses were saved, but everything else was swept away. All the flumes, chicken-coops, door-steps, hatching-troughs, filter-tanks, everything that was on the ground that would float, was carried off.

Baird was able to provide $2,500 for repairs.[8] In May 1878, Stone arrived on the McCloud to initiate repairs and oversee construction of a structure he referred to as a "rack," which was built across the river to impede the passage of salmon upstream, thereby making them more accessible for capture.[44]

In his report to the Commission for 1878, Stone mentioned a few incidents that provide additional insight into camp routine:[44]

> On Sunday, May 26 . . . [at about] midnight we were awakened by the dogs barking violently in the direction of the hill behind the house. Upon sending them out to see what was the matter, they went about ten rods to some thick brush, and returned yelping. At the same time we could distinctly hear stones being thrown at them. It was dark. There was only one man in the house besides myself, and we only had one gun between us. With the exception of the hostler at the stage station, a mile distant, there was not a white man within three miles. We were in a country which we knew was often frequented by desperadoes, and where the stage had been robbed six times in a month, and where murders are not of unfrequent occurrence. It might be only one or two burglars in the bushes, but how did we know that they were not a gang of cut-throats . . . ?

And a week later:

> About nine o'clock one evening we heard a great deal of noise, accompanied with some quarreling among the Indians about a quarter of a mile below the house. The noise continuing, two of our men started down the road . . . and on arriving at the fishery station found one or two men engaged in robbing a teamster who was stopping for the night. One or two shots were fired by our party, but the robbers escaped. We found, however, that the rascals had not only robbed the teamster of his money, but had also taken from his wagon twenty demijohns of whisky, which they had distributed indiscriminately among the Indians. The result was such as no one can realize who has not been in an Indian country. The Indians were all more or less intoxicated, were very noisy and quarrelsome, and were inciting each other to make a descent on the fishery, and, as they expressed it, "to sweep it clean with the ground." Our men . . . armed themselves for the occasion and determined to give chase . . . that very night. They found them about daylight at an Indian lodge, and placing the muzzles of their revolvers close to the robbers' heads, they captured them without resistance. One is

now in the State's prison. . . . The other was discharged for lack of sufficient proof of his guilt.

The fish began showing up on July 11. The week before that, Stone was approached by an Indian named Chicken Charlie, who wanted a coffin made for his father. Things didn't quite work out as planned, however:[44]

> We made the coffin, and after a while, when they supposed the Indian was dead, they put him in the coffin and proceeded to bury him; but before they had finished burying him he came to life again, and they took him out and waited a while longer. The next time he really died, and the following day he was buried over again.

Stone's rack required repair soon after the salmon began arriving. It seems as though the persistent jumping of the fish against the structure eventually weakened it and provided a place where the fish could escape. Stone assigned some of the Indians to effect repairs, and he praised them for their ability to work in the cold water and complete a job that he felt would have been ". . . difficult for white men, unless experienced divers. . . ."[44]

The temperature reportedly reached 149°F in the sun at 4:00 P.M. on July 26, though that was a bit cooler than on July 22, 1875, when Stone had recorded 153°F.[44] On July 29, there was a solar eclipse, which Stone said the Indians described as the grizzly bear eating the sun. Stone didn't mention whether the temperature was moderated by the event.

Things were otherwise fairly routine, except for the Indian scare. It seems as though reports were circulating that a large number of Indians from farther north were massing in the McCloud River area and were contemplating an attack on Copper City. Another rumor had it that Pit River Indians were acting in a hostile manner. A messenger from Copper City showed up at Stone's camp to ask about the possible attack. Stone sent back a message that he didn't think there was any danger. He might have acted a bit hastily.[44]

> The next morning, however, an Indian squaw told us that the Yreka and Upper Sacramento Indians were coming down to the McCloud to kill the McCloud Indians and what white men there were on the river, meaning ourselves. . . . We heard farther that Outlaw Dick, who murdered George Crooks here in 1873, and Captain Alexander, an Indian of very warlike disposition, had urged the northern Indians at a recent council to make a descent upon the McCloud and "clean out," as they expressed it, all the white men and McCloud Indians on the river. To add to the excitement, a Piute chief had visited our Indians the past week to stir them up to make war on the whites.

Having only one rifle available and with the reports of possible attack growing daily, Stone telegraphed for additional arms and ammunition:

> The excitement, however, gradually died away. The Piute chief returned to his own tribe; the Oregon Indians began to surrender and come in to deliver themselves up to the soldiers; the McCloud Indians recovered from their alarm, and . . . Captain Alexander and his Indians had changed their minds and were not coming.

Incidentally, the egg collecting in 1878 went pretty well. Stone estimated that there were as many as 14 million eggs in the hatchery when operations were suspended. Had there been sufficient demand, he estimated that another 4 million or more could have been collected.

Another Indian uprising was anticipated in 1879, though none materialized. Eggs taken that year were distributed to the following places:[45]

Iowa	Kansas	Maryland	Minnesota
Nebraska	New Jersey	New York	North Carolina
Ohio	Pennsylvania	Utah	Virginia
West Virginia	Wisconsin	Canada	France
Germany	Netherlands	New South Wales	

The fact that Pacific salmon die after spawning was apparently not completely understood in 1879. Baird wrote the following:[46]

> As the custom prevails of turning fish back into the river after the eggs have been taken, this device of obstructing the river has no doubt been beneficial in preserving many of the adult fish which would otherwise have died from the exhaustion conse-quent upon any further ascent of the river.

Baird's comments were undoubtedly influenced by a letter he received from Horace Dunn:[47]

> No doubt great numbers [of Sacramento salmon die after spawning], but a very large portion of the run return to sea again, as before the close[d] season between August 1, and November 1, was established it was a common occurrence to find spent salmon in market between the dates named.

Dunn apparently had the opinion that the fish that he saw in the markets, though in obviously terrible shape, were on their way back out to sea where they would recover. He thought the condition of the fish was associated with rough handling, not natural processes:[47]

> These salmon [the ones in the markets] were very much emaciated, had no scales, and varied in color from a rusty black on the backs to a faded brown on the belly. Some were of a dirty white color all over, as if they had been parboiled.
>
> In regard to the quality of the Sacramento salmon, I think they compare favorably with those caught in Maine. The mode of treatment here of salmon is simply barbarous. The fish are caught in drift-nets. . . . They, as a rule, lie in a boat for several hours exposed to the sun before being brought to the steamer's wharf. There they lie in large heaps for several hours more, and are dragged on board and put in large heaps again. At San Francisco the fish are dragged ashore and roughly thrown into wagons, and on arrival at the markets experience the same treatment again. . . .

Dunn stated that in earlier published reports Livingston Stone had indicated that all the salmon in the Sacramento River died after spawning (an observation that was correct). In responding to Dunn's letter, Stone indicated that he had been misquoted: ". . . I did not say that *all* the Sacramento salmon die after spawning, but limited my statement to the salmon of the McCloud River."[47] Stone could have extended his observation to include all the salmon not only in the Sacramento River but throughout the North Pacific. Many of the fish Dunn observed in the markets were obviously moribund when captured. Given the condition of the fish when they reached the market, one can only imagine that they were tasty and undoubtedly brought a high price.

While flooding in 1878 had caused problems for the Commission's hatchery crew at the McCloud River salmon station, it was nothing compared with the problems that occurred in 1881. January brought unprecedented rainfall to northern California. Stone cited a weather observer, one James E. Isaacs at Shasta, who had recorded 47 inches in January and 17.5 inches in February. The result was that by the evening of February 2, 1881, the McCloud River was over 16 feet above normal and rising. Livingston Stone reported the result of the subsequent flood:[48]

> When the day dawned [February 3] nothing was to be seen of the main structures which composed the United States salmon-breeding station on the McCloud River. The mess-house, where the workmen had eaten and slept for nine successive seasons, and which contained the original cabin, 12 feet by 14 feet, where the pioneers of the United States Fish Commission on this coast lived during the first season of 1872; the hatching-house, which, with the tents that preceded it, had turned out 70,000,000 salmon eggs, the distribution of which had reached from New Zealand to St. Petersburg; the large dwelling-house, to which improvements and conveniences had been added each year for five years— these were all gone, every vestige of them, and nothing was to be seen in the direction where they stood except the wreck of the faithful wheel which through summer's sun and winter's rain had poured 100,000,000 gallons of water over the salmon eggs in the hatchery, and which now lay dismantled and ruined. . . .

The river ultimately reached 26 feet, 8 inches above normal summer level. Stone estimated that it had been two centuries since the river had reached that level of flooding. He came to that conclusion because a previously undisturbed Indian graveyard that had been in existence for at least 200 years was impacted by the flood—a third of the graveyard was destroyed.

The loss was financially devastating:[48]

> The inventory showed that over $4,000 worth of hatching apparatus, house furniture, tools, and other articles were lost or destroyed by the flood, besides the buildings themselves. The whole loss could not have been less than $15,000.

Myron Green was dispatched on February 6 to telegraph Spencer Baird and request help. It took Green three days to go 25 miles because of the flooding. After telegraphing Baird he sent another telegraph to B. B. Redding (the California fish commissioner) in San Francisco and asked that Redding telegraph Senator Booth for an appropriation to rebuild the hatchery. Booth was able to obtain $10,000 for that purpose.

Stone showed up on the site on May 19. Some 30,000 board feet of lumber had been delivered, and there were over 20 whites and a dozen Indians present to rebuild the facility. Stone was faced with a $15,000-or-more costly job to be accomplished with only $10,000. He decided to complete the work and trust that Congress would make up the difference in funds.

While construction was progressing, Stone noted unusual mortalities within the salmon population beginning in late June. He examined dying fish and saw no external lesions or other signs of disease. Internally, the viscera were[48]

> . . . very much congested with dark blood, and the spleen was very much enlarged. Later in the season, those that I examined all had unhealthy gills. The gills in these cases were very much abraded on the outer edges, and were almost stuck together by slimy or gummy substance. . . .

He sent samples to the National Museum in Washington, D.C., for examination.

Even with the apparent disease problem, large numbers of fish were returning and it was incumbent on the workers to get the facility back into operation as quickly as possible. Stone described the results of their labor:[48]

> On the first of September we had on the fishery grounds a mess-house, hatching-house, and stable. We had also built a bridge 150 feet long across the river, and had added to it as usual a firmly built fence or rack that allowed the river to pass down but prevented the salmon from going up the river.

The 30 × 80 foot hatching house contained 40 hatching troughs, each 16 feet in length, and each containing seven hatching baskets. Each basket could accomodate 35,000 eggs, so the facility could handle up to 10,000,000 salmon eggs each year. The wheel and flume used to carry water to the hatching house was just completed during the last week of August when the first ripe salmon were collected. Stone again:[48]

> This year, on account of the extraordinary abundance of the fish, we frequently had to make but two or three [seine] hauls a day, and even at this rate we took all the eggs needed (7,500,000) before the spawning season was half over. . . .
>
> I may add here that this vast increase in the number of salmon in the river is the direct result of the artificial hatching of young salmon at this place. For several years past the United States Fish Commission has presented to the State of California 2,000,000 salmon eggs or more each year. These eggs the State fish commission has hatched each year at its own expense and has placed the young salmon in tributaries of the Sacramento. This artificial stocking of the river has resulted in a wonderful and wholly unprecedented increase of salmon in this river. . . . [The] annual catch of salmon in the Sacramento River is worth nearly half a million dollars more than it was seven years ago, before hatching operations were resorted to.

Stone was obviously pleased with his efforts, but as we shall see, the large returns of fish to the Sacramento River system were not going to last.

Egg taking ceased on September 8 with 7,500,000 eggs in the hatchery and fish still teeming in the river. Stone estimated that, had they been able to accommodate the eggs, the take could have reached 20 million.

While the eggs sat in the hatchery, the men prepared the shipping boxes. On September 18 the wheel that lifted water to the hatchery began to break up. Stone reported the crew's response:[48]

As soon as the accident was discovered not a moment was lost in establishing a line of buckets from the river to the hatching house to supply water to the eggs. Every white man and Indian that could be pressed into the service was employed, and in less than ten minutes we had three lines, of eight or ten men each, bringing water from the river in buckets, tubs, watering-pots, and anything that could be found, that would hold water.

Stone turned his attention to the wheel and found that seven of the paddles had broken off. He speculated that some debris in the river had precipitated the disaster. Repairs took from 11:00 A.M. one morning until 4:00 A.M. the next, with the bucket brigade continuously in operation. Stone had high praise for the diligence of the Indians, in particular. The men worked 17 hours straight,[48]

. . . with two very short interruptions, when I allowed them, three at a time, to run to the house to get something to eat. During all this [sic] seventeen hours some of them were carrying buckets of water that weighed sixty or seventy pounds each.

Things returned to normal, and beginning on September 24, egg packing for shipment was initiated. A railroad car containing 3,600,000 eggs packed in boxes with moss left California on September 28 and arrived in Chicago on October 3. From there the eggs were distributed to the states of Maryland, Minnesota, Nebraska, New Hampshire, Nevada (note the wrong direction), Pennsylvania, South Carolina, West Virginia, and New Jersey and to Ontario, Canada. Some 50,000 eggs were sent to New South Wales, Australia. The remainder stayed in California.

The following year (1882) was much less traumatic on the McCloud River, or more correctly, at the Baird salmon station. Demand was for 4 million fish that year, and the goal was met between September 2 and 25, which was good because after that the number of returning fish decreased considerably and Stone didn't think he could have met a target of 8 or 9 million eggs.

Stone reported on the contents of a letter written by B. B. Redding, just before his death, to Spencer F. Baird. Stone said Redding indicated[49]

. . . that several hundred thousand dollars had been invested in cannaries [sic] on the Sacramento River, that 1,600 men were employed in these canneries, and that this capital and these men would be ultimately thrown out of employment if the salmon hatching at this station [on the McCloud River] should be given up. He also stated that the hatching of salmon here had increased the annual salmon catch of the Sacramento 5,000,000 pounds in a year, and that the canneries on the river were dependent upon the salmon hatching of this station for their maintenance.

Frequent cross-country trips (one featuring a train wreck), floods, and Indian uprisings apparently weren't enough for Rev. Livingston Stone (or perhaps it wasn't enough for Spencer F. Baird), because the good reverend was also charged with spawning and hatching trout on the McCloud River. We'll look at the trout-rearing activities there and elsewhere later, but for the moment, it seems appropriate to let Livingston Stone rest awhile.

The Carp Saga

Commissioner Baird was a strong proponent of the introduction of common (also known as European) carp to the waters of the United States. He believed that the fish was one that had been domesticated for thousands of years and that it was a staple in the diet of the Chinese. He was also convinced that common carp were strict herbivores and proposed introducing them to the warm-water regions of the United States, particularly ". . . mill ponds and sluggish rivers and ditches of the South."[19]

The reality of the situation was that the various species of carp that have, indeed, been raised in China for some 4,000 years are quite different from the common carp. The original range of common carp was probably restricted to the watersheds of the Aral, Black, and Caspian Seas.[50] It was introduced into Greece and Italy when the Roman Empire was dominant. Europeans were familiar with the fish and there was undoubtedly a good deal of sentiment in favor of introducing carp to the United States by immigrants from both eastern and western Europe.

While Baird is often credited, or perhaps blamed, for the successful introduction to and propagation of common carp in the United States, he was not the first to do so, though he was the most vocal advocate of the idea. The history of attempts that preceded those of Baird is a bit cloudy.[51] Several references credited Captain Henry Robinson of Newburgh, New York, with the first introduction in 1831 and 1832. Supposedly, some of the fish escaped from his pond, creating a carp fishery in the Hudson River. However, when Professor Baird examined some of the fish, he was convinced that they were actually goldfish.

Robert Poppe described a successful introduction of carp into California by a Mr. J. A. Poppe (a relation perhaps?) in 1872. J. A. Poppe went to Germany and obtained 83 carp of various sizes.[52] Only the smallest survived the trip, which involved travel by steamship from Europe to New York and by rail from New York to California. A total of five, reportedly nearly dead, carp were ultimately stocked in Mr. Poppe's ponds in the Sonoma Valley. The surviving fish reproduced the following year and Mr. Poppe began selling the offspring to his neighbors. The efforts of Mr. Poppe were soon to be overshadowed by the U.S. Commission on Fish and Fisheries.

Carp became a focus of attention for Baird almost immediately after the Commission was established: "Sufficient attention has not been paid in the United States to the introduction of the European carp as a food-fish, and yet it is quite safe to say that there is no other species that promises so great a return in limited waters."[6]

Baird was apparently convinced that the carp was an excellent foodfish:[19]

> After considerable inquiry and investigation we are disposed to believe that there are varieties of the European carp of superior value because of their table qualities, and that the idea entertained by many that the carp is a very inferior food-fish has arisen from the testimony of those who have been so unfortunate as to have eaten only those of inferior quality.

Baird also acknowledged that there were some types of carp that were not good to eat, but was assured that the fish he wanted to import did not have that problem:[6]

> Mr. [Rudolph] Hessel informs me that there is the greatest imaginable difference in the taste of the so-called carp in the European ponds, and that a species very closely

allied to the carp . . . differs from it in the greater abundance of bones and its muddy flavor.

Not everyone agreed with Baird, of course. Among those who were not convinced was Seth Green: "A good deal has been said about the carp for stocking some of our waters; but if I am rightfully informed it is a coarse fish, about equal to our mullet."[11]

Baird discussed various "species" of carp in Europe, including what he called hybrids. One of the more interesting mentioned was the "naked carp" (undoubtedly what is now called the mirror carp, a strain that has only a row or two of scales. It is worth violating my policy of avoiding scientific names to note that Baird classified the naked carp as *Cyprinus nudus*.[6,19]

As evidence that carp was a desirable fish for introduction, Baird listed the following positive characteristics:[19]

1. Fecundity and adaptability to the process of artificial propagation
2. Living largely on a vegetable diet
3. Hardiness in all stages of growth
4. Adaptability to conditions unfavorable to any equally palatable American fish and to very varied climates
5. Rapid growth
6. Harmlessness in its relations to other fishes
7. Ability to populate waters to their greatest extent
8. Good table qualities

The desire of Baird to bring carp into the United States under the banner of the U.S. Commission of Fish and Fisheries was fulfilled in 1877 when Mr. Rudolph Hessel delivered more than 300 carp to ponds at the Druid Hill Park hatchery in Baltimore, Maryland.[8,18] Baird wanted to place a portion of the fish in Washington, D.C., and arranged for a site that already had on it two small lakes, known as the Babcock Lakes. The location was on a site called Monument Lot,[8] and the facility became known as the Monument Station.[46] Congress approved $5,000 (later supplemented by a second appropriation of $2,400) for pond renovation and other modifications. Two ponds, each 6 acres in area, were created.

In the spring of 1878, two-thirds of the Druid Hill Park carp were moved to the nation's Capitol. Those remaining in Maryland spawned, but apparently hybridized with goldfish, so the offspring were destroyed. In 1879, a total of 6,000 fry were distributed from the Monument Station ponds. Nearly 3,000 of them were spread around Maryland with the remainder being distributed in lots of 12 to 16 to various applicants.

Having carp in Washington, D.C., undoubtedly helped Baird promote the value of the species and provided an additional point of interest for visitors. Islands were constructed in the ponds, and water lilies were planted to provide additional amenities. The public was also accommodated during winter.[29] When sufficient ice cover was present, the public was allowed to skate on one of the ponds.[29,46] The place was a popular attraction. In 1882 the President and several members of Congress visited during harvesting.[29,53]

April 1879 was a big month for the Monument Station, as a telephone was installed. Baird reported that the importance of the telephone was that it gave ". . .

the superintendent and watchmen the means of instantaneous communication with the offices of the Commission and with the police headquarters, this latter advantage greatly adding to the safety of the property."[46]

Baird wanted to obtain an additional supply of carp and arranged with Dr. Otto Finsch, a German naturalist who was planning a trip to the United States, to bring along 100 fish.[46] The fish were transported in lots of 25 packed into "coal-oil barrels." Even though Finsch carefully monitored the fish, changed the water, aerated the barrels, and added ice to keep the water cool, only 23 fish survived the trip, which began in Hamburg on April 23 and ended in New York on May 6.[54] The fish were taken to Washington and introduced into the Monument Station ponds.[46]

Applications for carp had been solicited even before the first fish arrived from Europe. Mr. B. B. Redding, who (you will remember) was the fish commissioner for the state of California, was the first to place his order. His request was made in 1876 and filled from among the first 6,000 fish that were produced in 1879.[29] The following year,[55] 66,165 young carp were reportedly distributed.* The number of applications increased exponentially from 3 in 1876 to 20 in 1877, 98 in 1878, 324 in 1879,[55] and nearly 10,000 in 1882.[29]

Meeting the demand during the 1881 season required the distribution of 160,000 carp over a broad geographic area. Shipping was initially in ". . . wood-bound tin cans, holding about eight gallons of water, and making a shipping package weighing about 65 pounds."[56] The shipping companies accommodated the Commission by arranging for reduced rates (with fees paid by the applicants) ranging from $1 to $12 depending on distance shipped. Later in the year, smaller cans were used and the price of shipping was reduced. Recipients of fish could keep the cans for a price or return them. (The shipping companies provided free return freight.)

Experiments were conducted with the objectives of finding more optimum-sized shipping containers; the type that was settled on was a ". . . covered tin bucket having a capacity of 4 quarts. For facility of aeration several holes were punched in the cover of each bucket."[56]

Mr. S. C. Brown devised a shipping tag with the name and address of the recipient on one side and directions on the other. The tag also contained a postal receipt to be filled out and returned by the recipient. Buckets were returned by the applicants. The small buckets were used for shipments within 500 miles of Washington, D.C. Those shipments were about 8 pounds.

After fish had been delivered to the deep South, Marshall McDonald accompanied the fish into the mid South and Indian Territory (Oklahoma had yet to obtain statehood) on hauling car no. 1:[56]

> Texas, with 950 applicants, Arkansas, Indian Territory, Western Louisiana, and Missouri, with an aggregate of 150 widely scattered applicants, still remained to be supplied. . . .

There are discrepancies in the number of carp delivered in 1878 and 1879 when the reports of Baird, Smiley, and McDonald are compared, though the numbers are similar. Smiley[55] presented detailed records that the Commission kept of who received fish, how many they received, and the dates and locations of delivery.

Arrangements were made to have the fish rested and the water changed in Saint Louis. As these methods were novel, and the results considered doubtful by the most experienced messengers, it was thought best that I should accompany the expedition in order to enforce the observance of the necessary conditions of success and to take the responsibility of whatever failure there should be. The compliment of the car was . . . as follows: 40 large cans containing each 100 carp; 7 large cans containing each 150 carp; 18 crates containing each 320 carp; 3 crates containing each 400 carp.

The train left the East Coast with its load of fish and soon arrived in St. Louis, where the water was changed except for the crates that contained 400 fish. McDonald thought the procedure really hadn't been necessary as the fish were in good condition. The train left St. Louis on January 5, 1882, and arrived in Texarkana the next day, where the water was again changed. There were some mortalities in the three crates in which the water hadn't been changed in St. Louis and the fish were weak—water was changed in them (the fish had been in the same water for three and a half days). McDonald gave the fish a 24-hour rest in Texarkana;* then it was on to Sherman, Dallas, and Ft. Worth, Texas.

From Texas, the train entered Indian Territory, where fish were delivered to applicants in the Choctaw Nation. After returning to Dallas and picking up messengers, the train went to Austin via Hearne, Texas, supplying applicants along the way. A 12-hour layover was required in Corsicana to take care of the applicants. In Austin, Mr. R. R. Robertson, the Texas fish commissioner, took responsibility for delivering the fish in the immediate vicinity. Then it was on to San Antonio and Laredo. Of more than 800 applicants, ". . . not more than 7 were unsupplied."[56]

McDonald was particularly impressed with the prospects for carp in Texas: "The State of Texas seems to possess extraordinary facilities for raising carp, and as many of the recipients went to great expense to prepare ponds it is believed that carp-raising will soon become a valuable industry in that State."[56]

The 1882 carp deliveries involved a total of 259,000 fish distributed to 9,872 applicants in 298 of 301 congressional districts and 1,478 counties. The average distance traveled was 916 miles from Washington, D.C. Total mileage from Washington, D.C., to the final destinations was 90,450 miles. Most of those miles were covered by the various messengers employed by the Commission to ride along with the fish that were carried in baggage cars. Railroad Car No. 1 covered 20,601 miles and Car No. 2 traveled 13,901 miles.[57]

In 1883, a questionnaire was sent to recipients of carp.[58] Information requested included when and how many carp were received, a description of the ponds or other types of culture chambers into which the fish were stocked, type(s) of food provided, how many of the second generation were produced, incidence of diseases, and the eating qualities of the fish. Over 1,000 written statements were received. Some of the comments included:[58]

Food
I fed my fish regularly upon corn-bread twice a day . . .

I feed them with bread and truck from the garden—not regularly; only enough to keep them gentle.

---

*That's what he said. I suspect that it was the men that needed the rest, not the fish.

The fish are fed with curdled milk.

I feed them with corn and bread.

From May until winter I feed the carp with garden refuse, bread, rye, and dough. In winter I give them nothing.

I have never fed them and have never seen them since putting them in, but have no reason to suppose but that they are all alive.

I did not feed nor see them for a year.

Little or no attention has been given to them, except to occasionally throw in a handful of stale bread or cake made of corn-meal and flour.

I have daily given them bread made of one-third wheat-bran and two-thirds corn-meal.

I give them refuse fruit, vegetables, bread, corn-meal, and mulberries and blackberries, of which they are very fond. They are also very fond of grits, cow peas, and Irish potatoes cooked.

We fed the fish with cracked Indian corn and with plain corn bread.

I intend to give them a piece of boiled potato in a day or so, which I will slightly mash before putting in.

The carp come for their food every evening like chickens.

We feed them with stale bread about every day.

I give them bread and unground wheat 2 or 3 times a week.

The sound of my voice is sufficient to bring them to the surface of the water, and a whistle causes them to come for food. For this they scamper through the water like so many pigs. They disappear as suddenly at the voice of a stranger.

My little boy, 10 years old, now feeds them every day, and has them so tame that they will come like pigs when he calls and eat out of his hand; fine sport for boys. I give the carp corn, apples, potatoes, and the refuse of the kitchen. They are enormous eaters.

We feed them seldom, but sometimes on cabbage and young corn.

I feed my carp on corn and vegetables. They thrive on any kind of food.

The green scum on the surface of the pond and the large quantity of vegetable matter therein afford the carp a sufficient quantity of food.

I give the carp bread, corn-meal cakes, wheat, and rye.

We feed the carp irregularly with bread, corn, and sometimes cabbage.

I have been feeding them daily . . . with a peck of dough composed of two parts of corn meal and one part of wheat middlings stirred up in sour milk. The young now eat from my hand.

At first we fed the carp on different kinds of bread, and refuse from the kitchen, but we have not paid much attention to them of late.

I have given them corn and wheat bread, cabbage, beef, and worms.

I give the carp baker's bread, oats, and corn-crackers every day.

Having taught my carp to come to a certain place for food, [they came when] their favorite food—cracker crumbs—was thrown to them, till December 9, when they went into winter quarters.

I ascertained from experiment that carp disposed of oatmeal dough and a dough of rye meal mixed with chopped cabbage more quickly than any other kind of food given them. If carp are not fed in December, January, and February, they will go in the mud where they will be safer than were they regularly fed.

I have given the carp cabbage, clams, lettuce, and bread. They are very fond of minnows when scaled, and, in fact, eat almost anything given them.

. . . in late spring, I . . . give them lettuce leaves, turnip tops, and kitchen scraps. Later, I feed cabbage leaves, melon rinds, sweet corn, squash, &c., and in the fall pumpkins, boiled turnips, cabbage, potatoes, &c. They seem especially fond of pumpkins.

I feed the carp with green corn.

Daily I give the carp potatoes, corn, bread, scalded milk, and the refuse from the table in small quantities.

We give the carp bread and cheese daily, and potatoes weekly.

They seem to prefer fish-worms and watermelons to anything else.

I feed the carp largely on stale baker's-bread. And while curd of milk and spoiled cheese are excellent for them, they will eat almost anything that a pig will eat.

I train my carp to come for food at the sound of a bell.

I put in my pond the solid food of the refuse from my table and kitchen. The carp eat and thrive like pigs. The success of carp culture is in knowing how to feed them.

I have fed them with bread, but irregularly on account of the distance of the pond from the house.

I fed them during the winter on boiled rice, corn-meal, and flour . . . baked into cakes. . . . I saw nothing of these fish until May 10, and began to fear that some accident had befallen them, yet I continued to deposit food. I now have them so that they will come to my call and feed right under my feet. They rise to the top of the water and suck in a piece of biscuit as quickly as a hungry dog would devour it.

Once a week I throw in cabbage, lettuce, scraps, &c. In addition we give them a little bread in the evening, when we call them with a bell.

I feed a little corn-bread and a little loaf-bread every evening at 6 o'clock.

I find . . . boiled hominy to be the best food for carp. . . .

I feed the carp on wheat kernel, boiled potatoes, and millet-seed.

Eating Qualities

I have . . . eaten a few fried, boiled, and dressed with egg and butter sauce and parsley. The boiled were the best, their flavor being fine and next to shad. They are probably like our red-horse. As table fish, they are good.

They are not full of bones, and do not taste of mud, as some would have us think, but, on the contrary, are very free from small bones, and are a most excellent table fish. . . .

. . . every person who had the good fortune to be present [when carp was served] pronounced it superior to any of the native fish prepared for the occasion. . . . the meat was of a rich gold shade, very hard and firm, being quite similar to salmon.

The 2 fish taken were found to be good on the table.

To prevent the muddy taste that some complain of in carp . . . kill the fish as soon as caught, clean directly, soak in ice water a few minutes, then sprinkle with salt slightly, and hang up to dry. The above will make the fish *firm, sweet, and good.*

The small ones, split open and fried, are excellent. The opinion of every one is that they never ate better fish. I expect to have fish every day during the season.

It was cooked . . . by frying. We found it fat, succulent, flaky, and entirely free from troublesome bones. In solidity and flakiness it reminded one of the salmon.

The carp is the best fish I know of for workingmen and mechanics, who rarely lack an appetite, and who will always consider the fish good when they can get it. My personal opinion is that it is a very superior fish, and I will even go so far as to say that I prefer it to trout.

I tried them baked and fried, and found them far superior in flavor to the mullet or perch, and very nearly, if not indeed quite, the equal of the shad.

The carp is as free from bones as any fish I know of. Carp, however, should not be eaten during the spawning season, or immediately thereafter, as at that time . . . the flesh is soft and of an unpleasant taste. I do not believe, however, that carp taken from stagnant and offensive pools of water would be fit to eat; . . . the flesh of the fish will undoubtedly be affected and rendered unfit for table use, having a moldy and muddy taste.

Carp are palatable, nutritious, and healthy. Its flesh is of excellent flavor, and, like the shad, combines the qualities that go to make a perfect whole.

I . . . find it to be a very nice table fish, I believe the very best fish I ever tasted.

. . . it was a good table fish, and . . . reports to the contrary are without foundation. . . .

The quality of its flesh depends upon the character of its food. Carp in running streams or in ponds, where limited to mud or rank weeds for subsistence, are no better than the salt-water mullet; but it has been demonstrated by pisciculturists that when properly fed they are little if any inferior to salmon.

We had it fried for breakfast and 7 persons tasted it; all pronounced it excellent.

I tried the eating qualities of the carp again last November. A 4-pound carp was boiled, and the 8 persons, my neighbors and my family, who ate of it considered it very good. . . . The 2 small fish weighing 1/2 pound were fried, and the opinion as to its merits equally divided. About one-half considered them very good, the others detected a strong taste.

Enemies
There are some frogs; I sometimes kill a mud-turtle, and I am not altogether rid of suckers. A fish-hawk took one of the original fish, and one died from wounds the cause of which is unknown.

I know of a few snakes, one large and one small turtle, and 3 large bullfrogs in my pond. I have killed 8 fish-hawks and 4 white cranes. I examined the intestines of a small white crane . . . but found nothing resembling . . . fish, tadpoles, or anything else save a quantity of fine black mud.

There are in the lake sunfish, called by some rock bass, which are sweet and solid all summer and bite a hook quickly; small cat-fish, red horse, shiners, and the finest suckers I ever saw. . . .

During the first winter muskrats let out the water, and I suppose the young carp perished, as I have seen nothing of them since.

There are a few pike, which we have been trying to catch out.

There are turtles, terrapins, pike, catfish, mullet, sunfish, and eels in the pond.

A white heron was seen to catch the large carp, and the little fellows may meet the same fate when they get large enough.

The pond contains a destructive little minnow called top waters, large green frogs, an occasional turtle, and a few water moccasins.

I lost thousands of fine fish the first 2 years in my large pond, they being destroyed by turtles.

I find that the most destructive enemy of the carp is the cray or crawfish, found in low marshy lands.

My pond appears to be alive with bull-frogs and tadpoles. It also contains two mud turtles, which with the frogs and tadpoles I fear will destroy the carp.

A few frogs will slip in. The sun-fish have escaped all our efforts at extermination.

The carp you furnished were all destroyed by snakes and black trout.

There are sun-perch, a few mud-cats, and millions of minnows.

There are small perch and frogs in abundance, and it is impossible to get rid of them.

Miscellaneous

The dead carp had 4 wounds on its side and abdomen, having the appearance of being bitten by a small dog.

I fear the sunfish have eaten up all the little fellows, and I should much like to have 25 carp about 6 inches long, that the other fish could not master.

I have not seen any carp since I put them in, though I have looked a number of times. Still, I believe there are carp there. . . .

In a short time after we received them they disappeared. . . . I never found any of them dead. I think the water was too cold and too clean.

Some of the carp . . . had sores on them, and something like hair or long mossy stuff adhering to their backs.

Musk-rats destroyed my dam, leaving so little water that the heron and minks gobbled the young carp before they were 2 inches long.

The pond dried up the first summer after the carp were put in and they all perished.

Poachers visit the pond.

At least two major conclusions can be drawn from the information presented above. First, carp make excellent "garbage disposals." Secondly, carp must have changed considerably since they were first introduced into U.S. waters. The bony fish that people complain about today must have quickly evolved from the succulent animals introduced by the Commission in the 1870s and thereafter.

One report on the ability of carp to survive arduous conditions involved a group of fish shipped to Birmingham, Alabama, in January 1884. The water containing the fish froze solid during the trip from Washington, D.C. When the ice was thawed, 1,300 of 2,100 fish in the shipment revived.[58]

Shad

Several species of shad can be found in U.S. waters, but it was the American shad, an anadromous species, that was of interest to the Commissioner. As we have seen, in 1867, prior to the time that Baird asked Seth Green and W. Clift to produce shad for the U.S. Fish and Fisheries Commission, the commissioners of fisheries from four New England states had approached Green and asked him to determine if shad hatching was even possible.[10] Green, who had previously worked only with the large eggs of trout, amazed everyone by actually succeeding in spawning shad at his Hadley Falls, Connecticut, hatchery during the same year the challenge was presented to him. Green wasn't successful on his very first attempt, however. He first introduced fertilized shad eggs into trout-hatching troughs, but all the eggs died. Clift, reporting on the event at the first meeting of the American Fish Culturists' Association meeting in 1872, quoted Mr. Lyman of the Massachusetts Fish Commissioners, who described Green's activities as follows:[59]

> "Nothing daunted, he examined the temperature of the brook, and found, not only that it was thirteen degrees below the temperature of the river [from which the shad came] . . . but that it varied twelve degrees night to day. This gave the clue to success. Taking a rough box, he knocked the bottom and part of the ends out, and replaced them by a wire gauze. In this box the eggs were laid, and it was anchored near the shore, exposed to a gentle current, that passed freely through the gauze, while eels or fish were kept off. To his great joy, the minute embryos were hatched, at the end of sixty hours, and swam about the box like the larvæ of mosquitoes in stagnant water. . . . the contrivance was still imperfect; for the eggs were drifted by the current into the lower end of the box, and heaped up, whereby many were spoiled for lack of fresh water and motion."

Never fear, Green eventually solved the problem.[59,60] The description of Lyman quoted by Clift provides the most colorful description:[59]

> "The spawn-box he at last hit upon and is as simple as it is ingenious; it is merely a box with a wire gauze bottom, and steadied in the water by two float-bars, screwed to its sides. These float-bars are attached, not parallel to the top line of the box, but at an angle to it, which makes the box float with one end tilted up, and the current striking the gauze bottom at an angle, is deflected upwards, and makes such a boiling within as keeps the light shad eggs constantly free and buoyed up.
> "This simple contrivance of Green's is one of the most important discoveries of modern times. Its grandeur will be much better understood ten years hence, which it shall have been applied to all our shad streams, and the yield shall have been increased some thirty, some sixty, and some a hundred fold."

Taking nothing away from Green's contraption, the suggestion that it was "one of the most important discoveries of modern times" seems a bit of a stretch, even for the 1870s. Lyman was obviously pretty easily impressed.

The details of stocking shad produced by Clift and Green in 1872 have already been disclosed. By that time, according to Baird,[6] Green had produced nearly 7 million shad for release in New York, while more than 92 million young shad had reportedly been released in Connecticut.

While apparently hell-bent to liberally sprinkle fish in both the marine and fresh waters of the United States, Baird showed some pragmatism when he questioned whether shad could, indeed, proliferate in all the nation's waters:[6]

> Whether shad can live permanently in fresh water, and maintain those characteristics of flavor and size which give them such prominence, and whether they can be established in the Mississippi Valley are problems not yet solved; but the results to be obtained . . . are of such transcendent importance in relation to the food-supply of the country, and the cost of the experiment so trifling, that it would be inexcusable not to attempt it.

In 1874, shad distribution was extended beyond the New England states. Shad from the Hudson River were hatched and the fry distributed in tributaries of the Mississippi River, as well as in the Brazos and Colorado Rivers of Texas. Fish from the Hadley Falls, Massachusetts, spawning station were distributed into the Mississippi River drainage, the Great Lakes, Lake Champlain, and some New England rivers. More than 3 million young shad were stocked between June 25 and August 15. Stocking numbers ballooned to 12,500,000 the following year.[19]

Seth Green had taken shad to California in 1871 before the U.S. Fish and Fisheries Commission was formed.[9] Livingston Stone, after returning to Washington, D.C., following the train wreck of the first aquarium car in 1873, also took a load of shad to California.[27] Stone cited reports that two adult shad had been caught in California in 1877, confirming some survival.[22] He felt that if two shad had been turned in, it was likely that hundreds more had been caught but not reported.

Shad spawning was not, on the surface, a particularly hazardous activity, but in 1874 there was a casualty. That year, Dr. J. H. Slack, a Deputy Commissioner and early fish culturist, was dispatched to Point Pleasant, Pennsylvania, to collect adult shad for spawning. He arrived on June 25 and remained there until the afternoon of July 2. The following day he[61]

> . . . was prostrated by a sudden attack of pleuro-pneumonia from which he never rallied, and which terminated fatally on the 27th of August. There is no doubt that exposure on the river at night, in the persecution of his work, induced the disorder.

One of Slack's claims to fame was publication of a book entitled *Practical Trout Culture*. The *Progressive Fish-Culturist* gave some insight into how Slack approached his work by quoting some lines from the book relating to selection of a fish farm site:[62]

> . . . the immediate vicinity of a large town is to be avoided, as the roughs, that class of population to be found in every city, have a fondness for trout; and a nocturnal visit from individuals of this stamp is generally attended by results far from pleasant. In fact, the stealing of trout from a private pond is too frequently regarded by even the so-

THE FISH COMMISSION GOES TO WORK

called better classes as a venial offense. In many . . . states it is considered in law only as a trespass. . . .*

Protecting one's fish from vandals could, in fact, get the fish farmer in trouble:[62]

> . . . Mr. Ward, of Mumford, N.Y. was obliged to suffer an imprisonment in the county jail . . . for peppering with shot the carcass of a scoundrel whom he detected in the act of stealing his fishes. . . . it is advisable that the ponds should be so located as to be in full view of the residence of the proprietor, and a good dog, or better still, a pair, will generally give notice of the approach of a nocturnal visitor.

In his 1872 book, Slack also provided advice on feeding fish. Part of the following was reprinted by the *Progressive Fish-Culturist*: "As the weather becomes warmer, maggots should be fed, these most disgusting, but to the fish culturist most valuable, creatures being the nearest approach to the natural food of the trout that can be obtained."[63] Slack described a feeder box with a drawer in the bottom. Offal would be placed on a screen above the drawer, and the maggots produced would fall into the drawer:[63]

> . . . emptying of the drawer is by no means a pleasant task, and the old-fashioned method of suspending offal from a wire directly over the pond, thus allowing the maggots to drop, as it were, directly into the mouths of the fishes, is perhaps preferable. . . . even if [there is] some inconvenience to the eyes and nostrils . . . [insects] must be in some way supplied. During one season, disgusted by the odor, we depended upon an increased supply of meat as food for our young trout, omitting entirely the maggots; but the small size attained by the fishes during the year, less than half that of the preceding, when less meat and more maggots were fed, warned us not to repeat the experiment.**

Not long after Slack wrote his book on trout culture, Charles Griswold tried shipping some shad of 1.5–4 inches in length in cans and stone jars.[64] To some of the containers he added gravel, which he thought harbored food that could be used by the fish during transit. He ultimately concluded that there was no advantage to transporting larger fish in preference to fry.

Stone's report for 1877 indicated that shad were probably becoming established in California.[22] In 1878, Baird indicated that a number of adult shad had been taken in California.[8] He felt that after a few more shipments from the East were made, the

---

*The situation has not changed a great deal since the 1870s. A few years ago, a colleague of mine in Texas was hauling a load of channel catfish in a tank on the back of his pickup truck when a car pulled out of a side road onto the highway immediately in front of him. Unable to avoid a collision, he broadsided the car, at which time water and a large number of fish entered the pickup cab through the sliding rear window, which happened to be open at the time. Insurance covered the damage to my friend's truck, but Texas law did not consider fish to be livestock, so he was not reimbursed for their value.

**In about 1980 I visited a commercial fish farm in Texas where various species were being produced and sold for stocking in farm ponds; I saw a big chunk of meat hanging over one of the ponds and asked about its purpose. "That's how we supplement our feeding program," said the farmer. He explained how maggots produced on the rotting meat fell into the pond. I wonder if that farmer knew he was practicing a technique that was over 100 years old.

population in California would become self-sustaining. Similarly, reports from Louisville, Kentucky, indicated that several hundred shad that had been stocked in the Ohio River were appearing in the markets. Shad production in 1878 was 15,500,000, and increased to 16,062,000 in 1879.[46] Between 1872 and 1881, over 172 million shad were hatched. Statistics compiled by the Commission in 1881 showed the numbers stocked during selected years and the locations where they were stocked:[65]

Number of shad stocked in waters near the hatching stations:

1879	5,587,000
1880	7,864,600
1881	46,518,500

Number of shad transported to other waters:

1879	10,002,500
1880	20,761,400
1881	23,516,500

Places receiving shad and the numbers received during 1881:

Connecticut	1,000,000
Delaware	940,000
District of Columbia	205,000
Georgia	1,800,000
Iowa	1,100,000
Kansas	200,000
Kentucky	707,000
Maine	1,150,000
Maryland	24,705,000
North Carolina	4,357,500
Ohio	1,020,000
Pennsylvania	3,500,000
Rhode Island	500,000
South Carolina	620,000
Tennessee	400,000
Texas	277,000
Virginia	24,280,000
West Virginia	175,000

Interest in the distribution of shad was not confined to the United States. There was also a demand for shad in Europe. In 1879, experiments were conducted to determine if shad eggs could be shipped successfully across the Atlantic Ocean.[66] H. J. Rice looked at controlling temperature and limiting water use to find a means of keeping shad alive for transatlantic voyages. He described what may be one of the first (at least in the United States) recirculating systems.* A reservoir provided water to the conical hatching chambers, which overflowed into another reservoir from which it was pumped (by a steam pump) back to the first reservoir. Some aeration took place when the water flowed from the cone to the lower reservoir, though treatment of the water was, at best, minimal. The system did increase the water volume to which the eggs were exposed, thereby diluting metabolites. Temperature was controlled by adding ice to the upper reservoir.

The first batch of 50,000 eggs was stocked into the novel system on June 8, 1879. All of them were dead by June 11, so the system was cleaned and restarted on June 12. Rice described what happened during his second attempt:[66]

> [The developing eggs] appeared to be healthy and in good condition; but gradually the water became filled with sloughs and decomposing animal matter, and early in the fifth day, or by the morning of the 17th of June, the fish were all dead. The eggs were accordingly thrown away and the apparatus again cleaned and placed in readiness for a third trial.
>
> It had become pretty evident that the trouble was in the water, and we determined to try the next time the effect of more thorough aëration upon it.

The third trial also employed 50,000 eggs. Water temperature was initially 70°F, but was eventually reduced to 53°F. A second cone was added to the apparatus, immediately after the first. Aeration was provided by agitating the water in the second (apparently eggless) cone for 5–10 minutes every 2 or 3 hours. This was done with a dipper in the following manner:[66]

> . . . running the dipper down deep in the vessels and getting the water from near the bottom, then lifting the dipper high above the vessels before pouring it back, so as to give the water as much of a fall as possible. In addition to this method of purifying the water a certain quantity was taken two or three times each day from the surface of the hatching-cone. In this manner it was intended to take of that water which had just passed over the eggs about the same quantity that would be added to the supply-tank as fresh water by the melting of the ice, and in taking it from the hatching-cone any sloughs or dirt which had accumulated around the top of the cone could be included.

The eggs remained in good condition for a day longer than in the earlier trial, though the temperature was lower in trial 3, so the metabolic rate of the developing embryos was slower. A fishy odor was noted on the fifth day, and foam was forming on the surface of the water. On the sixth day the foaming increased. Some of the water was removed from the system by taking it from the hatching cone overflow, and a charcoal filter was placed under the supply tube of the supply reservoir. That didn't seem to help, so on the sixth day new water at 73°F was added in an attempt to save the eggs. The damage had already been done and the eggs succumbed.

The spawning season was ending, so further experiments were not possible. Rice indicated that his next move would be to have sufficient water available to exchange the volume of the system completely every 3 days.

---

*A recirculating system is one in which the water is not continuously discharged and replaced. Instead, water passes through a chamber containing fish and is then treated in some manner to maintain its quality, after which it is reintroduced to the fish tank.

In 1884, Frank Clark reported on studies he had conducted to determine the minimum quantity of water that could be used for hatching shad eggs. Aeration was used to keep the eggs in motion. Eggs without water flow died, so Clark concluded that at least some water flow was necessary in the presence of aeration. He also felt that air should be introduced from the bottom to keep the eggs in suspension:[67]

> In every case where an aërator was in use, and they were kept until the fifth day, they commenced dying, and in twenty-four or forty-eight hours all were dead. My opinion is that the violence of air-bubbles and water weakened or wore them out.

Commercial fishermen were the source of adult shad for spawning. The Commission purchased ripe shad, spawned them, hatched the eggs, and stocked the young fish to enhance the commercial fishery. The fishermen did not always see the activities of the Commission as a positive influence on their livelihood. William Babcock ran afoul of the fishermen's negative feelings in 1883:[68]

> I found a decided prejudice existing among the fishermen because the elements had not been favorable last year. As it could not be attributed to anything else, in their minds, the fishermen seemed to have selected the Commission to bear all the blame, and therefore demanded exorbitant prices for the right of collecting fish eggs. . . . In one instance the owner of a fishing-shore declined to sell any fish to the employés of the United States Fish Commission for the full market-price at his shore. Matters were finally arranged in a satisfactory manner, and, as everything has been done to aid . . . the fishermen, even they in time must be convinced that the efforts of the Commission are intended for their benefit. Such antagonism, however, should lead the Commission to establish its own stations for seine-hauling and collecting fish eggs. . . .

Whitefish

A glimpse of the early activity associated with spawning and distributing whitefish was described earlier in conjunction with some of the first activities associated with U.S. fish culture. The U.S. Commission on Fish and Fisheries was quite interested in whitefish culture in the early years. In 1872, Mr. N. W. Clark, of Clarkston, Michigan, obtained some 500,000 eggs from George Clark of Detroit and placed them in hatchery troughs.* In January 1873, about 200,000 eggs were shipped by the Commission to California.[6] The shipment arrived dead, so a second shipment of about 100,000 eggs was subsequently sent. Those eggs were successfully hatched after arrival and were placed in Clear Lake. That was the first time the species had been shipped across the Continental Divide.[39] It soon became clear to Baird that the intense interest in whitefish production in various states meant that the Commission could allow the states to raise that fish, freeing the federal program to concentrate more heavily on other species.[8]

Seth Green had successfully hatched whitefish in 1869,[10] and kept up his activity with that species. In a report to the American Fish Culturists' Association in 1876, he indicated that Mr. Oren Chase, an assistant of Green's who was operating a hatchery

---

*A lot of Clarks are in that sentence, I'm afraid, but those are the facts.

in Detroit, Michigan, had 8 million eggs on hand.[69] The activity had been arranged for by the Michigan fish commissioners.

Even though the Commission was supposedly downplaying whitefish, Baird acknowledged that culture of that species had not been abandoned:[8]

> Mr. Frank N. Clark usually collects several hundreds of thousands of eggs, and develops them at his fish-culture establishment at Northville, Mich., for any desired assignment. These, for the most part have been sent to the commissioners of California, and also to various parties in Pennsylvania, New Jersey, Wisconsin, &c.

In 1881, the Commission shipped a total of over 2 million whitefish eggs to California, Connecticut, the District of Columbia, Iowa, New Jersey, France, and Germany. Fry whitefish were shipped to Michigan, New York, and Wisconsin. The total number of fish involved in those shipments was 17,750,000.[28]

Whitefish became a species of particular interest to states bordering the Great Lakes. James Milner had documented the general decline of foodfish species in the Great Lakes by 1874 and suggested two solutions: protective legislation and artificial propagation. He indicated that protective legislation was the least effective of the two and promoted culture as the best approach.[70] A whitefish stocking program was immediately launched.

In 1885, Frank Clark reported on his attempt to determine if the program had produced a positive effect in Lake Erie.[71] He could not document all stocking episodes, but indicated that fish had been planted by the Canadians along with the U.S. Commission on Fish and Fisheries, and the fish commissions of Ohio and Michigan. Numbers of whitefish and the year in which they were planted by the various U.S. agencies are as follows:

1875	150,000	1876	300,000
1877	450,000	1878	12,000,000
1879	7,000,000	1880	13,000,000
1881	12,000,000	1882	42,000,000

Clark went on to provide statements from commercial fishermen who indicated that their catches had generally improved in 1883 and 1884. Most attributed the enhanced catches to the stocked fish. To give an indication of the volume of fish being taken from Lake Erie, the comments of just a few fishermen follow:[71]

> My catch of whitefish . . . is as follows: 1883, 175 tons; 1884, 225 tons.

> For the past five years our annual catch of whitefish has been about 150 tons, until 1884, when it was 200 tons.

> I . . . think the whitefish for the past three years have been increasing. . . . my catch . . . for the past two seassons [is] as follows: 1883, thirty tons; 1884, fifty tons.

> My figures for the last two seasons are: 1883, thirty tons; 1884, fifty tons.

> We [caught] about 132,000 pounds of whitefish [in 1883], and in 1884, 170,000, an increase of nineteen tons for 1884.

One fisherman summed up his view with respect to the fish stocking program: "'I have not thought much about the hatching business, but it must do some good; at least there seems to be an increase in whitefish that past two or three years.'"[71] Clark, who had conducted the interviews, was obviously convinced that the stocking effort was successful:[71]

> . . . sufficient is shown to prove beyond a doubt that the aggregate catch was greater than for several years, that whitefish are decidedly on the increase in Lake Erie, and that the increase is simply the legitimate result of the work of the hatcheries.

While Clark could not have known it at the time, the battle to maintain fish stocks in the Great Lakes had only just begun. Clark was working several years before the sea lamprey invaded the lakes and wrecked further havoc, but that's a story for another book.

Trout

We've given Livingston Stone a bit of a rest, so perhaps it's now time to revisit him with respect to his activities in trout rearing. Stone had been raising trout for a few years before being tapped by Spencer F. Baird to head up salmon and trout spawning activities in California.[5,9] In fact, in 1872, he presented a paper on the economics of trout farming to the American Fish Culturists' Association.[72] He placed the cost of feed at about 3¢ per pound. Sale prices of the fish ranged from about $0.50 to $1.25 in Boston and New York. He figured four years to market, a food conversion ratio of 5:1 with two and three year olds.*

Stone's balance sheet for an inventory after four years of 4 million fish (1 million each of one, two, three, and four year olds) showed the following:[72]

Cost of buildings and equipment	$6,000
5,000 spawners	2,500
Three men's labor for 4 years @ $300/yr	3,600
Feed costs	35,000
TOTAL	$47,100
Value of crop at end of 4 yrs	$542,000
PROFIT (all fish at $0.25/lb)	$495,000

A half million dollars was a prodigious amount of money in 1872 (and it's not peanuts now). The enormity of the profit potential becomes even more clear when one compares wages then and now. Stone acknowledged that risk was a factor that could not be overlooked. He thought that cutting the size of the operation by 10 would make a reasonable operation.

So far we have spent some time looking at the trials and tribulations associated

---

*Today we look at food conversion ratios by dividing the dry weight of feed offered over a given period of time by the wet weight of gain shown by the fish over the same time period. So, if a fish gets fed 1.5 pounds of dry feed to gain 1 pound, the food conversion ratio will be 1.5/1.0 = 1.5. In Stone's day, wet feed (usually organ meats such as liver and kidney) was offered. The conversion ratios were based on wet weight of feed to wet weight of fish. So it took 5 pounds of food to produce 1 pound of fish, which wasn't too bad, considering it takes something like 15 pounds of wet grass to put a pound on a cow.

with the McCloud River salmon spawning station that Stone established in 1872. He also set up a trout spawning station not too far away, though that station was not established until 1879.[28] At the trout station, Stone had ponds installed where broodfish could be retained from year to year, unlike the salmon which, depending on whom one believed, either had to return to sea or died after spawning. One advantage of the trout station was that flooding was not a worry because the ponds were on high ground. At least it wasn't a problem until the flood of 1881: "The mischief that was done proceeded from an entirely unexpected source, which well illustrates the fact that in the new country like this when trouble begins no one can tell what will come next."[73]

The problem was from mud. The creek that fed the trout ponds filled up with suspended sediments, which were deposited in the ponds, nearly filling them. The men had to get in the ponds and shovel out the mud by hand. Gills became clogged and many fish died immediately. Others died over a period of months, leaving only about a thousand healthy fish in the ponds by July 1.

Trout spawning had begun before the peak of the flood. Myron Green started out for the town of Redding on January 26 with 75,000 trout eggs that had been packed in boxes. Green was supposed to catch a train for the East Coast, but when he reached the Pit River about 7 miles south of the trout ponds, he found that it was running very high. Stone explained what happened:[73]

> . . . Mr. Green succeeded, at considerable risk, in getting the eggs across the Pit. By the time they reached the Little Sacramento at Reid's Ferry this river had become all but impassable, and no one could be found who was willing to venture to cross it. The eggs consequently lay there several days. In the meantime the floods had spread over the whole country, and the California Pacific Railroad for a hundred miles below Redding was more or less under water. The consequence was that the trout eggs spoiled in the crates long before they could be started on their eastern journey. I know that some dissatisfaction was felt by the eastern consignees of these eggs, but if they could realize the difficulties which had to be encountered at the other end of the route in shipping the eggs they would not want to attach any blame to any one.

Ultimately, the McCloud River trout station was able to produce 179,000 eggs in 1881, 40,000 of which were sent in February to the Honorable B. B. Redding, secretary of the California Fish Commission, in San Francisco. The 75,000 eggs that were lost had been destined for Maryland, Iowa, and Illinois. Between late February and the end of April, shipments were made to Illinois, Iowa, New Hampshire, Maryland, Michigan, Minnesota, Wisconsin, Kentucky, Pennsylvania, and New York. Most batches sent East to particular recipients contained fewer than 10,000 eggs.

Aside from the filling of ponds with mud during 1881, activities at the trout station were relatively routine, though there is an indication that another aquaculturist's health was adversely affected by the rigors associated with fish spawning:[74]

> Mr. Loren Green took most of the eggs this season, and is entitled to much credit for the endurance and perseverance that he exhibited in accomplishing his work. I may add here that Mr. Myron Green, who has had charge of the trout-ponds ever since they were started, resigned his position this summer, owing to his health having become impaired by exposure during the rainy seasons of this climate.

Stone shipped fish from the McCloud River trout hatchery throughout the country, but he was not insensitive to the need of maintaining the local trout population:[74]

> Up to the present season no systematic attempt has been made at the McCloud River trout-ponds to rear young trout, the surplus of young fry at the end of each season having been turned into the river, and I may add here that it is very proper, and perhaps indispensable, that a considerable number of young fish should be put into the river each year in order to keep up the river's stock. . . .

Holy Mackerel! (and Other Species of Interest)

A number of marine fish species were the targets of culture during the first few years of Commission activity. Experiments pursuant to determining the feasibility of artificial propagation were being conducted prior to 1880 on herring, pollock, and haddock.[75] Other species of interest during the next several years were winter flounder, tautog, cunner, scup, sea bass, and Spanish mackerel.[76] Striped bass eggs were successfully hatched by Mr. M. G. Holton (a former assistant of—who else—Seth Green) at Weldon, North Carolina, in about 1873.[6] Subsequent success with striped bass culture was apparently not so promising, though some success was achieved with egg hatching in 1879.[46] Sole were brought from Europe in 1878, but they all died during the voyage.[8] Some of the work was not pursued to any extent, though with others, huge numbers of eggs and fry were produced.

Cod was a species that drew early interest, though captive hatching didn't begin until the winter of 1878–1879. Initial activity was associated with a facility at Gloucester, Massachusetts,[75,76] but the effort was soon expanded to Wood's Hole.[46,75] James Milner and Frank Clark were involved with setting up the facility at Gloucester. Milner left because of illness, and Clark was called away for other Commission business. They turned the effort over to Capt. H. C. Chester.[76]

At first it was not known if cod eggs floated, sank, or were adhesive in nature, so the approach to hatching them was initially unclear. It turned out that the eggs floated, and working out the process of hatching them involved a good deal of work.[46] Modifications of cones that had been developed a few years earlier for use in hatching shad eggs ultimately resolved the problem.[75,76]

To examine whether cod fry could survive hauling, some of the young fish were shipped from Gloucester to Washington, D.C., in 1878: "These were placed on exhibition in the rooms of the Committee on Appropriations in both House and Senate, and were also exhibited to the President and Cabinet."[46]

Three methods of taking spawn were attempted during the first two years of activity.[46] The first method was to examine fish landed at the wharf. Because they had been dead several hours when landed, viable eggs were usually not obtained. A second method was to send out men in the Commission's vessel to take eggs on the fishing grounds. Hand lines were used, but most of the fish caught were not ripe. Commission personnel then tried fishing off commercial hand-lining vessels. While some ripe eggs were obtained, it became clear that much better success could be obtained if the fish were taken by trawling.

In the meantime, live boxes were constructed, and some of the undeveloped fish were held. The fish were inspected every two or three days until they were ripe and could be stripped of their eggs.[46] Finding running males did not appear to present a problem.

The method of collecting spawners that was ultimately adopted in later years involved the placement of Commission personnel on commercial fishing vessels:[76]

> . . . permission to place spawn-takers aboard the fishing vessels is obtained, with the understanding that they will be allowed to take eggs from the fish secured, that they be given the freedom of the vessel in order to properly take care for the eggs, and that no charges be made against the Commission except that 25 cents be paid for each meal furnished to the spawn-takers.

By the time the Commission published a fish culture manual as part of its 1898 annual report, cod were ". . . being "propagated artificially on a more extensive scale than any other marine fish."[76]

Marshall McDonald and Mr. R. E. Earll found Spanish mackerel spawning in Chesapeake Bay. Earll found that the eggs were easy to obtain and that they could be readily hatched ". . . by methods requiring little apparatus or attention on the part of the observer."[77] Not everyone seemed to share that view, however, as the *Fish Hawk* personnel reported in 1885.[78]

McDonald set up a hatching facility at Cherrystone, Virginia, in July 1881. Fish were obtained from commercial fishermen operating pound nets. Most were either spawned out or immature at capture, causing McDonald to speculate that the fish spawned at night while the commercial fishermen checked their pound nets in the early morning. McDonald concluded that if the Commission wanted to pursue spawning of Spanish mackerel, its own pound-net site would have to be established and a crew would have to be assigned to operate the net.

Spanish mackerel first became a focus of personnel aboard the steamer *Fish Hawk* during the summer of 1881.[78] Eggs were obtained and placed in various types of hatching devices, but all the eggs died. In 1883, the *Fish Hawk* crew again focused some of their attention on Spanish mackerel. The crew worked with commercial fishermen to find ripe fish but found only a small number that were not spawned out. One explanation, which was similar to the one McDonald had come up with a couple of years earlier, was that the fish expelled their eggs while being trapped in a net for long periods.[79] Some eggs were obtained and hatched, but the fry did not survive.[78,79] Dr. J. Alban Kite developed the hatching jar used in the abortive 1883 work. Kite later came up with a modification of his design, which he believed would allow for the successful hatching of not only Spanish mackerel, but also cod.[78]

Today's aquaculturists always seem to be looking around for new species to culture. Some of those of contemporary interest were targeted by the U.S. Commission on Fish and Fisheries over a century ago. Included are many we have examined, along with higher vertebrates. While we may think, for example, that interest in frog culture is a relatively recent phenomenon, good old Seth Green had already been thinking about it in the 1870s:[80]

> There are many stagnant pools about the country, useless in their present state, and believing that they should be utilized, I cannot think of any better use for them than to make them into frog-ponds. I believe that the man who could raise a million frogs, and

get them to market, would be a rich man. He will find many difficulties to overcome;
but allowing him two years for experimenting, good results might be anticipated.

After nearly 125 years, the experimenting continues. Green did provide some point-
ers as to the state of the art in 1874. He described how to collect egg masses and how
to take care of the tadpoles. Green recognized that once they metamorphosed into
frogs, they ate insects. He had worked out ways of producing maggots (the same
disgusting topic that we've already discussed), but he thought it might be possible to
train frogs to eat meat. The question was how to do that on a massive scale:[80]

> I have many times tied a small piece of meat on a fine thread and attached the thread to
> a long fish-rod and moved it about near the frog's nose, and he would take it very
> quick. But you could not afford to teach a million in this way.

A review of edible frogs in the United States, and on their culture, was published in
1898,[81] and the subject has come up frequently since that time. Providing sufficient
numbers of live insects to feed a million frogs continues to be the major problem.
Having that many fogs around could be a bit distracting if even a small percentage of
them started croaking.

The Commission did not accept all challenges. They sometimes deferred to the
states and even to the private sector with respect to the culture of some species. An
example of where the Commission deferred to the private sector involved the diamondback terrapin, a turtle that was (and in some places continues to be) a prized
human food item. By the early 1890s the diamondback terrapin was already becoming scarce and the price had risen sharply. The Commission received numerous
inquiries about terrapin culture. Those inquiries were passed along to culturists who
had established terrapin farms. It was acknowledged that ". . . sufficient time has not
elapsed to fully demonstrate the feasibility of rearing terrapins for market from the
egg."[82] Slow growth and a high incidence of reproductive failure of captive turtles
were cited as major problems.

There wasn't anything about alligator culture discussed in any of the early aquaculture literature. Interest in those beasts came along later and, incidentally, seems to
have been an initiative of the private sector.

Shellfish

Invertebrates were not overlooked by the early American aquaculturists. Lobsters
and oysters were early favorites and the subject of a good deal of study. Surveys of
where natural populations could be found, the extent of their fisheries, their life
histories, and their culture potential were subjects of interest to the Commission.
You may recall that adult oysters and lobsters were both part of the shipment that
flopped into the river along with Livingston Stone and his party in 1873.[19]

Lobsters. The taking of egg-bearing female American lobsters* was recognized as a
reason behind declining populations, and led John Brice to conclude that if the

---

*This name refers to the lobsters found in New England which have the big claws (chelae). Florida
lobsters lack them.

practice did not occur, there would be no need for artificial propagation.[76] Other means of protecting lobsters from overfishing included a suggestion by S. M. Johnson in 1883 to design lobster traps with slats wide enough apart to allow small animals to escape. Even more radical was his idea of imposing a closed season on lobster fishing.[83] The notion of prohibiting the taking of egg-bearing females had apparently not been developed, and had it been put in place, there was no enforcement agency in place to police the policy.

While no attempts to reproduce lobsters in captivity had been made prior to the establishment of the Commission, a technique known as "parking" had been developed; this involved holding injured, newly molted (soft-shell), and submarketable lobsters in enclosures until they became marketable. So-called parks had been established in Massachusetts in 1872 and in Maine about 1880.[76] Experiments with artificial hatching began in 1885 upon completion of a new laboratory and hatchery at Woods Hole, Massachusetts. Development of the technology for rearing lobster larvae for long periods was beyond the scope of Commission interests or ability. It was recognized that even the early stages are cannibalistic and that crowding them in a hatchery only increases the opportunities for them to consume one another. The idea was to hatch as many eggs as possible and stock the larvae in the sea, where they would, undoubtedly, thrive and enhance the fishery. There is little evidence that such enhancement ever really occurred; more likely, the lobster larvae provided a good food source for young fish. In any event, the Commission can certainly not be faulted for underachievement in terms of lobster larval releases.[76]

In 1886 the experiments had progressed so successfully that several million eggs were collected and hatched at Wood's Hole, the fry being deposited in Vineyard Sound and adjacent waters. From 1887 to 1890, inclusive, the number of eggs collected was 17,821,000.

From the eggs collected up to 1889 the average production of fry was about 54 percent. . . . in 1890, from the 4,353,000 eggs collected, over 81 per cent yielded fry. Work was continued at Wood's Hole on about the same scale until 1894, when the collections aggregated 97,000,000 eggs. In the same year lobster propagation was undertaken at Gloucester and a collection of 10,000,000 eggs was made there.

In the fiscal year 1896 the number of eggs taken by the Fish Commission was 105,188,000, the resulting fry liberated numbering 97,579,000, or about 93 per cent; and in 1897 the collection amounted to 150,000,000 eggs, of which 135,000,000, or 90 per cent were hatched.

Once hatched, the larval lobsters were transported back to the sites of adult capture. The technique involved placing larvae in 10-gallon cans: 200,000 larvae for short trips and 125,000 for longer ones.[76] The Commission had learned that the larvae could be transported for up to 24 hours and sometimes used the railroads to expedite distribution. When it was not possible to return the larvae to their paternal waters, ". . . they were liberated in Vineyard Sound and Buzzards Bay with an outgoing tide, so as to insure [sic] their wide distribution."[76] While the majority of the effort was expended on producing lobster larvae for stocking, some attempts to move adult lobsters around the country were made during the early years of the Commission. Once again, you might recall that there were lobsters aboard the ill-fated aquarium car that ended up in the Elkhorn River, Nebraska.[19] A second load of lobsters was shipped by rail from Charlestown, New Hampshire, to California in 1874. Marshall L. Perrin was responsible for the shipment and presented the details in his report published in 1876. The 150 lobsters were delivered to Charlestown on June 3, packed in seven pine boxes. The lobsters were lying on wood with sponges over and around them. Six 149-gallon containers of seawater were obtained the morning of the trip. Most of the water was put in two saltwater tank in the aquarium car.* The aquarium car tanks were constructed of wood and were sealed with a mixture of resin and tallow. Saltwater fish were put in those tanks.

On June 4, the lobsters were repacked aboard the aquarium car in boxes that had 12 compartments (which Perrin refered to as "apartments," but what we would now refer to as "lobster condominiums"). The lobsters were laid on a bed of straw and covered with sponges soaked with salt water. The boxes had holes in the bottom for drainage. Each day during the trip the boxes were inspected, and water from sponges was squeezed out over each lobster, ". . . which, if alive, [would] respond . . . by blinking its eyes and stretching its claws, perhaps moving its body a little."[84] Pieces of ice were added to each box to keep the lobsters cool.†

On Saturday, June 6, at noon, 60 lobsters were put in one of the large saltwater tanks with striped bass and some other fish. Those 60 lobsters reached Chicago alive by Sunday morning, but they, along with the fish, were all dead by Sunday evening—the lid having fallen into the tank. Whether the lid in the tank was really the cause of the problem isn't known, but Perrin speculated that the air pressure built up under the lid and killed the lobsters. (Air was being continuously pumped into the tank.) That's an interesting but highly unlikely theory. Another possibility presented was that the air was trapped for a long time and not properly renewed (which seems like a more likely explanation). There was also suspicion that the resin-and-tallow mixture, the wood from the tank, the mixing of fish and lobsters, and the stocking density could have contributed to the mortality.

Mortality was also a problem with respect to the lobsters in the boxes. By June 8, there were only 20 alive. Perrin felt that five days out of water in shipment was about all lobsters would tolerate: "After the fifth day crowds of lobsters take offense at something, and revenge themselves by dying."[84]

Water was being used more rapidly than Perrin thought would be necessary. He didn't want to use water form the fish tanks, so a 149-gallon reserve cask was opened for use. Perrin telegraphed ahead to California and asked to have some Pacific Ocean water shipped east to meet the train. Perrin was concerned that the melting ice being used to keep the lobsters cool was diluting the seawater excessively, so he left ice off a few times only to learn that the cold was more indispensable than high salinity.

There was also the so-called destructive air problem:[84]

> Every time the car-doors were opened or the atmosphere around the lobster-boxes disturbed, there inevitably rushed upon them a draught of warm and dry but injurious

---

*This was a second aquarium car. The one that dropped into the river in Nebraska was defunct.
†I don't think the lobsters of today can blink their eyes, which may be an indication of extremely rapid evolution.

air, fatal at once to a lobster in case the current strikes it. There must be some medium, as a wide or at least constant stratum of moist atmosphere, to guard the lobster against this destructive air; and at the same time that it would prevent this evil, it should produce the needed low temperature.

On June 9, Perrin removed the straw and packed sponges around the few remaining lobsters. Mortality diminished, and he was convinced that the action he had taken was responsible. He felt that the lobsters got their claws entangled in the straw and were "restrained" by it. Next, he removed the partitions in the boxes, allowing the lobsters more room to move about:[84]

> When in an apartment with partitions, they never staid [sic] in the middle, but worked themselves over to one side, and struggled against the wooden partition; in this way tiring themselves out, which is of course an evil. A lobster needs room to stretch all its limbs, if it wants to do so. For this reason they are better in boxes without partitions, provided they are not near enough together to bite each other. Rubber bands around the claws are an extreme case of restraint, and are extremely pernicious.

On June 10, a pair of lobsters were dropped off at Ogden, Utah, to be stocked in Great Salt Lake. The train passed out of Utah that night with eight lobsters and one pail of seawater remaining. On Thursday, June 11, a freight train from the west met the aquarium car in Nevada and delivered four barrels and four metal tanks of Pacific Ocean water. The water had been transported in metal containers was useless— Perrin was aware that such containers were toxic—but the water from the barrels "was welcome and immediately used."

The train reached San Francisco Bay on June 12 with four live lobsters. Those were stocked from the Oakland wharf during the afternoon. I don't think it's necessary to speculate on what might have happened to those lobsters.

Oysters. The history of oyster culture in the United States has largely been one of replacing substrate in existing beds for spat* settlement and moving spat-laden substrate into areas where oysters are not present or are depauperite. Even the most modern methodology, which involves the production of spat in hatcheries and may involve growout on strings suspended from rafts or longlines, depends on placing the young oysters in an environment where natural food is available at a level that will support rapid growth. While more control is exerted over the life cycle of cultured oysters in some places today, the techniques developed by the U.S. Commission on Fish and Fisheries and others during the nineteenth century continue to be in use today.

Interest in studying oysters and the potential of supplementing them arose from the same concern that led to the organization of the Commission in the first place, that is, the view that the fishery was in decline. In a paper on the oyster situation in

---

*The free-swimming larvae of oysters are called spat. The spat will settle within a few days after the eggs hatch, and if the substrate is suitable (consisting of mollusc shells or other hard substrates) and the environment is conducive for their growth (with proper water quality and food availability), they may survive and prosper.

New York, that view was expressed before the American Fisheries Society in 1885 by
Eugene Blackford a New York fisheries commissioner:[85]

> This work was begun on the supposition that there was danger of a failure of our
> oyster supplies in the near future, unless some steps were taken toward remedying
> certain practices and evils which were thought to be detrimental to the success of and
> continuance of the industry.

Also pointed out by Blackford was that the planting of oysters on depleted beds was
having a positive effect:[85]

> . . . the natural oyster areas of the State [New York] are in bad condition and very much
> less in extent than they were a score or more years ago, [but] the loss in the natural
> areas has been much more than made up in the formation of planted beds, some of
> which occupy the localities of natural areas, which have been exhausted of their natural
> supplies, and have been repopulated by artificial means. . . .

Blackford indicated that allowing the privitization of at least some oyster bottoms
would help ensure that production would continue to prosper. He also noted that
certain areas had been made unsuitable for production because of pollution and
recommended the passage of laws to control sewage releases into coastal waters.

It is clear that before 1885, oystermen were moving spat-laden cultch* around to
extend and renew overfished oyster beds.[86] At the same time, studies were underway
to learn more about oyster biology and culture. One such study was reported at the
American Fisheries Society Meeting of 1886 by Fred Mather. Mather had learned to
spawn oysters and obtain spat sets on cultch material placed in ponds and a wooden
tank. He cited John A. Ryder, of the Commission, as having laid down the following
principles with respect to oyster culture:[87]

1. Oyster embryos . . . affix themselves throughout the three dimensions of . . .
   [the] water.
2. The [spat] will adhere to smooth surfaces as well as to rough ones.
3. The surface upon which spatting occurs must be kept as free as possible from
   sediment and organic growths, in order that the tiny young molluscs may not be
   smothered and killed. . . .
4. Artificial fertilization of the eggs of the oyster is feasible, and will become an
   important adjunct to successful spat culture.
5. Water [containing] embryo oysters may be passed through a steam pump without injury to the embryos.
6. Oyster [spat] adheres to the under surface of shells or other collectors most
   abundantly, because the lower side is cleanest and most favorable to the survival
   of the animals.
7. The spat of the oyster will grow and thrive with comparatively little light.
8. The specific gravity of the water may range from 1.003 to 1.0235.
9. The most favorable temperatures . . . for spatting seem to be from 68 degrees to
   about 78 or 80 degrees Fahr.
10. Spatting will occur just as freely in ponds or tanks as in the open water.

Some interesting and important facts about oyster culture are contained in this list.

*Substrate material on which oyster spat are allowed to settle is known as cultch.

Fred Mather (1833–1900)

Ryder reviewed the history of oyster culture,[86] indicating that Prof. W. K. Brooks at Johns Hopkins University had been the first to study the development of the American oyster, using artificially fertilized eggs in the process. That had occurred in 1878 and 1879.

Ryder had worked with oysters at a hatchery on Saint Jerome's Creek, Maryland, in collaboration with Marshall McDonald. In 1884, Ryder reported on that work.[88] One of the studies conducted involved examination of the feeding habits of oysters. From that study it was concluded that algae were an important food. Ryder spawned oysters on several occasions and developed a filter system that allowed him to exchange water without losing the larvae.

Also reporting on oyster culture in 1884 was F. Winslow, who stated that while egg fertilization and development of oyster larvae was not a problem, keeping the young oysters alive for extended periods had not been possible. Winslow attributed the problem to an insufficient supply of water to aquaria in which the larvae were held.[89] He indicated that the larvae were so minute that removal of water without losing the larvae was not possible.

Winslow described the fertilization process and the procedure for collecting broodstock and for obtaining eggs and sperm. He felt that aeration of the water during incubation was important, and that it was critical that the air entering the water be the same temperature as the water.

Dr. W. K. Brooks, the first to spawn oysters in the United States, had added sources of calcium carbonate (as pieces of oyster shell and sand dollars) to the larval oyster changes and found that larval development was more rapid than when lime source was not made available.[89] That work had been conducted at Beaufort, North Carolina, before Winslow showed up. With Winslow present the lime didn't seem to have that much effect, so Winslow thought other, unknown factors had been the reason for the more rapid development.

Winslow recommended obtaining seawater just before its use and cleaning oyster shells thoroughly before putting them in the aquaria as a means of reducing the development of infusoria (protozoans). He recognized that providing food for high concentrations of oyster larvae in aquaria was a serious problem, compounded by the fact that no one knew what oyster larvae ate. (Winslow apparently was not aware of the food-habit studies undertaken during the same period by Ryder, or didn't make the obvious linkage.) Brooks tried adding starch to encourage bacterial growth. That didn't work, nor did the addition of ditch water thought to be a source of detritus, which some thought might be the source of oyster food. He then collected water and mud from the bottom in an oyster production area and dropped it in the aquaria. Some larvae were seen to consume organic particles released from the mud, though all the larvae eventually died. Winslow figured that was because there was insufficient food.

Winslow reported that Marshall McDonald had put together a recirculating water system which had siphons to move water from one chamber to another.[89] Dippers were used to remove water continuously from the aquarium with the oysters and place it into the upper vessel in the series. The efficiency of such systems developed by other workers in later years was greatly improved when pumps were added to lift the water from the lowest to the highest chamber, but apparently McDonald didn't have access to such equipment. Two closed vessels contained scrubbed oyster shell to add calcium carbonate, and another was stocked with seaweed (*Laminaria*) to remove carbon dioxide and add oxygen. Oysters were found attached to the walls of the system, the first time spat settlement had been reported. Winslow said he couldn't tell how the oysters caused themselves to become attached. He and McDonald did determine that flowing water would not wash the attached oysters off the glass. All of the attached oysters died within five days of fertilization.

The work of Mather at Cold Spring Harbor, New York, apparently attracted some positive response from oystermen. He indicated that some of them felt that ". . . seed oysters could be raised in quantities by any person who had an inclosed [sic] pond . . . where the water came in at times of high tide, and that they would be reasonably certain to get a fair 'set' on proper cultch."[87]

W. Ravenel conducted experiments on oyster spawning in ponds during 1885.[90]

He stocked adult oysters in ponds that received various rates of tidal flushing, but obtained successful reproduction only in ponds where frequent exchanges of water occurred. Reasons for the observed results were not provided, but it is certainly likely that food was limiting in ponds that did not receive frequent water renewal. Oysters depend on an adequate supply of phytoplankton for food. In some cases the new water passed through two or more of Ravenel's ponds before entering the last pond in a series, so there may have been little or no food available to the downstream oysters. Water quality in the mostly static ponds could also have been poor.

Studies on oysters continued, and a good deal of information on their biology was learned, though few changes in culture practices were made. As the nineteenth century drew to a close, J. Percy Moore summarized the situation. He indicated that to increase supplies of oysters, the method employed in the United States was known as "planting":[91]

> This consists in placing firm bodies [cultch] in the water for the purpose of catching the spat or in spreading young oysters upon the bottom in places suitable for their growth. Vast as our oyster-fields, but a small portion of the bottom available for the growth of this mollusk has been utilized by nature. This has arisen from the fact that in many cases where the other conditions are favorable the bottom is of such a character as to prevent the attachment of the young, though perfectly adapted to the rapid growth of the adults. If then the spat be caught on planted cultch, or partially grown oysters be placed upon such bottoms, the difficulty is overcome and nature has been assisted to the degree necessary. . . .

Planted beds were often under private ownership, and Moore contended that predation problems would be reduced as compared with public beds because private oyster growers would do a better job of tending beds that they controlled.

Some work on the pond culture of oysters continued. It was Moore's contention that modifications in contemporary techniques might make it possible to rear brood oysters in ponds located in regions where there was little or no oyster production.[91] The pond-reared oysters would provide the spat required to establish new beds.

CHAPTER 2

A New Science is Born

In a review of the first century of fish culture in America, E. M. Wood placed the beginnings of fisheries science at about 1880.[92] Prior to that time, it might be argued that U.S. aquaculture consisted mainly of observation, trial and error, and the persistence of talented individuals who had little or no training as scientists, but who were often acknowledged to be highly skilled naturalists. Beginning around 1880, scientists started to look at the chemical composition of fish, collect statistical information, investigate methods of preservation (including freezing), and seriously examine food habits, growth, and other aspects of fishery biology. Scientists from universities became interested in fish culture, and the American Fish Culturists' Association, which had been formed a decade earlier, would soon become a scientific organization renamed the American Fisheries Society.

The development of fish culture as a science was addressed by John Gay and William P. Seal in 1890, less than 20 years after the U.S. Fish and Fisheries Commission had been formed:[93]

> Looking backward, we can see the gradual evolution of a complex but systematic scientific organization grappling with questions absolutely new and untouched, the development of methods simple and efficient from those heterogeneous and crude, and the final creation of an enthusiastic body of trained experts thoroughly imbued with a faith in the latent possibilities to be achieved in the great future of fish culture.
>
> The fact that so great a scientific authority as Prof. Huxley should have expressed a doubt as to man's ability to diminish or in any way to control the harvests of the sea . . . has no doubt in the past had the effect of producing at least a very conservative feeling in the minds of many regarding the possible limits of the fish cultural attainment.
>
> The difficulties to be encountered and the influences to be overcome were fully appreciated by the great founder of practical fish culture, Prof. Spencer F. Baird.

Gay and Seal went on to discuss how the bringing of civilization to the wilderness was accompanied by severe reductions in wildlife and fishery resources, but indicated that production and restocking programs could correct the problem. They were convinced that the fish stocking programs in the United States had met with a good deal of success.[93]

> When we consider that in the case of the shad alone, the survival of five per cent. of the fry distributed by the United States Fish Commission will supply the entire shad-catch of the Atlantic coast, we begin to get some idea of the possibilities resulting from the work carried on on a scale of adequate magnitude. When we realize that in this fish alone, since 1880, the catch has been doubled, resulting in an addition of over $1,000,000 per annum to the food production of the country, the great economic value of fish culture becomes more apparent. The great influx of shad during the present season, completely glutting the market, shows a continued rapid rate of increase as the work of propagation is increased. The introduction and rapid spread of shad on the Pacific coast is a further proof of the beneficient [sic] effects of fish culture.

Gay and Seal went on to laud the success associated with the whitefish stocking program and with salmon restoration: "Fish culture, therefore, we may fairly conclude, no longer needs defence [sic], but may move forward serenely to a realization of the brightest conceptions of those who first conceived its value."[93]

Gay and Seal next put forth a list of things the American Fisheries Society could do to promote fish culture. Their list included advocating the removal of politics from the activities of the U.S. Fish and Fisheries Commission and the various state commissions, establishing state branches of the Society, promoting more cooperation among the state commissions and the U.S. Fish and Fisheries Commission, serving as a repository of fishery literature, and establishing awards in the form of medals for those who advance the science. They also suggested the establishment of a national fish culture school, which would be operated by the Commission.[93]

While fisheries science in all its aspects—including aquaculture—was developing rapidly, there was still a long way to go. Most of the early aquaculturists were good observers, and they made many breakthroughs by watching the behavior of the species in which they were interested. Once in a while, however, they got badly misled. For example, in the waning years of the nineteenth century, a debate was raging over where eels spawned and if, indeed, male eels even existed. No adult males had ever been seen, and the eggs of eels had never been collected or observed in nature. Some people believed that eels spawned at sea, but others, like Robert B. Roosevelt, felt the opposite was true and believed his observations proved it.

Since Robert Barnwell Roosevelt had a significant influence on United States aquaculture, let us take a look at his life. Roosevelt, whose nephew Theodore became President of the United States, was born in New York City on August 7, 1829, and died on June 14, 1906.[94] He was educated as a lawyer, and became interested in fish culture after meeting Seth Green. In 1850, he gave up the practice of law, having inherited sufficient money to do so, and helped finance Green's hatchery at Caledonia, New York.

In 1867, Roosevelt drafted a bill which led to the creation of the New York fish commission. He became a member of the commission and later a member of Congress. He was a dedicated sportsman and naturalist.

Roosevelt made his observations on eels in trout ponds on Long Island, New

Robert B. Roosevelt (1829–1906)

York, and concluded that it was not possible for them to have come upstream from salt water since they would have had to traverse a 2-foot outfall as elvers to get into the freshwater outflow. That and other observations led him to conclude that the eels had been spawned in the pond, not at sea:[95]

> If my present conclusion is right, it accords with the practice of all migratory fish, and brings eels into the ordinary catalogue of breeding in fresh-water and growing in salt. It seems to me impossible that I could be deceived.

Roosevelt was not aware of the ability of eels to "swim" through wet grass, but he was convinced that eels couldn't behave contrary to any other known type of fish found in fresh or salt water:[95]

The accepted opinion of their method of reproduction goes on the idea that they deposit their spawn in the salt-water . . . and that the young ascend the streams in May to find some suitable mud-endowed pond where they can live, luxuriate, eat, and grow fat. All this is contrary to the habits of every known variety of fish. . . .

In the discussion that followed Roosevelt's presentation of the above information to the American Fish Culturists' Association in 1878, Spencer F. Baird indicated that he had within the previous few weeks received eels with ripe ovaries, but acknowledged that males had not been found.*

The following year, in another paper presented to the American Fish Culturists' Association[96] Roosevelt revisited the issue. He had determined that the eels had been collected from fresh water during the fall—he thought they had been captured in Maine. The finding of developing females in fresh water supported Roosevelt's earlier argument:[96]

The received theories of the descent of the mature fish to the sea in autumn to spawn, and the ascent of the young in spring to the fresh waters to grow were discredited, and if not disproved, are now shown to be at least exceedingly doubtful. . . .

. . . many . . . gentleman insist that spawning eels descend the rivers in the fall to the salt water, and point to the construction of eel-weirs as proof; but it is possible that they are not descending, but are only roaming about looking for an appropriate place to spawn. They are caught in weirs late in the season, when the *ova* must be well matured, as was the case with those taken [in Maine], and just before they hibernate. They would scarcely get to the salt water before they would have to spawn, if they were to do so before hibernating, and yet it is a general rule that fish cannot change instantly from a fresh to a salt water element, or *vice-versa,* and never spawn immediately after making such a change.

Such controversies were stimulating for those involved in the newly developing science and led to lively discussions. Finding the answers to such intriguing questions sometimes involved years of study and required resources that were not available to academics. The government, in 1872, created the Fish Commission to examine the degree to which food fish were being overexploited, and that appointed a Commissioner who received no additional salary for his activities. By the 1880s, the Commission had established a fleet of vessels, constructed railroad cars for fish deliveries, established laboratories and hatcheries, and was becoming increasingly involved in an array of fisheries and oceanographic research topics that went well beyond the initial charge under which the agency had been established.

THE FISH COMMISSION MOVES TOWARD THE TWENTIETH CENTURY

Significant changes occurred in the Commission occurred in the waning years of the nineteenth century. There were major personnel changes, new facilities were opened

---

*The debate on eel reproduction was one that went back into ancient history. Roosevelt's views were discredited when scientists finally learned that the American eel and its European cousin both migrate to and spawn in the region in the middle of the Atlantic Ocean known as the Sargasso Sea.

and old ones expanded, production was increased, and the Commission became heavily involved in annual expositions where its activities were highlighted. Our attention will remain riveted on the aquaculture-related activities of the Commission, though other types of fishery investigations, studies of all types of marine life, and the development of U.S. oceanographic science were all being pursued.

Changes at Headquarters

Spencer F. Baird served the Commission well from its formation in 1871 until his death in 1887. President Grover Cleveland immediately appointed Dr. G. Brown Goode, who was assistant secretary of the Smithsonian Institution, to assume the role of Commissioner. That appointment lasted for about six months, after which the authorizing legislation for the Commission was changed.[97] The new act specified the following:

> . . . there shall be appointed by the President, by and with the advice and consent of the Senate, a person of scientific and practical acquaintance with the fish and fisheries to be a Commissioner of Fish and Fisheries, and he shall receive a salary at the rate of five thousand dollars a year, and he shall be removable at the pleasure of the President. Said Commissioner shall not hold any other office or employment under the authority of the United States of America.

The President approved the act on January 20, 1888, and Marshall McDonald was appointed Commissioner. For the first time, the job of Commissioner was seen to be sufficiently extensive that the individual in it was compensated.

Some of the exploits of Marshall McDonald with the U.S. Commission on Fish and Fisheries were given in Chapter 1 and will not be repeated here. His professional life was reviewed by W. C. Kendall in 1939, but little personal information was included in that biography.[98] Kendall did report on a few aspects of McDonald's life that have not heretofore been subjects of our discussion. McDonald was born in 1835. He received his first position in fisheries in 1878, when he was appointed fish commissioner for Virginia, which had been established in 1874. Two years earlier he had been placed in charge of a hatchery that was associated with the Virginia Military Institute.

In 1876 the Virginia fish commissioners sent McDonald to the Centennial Exposition in Philadelphia to study fishway models that were on exhibit. He was supposed to find a design that would be suitable for use with shad ascending the James and Rappahannock Rivers. Failing to find a suitable design at the exposition, he developed one of his own, which was apparently put into operation on the Rappahannock near Fredericksburg, Virginia, in 1881 or 1882. Another fishway of McDonald's design, built in 1882 on the Potomac River near Shepperdstown, Virginia, reportedly allowed the passage of bass and other species. In the following year a fishway was erected at Bosker's Dam on the James River upstream of Richmond. Whether McDonald's designs accommodated the passage of shad remains in question, but they were in sufficient demand that he apparently started a commercial enterprise, The McDonald Fishway Company, to meet the demand. In any case, several additional fishways of McDonald's design were established in the 1880s.

Marshall McDonald (1835–1895)

McDonald's ingenuity certainly didn't stop at the design of fishways. He was also heavily involved with the development of hatching equipment for the small eggs of marine species. He developed an apparatus for the hatching of cod eggs in which the water level rose and fell to simulate tidal activity. He called that one the tidal-hatching box.[98]

In 1883, McDonald described experiments that led to the development of an automatic fish-hatching jar, now known as the McDonald jar.[99] Patent number

263933 (U.S. Patent Office) was issued to McDonald and Stephen C. Brown in 1882.[100] McDonald jars, which are found, sometimes in modified form, in hatcheries all over the world today, were designed for the hatching of shad eggs, but they had soon been adapted for use with many other species, including whitefish, yellow perch, cod, and even lobsters.[98] Modern hatcheries have used the jars for many additional species.

McDonald also developed other types of hatching apparatus, and made a number of other valuable contributions during his career. He remained Commissioner from the time of his appointment in 1888 by President Grover Cleveland until his death on September 1, 1895, after a period of illness.[101,102]

Herbert A. Gill, who had been chief clerk of the Commission, was appointed Acting Commissioner upon McDonald's death. He held the office until March 30, 1896, when John J. Brice was selected to fill the vacancy.[103] Brice only held the office until February 16, 1898, when he was replaced by George M. Bowers.[104] Bower's tenure was considerably longer than that of his immediate predecessors, as we shall see.

McDonald began publishing financial information on the Commission in his report of activities for 1888 and 1889, the first report issued after he became Commissioner.[105] Of the total appropriation for the fiscal year ending June 30, 1889, of $257,580, the breakdown of expenses was as follows:

Compensation of Commissioner	5,000
Propagation of foodfishes	135,000
Office rent	2,500
Distribution of foodfishes	31,180
Maintenance of vessels	53,900
Inquiry respecting foodfishes	20,000
Statistical inquiry	10,000

Propagation and distribution of foodfishes were aquaculture activities. The distribution aspect of that category would have required a great deal more money had the railroads not cooperated by, in many instances, providing free shipment of the Commission's eggs and larvae. In addition, some ship time was still being devoted to the collection and hatching of eggs. Thus, the bulk of the funding was related to aquaculture. McDonald had a chief clerk and a disbursing agent (accountant) working in his office. It is not clear whether their salaries were paid from one of the other accounts or from some other source. The budgets for 1892 and 1893 were similar in magnitude and distribution of funds to that of fiscal 1889.[105,106] The Commissioner's salary remained the same at $5,000 for several years.[107]

The first Commission annual report issued after the death of McDonald showed a different breakdown and a significant increase in total expenditures.[101] Salaries were placed at $172,120.00—the Commissioner's compensation was not broken out. Office administration was $9,000. Propagation and distribution of foodfishes had been lumped together and totaled $92,181.94. There were also expenses associated with constructing and repairing stations operated by the Commission. Rent expense had fallen to $2,000.

The Commission was clearly growing. As a final example, the following is an account of expenses for fiscal 1898:[104]

Salaries	$195,620
Miscellaneous expenses	
Administration	9,000
Propagation of foodfishes	132,500
Vessel maintenance	30,500
Inquiry respecting foodfishes	10,800
Statistical inquiry	5,000
Ship purchase, modification, and repair	37,140
Rebuilding fish transportation cars	10,000
Establishment, construction, etc., at fish culture stations	53,516

What had started as a modest effort to look into the depletion of U.S. foodfish stocks in 1871 had become a half-million-dollar organization by the end of the century.

Expansion of the Hatching Stations

Baird had supported the establishment of hatching stations early in his tenure as Commissioner. Over the years, he expanded the number of hatcheries and related stations, and that expansion continued under later administrations. By 1883, the following 16 stations were in operation, a few of which were used primarily for research, though most were involved in propagation:[108]

Glouchester, Massachusetts	Baird, California, trout station
Wood's Hole, Massachusetts	Wytheville, Virginia
Saint Jerome, Maryland	Cold Spring Harbor, New York
Grand Lake Stream, Maine	Havre de Grace, Maryland [aka Battery Station]
Bucksport, Maine	Central Station, Washington, D.C.
Northville, Michigan	Fort Washington, Maryland
Alpena, Michigan	Monument Reservation, Washington, D.C.
Baird, California, salmon station	Washington Arsenal grounds

The situation did not change in 1884, though Baird was beginning to lay the groundwork for the establishment of new facilities.[109] By 1885, all of the 16 stations listed above were still in operation, and a temporary station had been added at Lambertville, New Jersey, for the hatching of shad.[110] In helping to set the stage for developing at least one new facility, Baird had sent Livingston Stone to the Columbia River in 1883 to find a suitable spot for a salmon station.[111]

Stone explored extensively and tried a suitable site that was convenient to the Northern Pacific Railroad so the eggs collected could be transported easily. The one site that appealed to him because it met his criteria was located on the Little Spokane River. The criteria were as follows:[111]

1. Abundance of breeding salmon
2. Accessibility of location
3. Adequate water supply
4. Convenience of location for obtaining water
5. Availability of location [relates to whether the site is already privately owned]
6. Facility for catching parent fish
7. Facility for arresting the upstream progress of the migrating salmon
8. Security from high water and other dangers

There were 15 stations active in 1888, and 17 in 1889. Some of the names of the players had been changed, so it may be best to begin anew with the list as developed for 1888 by Marshall McDonald:[105]

Schoodic Station, Maine
Craig's Brook Station, Maine
Gloucester Station, Massachusetts
Wood's Holl* Station, Massachusetts
Cold Spring Harbor Station, New York
Battery Island Station, Maryland
Fort Washington Station, Maryland
Central Station, Washington, D.C.
Fish Ponds, Washington, D.C.
Wytheville Station, Virginia
Sandusky Station, Ohio
Quincy Station, Illinois
Northville Station, Michigan
Alpena Station, Michigan
Baird Station, California

Repairs were being affected at the Baird Station, and three new stations were under construction:

Clackamas River Station, Oregon
Neosho Station, Missouri
Duluth Station, Minnesota

Both the Duluth and Clackamas River stations were in operation by the end of 1889.[105]

By 1892, the number of stations had reached 22. At that time the Baird stations (one for salmon and one for trout) were listed as a single entity. No longer listed were Fort Washington Station, Maryland, or Quincy Station, Illinois.[106] The Sandusky Station was a facility of the Ohio Fish Commission and had been temporarily used by the U.S. Fish and Fisheries Commission during the 1888–1889 season.[105] Quincy, Illinois was back on the list for 1886.[103] Added to the list for 1892 were Green Lake, Maine; Delaware River (which was actually associated with activities aboard the steamship *Fish Hawk*); Bryan Point, Maryland; Put-in-Bay, Ohio; Quincy, Illinois; and Leadville, Colorado.[106]

In 1893, new fish hatching stations were established in St. Johnsbury, Vermont, and near the St. Lawrence River in New York. Also in 1893, the site for a new facility in Bozeman, Montana, was selected.[107] The Commission was also looking for a site in Alaska. The Fort Gaston Station, California, had been added by 1894. There was also a temporary station established in conjunction with the Chicago World Fair that year.[112] The Schoodic Station in Maine and Cold Spring Harbor Station in New York were no longer listed, nor were the Fort Washington Station, Maryland, or the Sandusky Station, Ohio. New on the list was the Green Lake Station, Maine, which

---

*You may recall that Wood's Hole was sometimes spelled *Wood's Holl* or *Woods Holl* at that time.

may represent a name change. Plans were being made for establishment of a fish hatchery at San Marcos, Texas, contingent on construction of a dam across the San Marcos River by the San Marcos Water Company.[101]

New on the list for 1895 were Korbel, California.[113] That year J. J. Brice had been put in charge of West Coast operations for salmon production. He was charged with formulating a plan to bring back salmon to their original numbers. He recognized that rehabilitation of a depleted river was both expensive and labor-intensive. He had established the hatchery at Fort Gaston, which was a central hatchery. Smaller satellite hatcheries collected eggs and returned the fry obtained to the immediate area from which the eggs came. Brice recommended the establishment of hatcheries at four additional sites: Chilcat River, Alaska; Puget Sound; Colville Reservation on the upper Columbia River; and Eel River, California. His cost estimates for operations at the central hatchery and the establishment of four new main facilities plus satellites were:[114]

Four central hatcheries, including buildings and apparatus	$8,000
Twenty auxiliary hatcheries at $300 each	6,000
Five superintendents at $1,200/yr	6,000
Six laborers, at $40/mo for a year	2,880
24 laborers at $40 for 4 months	3,840
Miscellaneous expenses (five hatcheries)	7,500

The five hatcheries would not include expenses for Baird, but included Gaston.

Brice also recommended establishing one river and its tributaries as a government fish preserve from which seining or taking salmon in any way for commercial purposes would not be allowed. He suggested the Klamath River as the preserve.

In fiscal 1897, a team was sent to the Columbia River Basin to search for a hatchery site to supplement the Clackamas River hatchery. A number of streams were evaluated during August and September. Salmon in significant numbers were found in the Big White Salmon and Little White Salmon Rivers and in Tanner and Eagle Creeks.[115] A hatchery site on the Little White Salmon River was selected and was in operation by 1899.[116]

Redfish Lake, Idaho, has been in the news during the past few years because the sockeye salmon in that lake have been designated as endangered by the federal government. In 1992, sockeye returning to Redfish Lake were captured and spawned in captivity with the idea of using the offspring as the basis of developing sufficient numbers of fish to eventually reestablish the stock. Because of the small number of returning fish to Redfish Lake, crossing them with genetically similar fish from Alturas Lake, which has a larger fish run, is being considered. Research on salmon in Idaho is not a recent endeavor. Studies on the spawning habits of both sockeye and chinook salmon in the headwaters of the Salmon River and in both Redfish and Alturas Lakes were initiated by the Commission in 1894 and continued through 1896. As reported in 1897, some of the findings were:[115]

1. Both species reached the spawning grounds in excellent condition.
2. All of the fish of both species died after spawning.

3. The young of both species remained in the vicinity of the spawning beds for a year.*

4. Redfish Lake eggs laid in September hatched the following March.

By 1898, both the San Marcos, Texas, and Bozeman, Montana hatcheries were in operation. The San Marcos hatchery concentrated on black bass, rock bass, and crappie during 1898, while the Bozeman Montana facility produced brook trout, steelhead trout, and grayling.[104]

As a new century dawned, the number of hatcheries in operation by the U.S. Fish and Fisheries Commission had reached 35, including the Steamer *Fish Hawk* hatchery, which conducted its operations on the Delaware River. Since you may be confused (as I was) about which stations were still in operation and which may have been added over the years, the following list for 1900 is provided:[117]

Green Lake, Maine	Detroit, Michigan
Craig Brook, Maine	Alpena, Michigan
Grand Lake Stream, Maine	Sault Ste. Marie, Michigan
St. Johnsbury, Vermont	Duluth, Minnesota
Nashua, New Hampshire	Quincy, Illinois
Gloucester, Massachusetts	Manchester, Iowa
Wood's Hole, Massachusetts	Neosho, Missouri
Cape Vincent, New York	San Marcos, Texas
Steamer *Fish Hawk*	Leadville, Colorado
Battery, Maryland	Spearfish, South Dakota
Fish Lakes, Washington, D.C.	Bozeman, Montana
Central, Washington, D.C.	Baird, California
Bryan Point, Maryland	Battle Creek, California
Edenton, North Carolina	Clackamas, Oregon
Wytheville, Virginia	Rouge River, Oregon
Erwin, Tennessee	Little White Salmon, Washington
Put-in Bay, Ohio	Baker Lake, Washington
Northville, Michigan	

One of the early aquaculturists who has now been enshrined in the Fish Culture Hall of Fame was D. C. Booth, who operated the Spearfish, South Dakota, fish hatchery for a number of years beginning in 1899, and for whom the hatchery is now named.† Booth was born on August 5, 1867, at Palatine Bridge, New York. He was educated in the New York public school system and attended Colgate University. He became a government employee in the U.S. Customs Service in 1893 and worked the Columbian Exposition in Chicago that year. (His family had moved to Chicago in 1890.) He subsequently joined the Fish Commission, beginning his career at the Cape Vincent hatchery. He then worked at Woods Hole and Leadville before being

---

*Chinook salmon spawn in rivers, and the fingerlings migrate downstream to the sea within a few months after hatching.

†Booth was inducted in 1986, the second year that the Fish Culture Hall of Fame was in existence. His biography was provided by Arden Trandahl.

reassigned to Spearfish, where he became the youngest hatchery manager in the federal system.

Leadville had excess fish eggs that they wanted to send to someone. While Spearfish was still incomplete in terms of equipment, Booth told Spearfish he would accept the Leadville eggs. The Spearfish crew constructed hatching trays within two days and were able to accommodate the egg shipment when it arrived.

Booth's ingenuity was taxed further when two months after he assumed the manager's position the hatchery was flooded. A gulch running through the hatchery site was the source of the problem. Dry most of the time, the gulch became a significant threat during downpours. Booth and his crew immediately began their efforts to divert floodwaters by digging drainage canals and constructing retaining walls. The activity required several years of spare-time labor and resulted in the stone walls, culverts, and bridges that can still be seen at the hatchery. The work was sufficiently complete by 1904, when the hatchery was spared from damage in another flood.

There were no native trout in the Black Hills area, so until the Spearfish hatchery was opened, trout were shipped to the region from Leadville, Colorado. Trout hatched at Spearfish were first delivered on April 19, 1900. The fish, 25,000 brook trout, were sent by rail in iced 10-gallon milk cans to Englewood and Savoy, South Dakota. Over the next three decades Booth shipped fish throughout the Great Plains. During his tenure as superintendent at Spearfish, Booth hatched cutthroat, brown, rainbow, and even a few steelhead trout in addition to brook trout.

In 1901 Booth was charged with managing the fisheries in Yellowstone National Park. He and his staff arrived early in the year and, since no facilities were in place, lived in tents while constructing fish traps, hatching troughs, and other necessary structures from scrap lumber. Fish were spawned and their eggs shipped by horse-drawn wagon to Gardiner, Montana, where they could be loaded onto railroad cars. The Yellowstone activity was one of few sources of trout eggs in the West for a number of years. Booth continued the Yellowstone operation until 1910, after which he was replaced, though only after an intense argument with his superiors.

With the exception of two years he spent at Winona, Minnesota, Booth remained at Spearfish for the remainder of his career. He retired on November 1, 1933. During his career he pioneered many trout management techniques and trained a cadre of people who became hatchery superintendents or held other important positions in the fisheries profession.

Fish Distribution in the 1880s and 1890s

The U.S. Fish and Fisheries Commission and various state commissions distributed hundreds of millions of eggs and fry during the final two decades of the nineteenth century. Many pages in the annual *Report of the Commissioner* were devoted to detailing fish distribution. (It is certainly not my intent to inundate you with those details, but the magnitude of the activity is something that is interesting, if not truly mind boggling.) Fish were distributed to both public and private waters (with invertebrates undoubtedly restricted to public waters since they were all marine species). While fingerlings and even adults were distributed, the bulk of the animals were stocked as eggs or fry.* Numbers of fish stocked, a list of the species propagated and stocked, and an indication of the miles traveled in conjunction with fish distribution on the railroads for a few years from 1885 to 1900 (inclusive) are hereby provided. Readers desiring more details on the activities may consult references 101, 103–107, 109, 110, 113, 115, and 118–123.

Total Numbers Propagated and Distributed during the Fiscal years Shown

1885	173,666,083	1895	619,915,852
1887	210,628,413	1896	498,488,268
1888	238,986,117	1898	857,309,546
1889	322,795,830	1899	1,056,371,898
1892	305,918,346	1900	1,164,366,754
1894	450,310,543		

Some Species Mentioned as Having Been Propagated and Distributed during Those Years

Atlantic salmon	Pike-perch (walleye)
Black-spotted (cutthroat) trout	Quinnat (Chinook) Salmon
Brook trout	Rainbow trout (including steelheadtrout)
Brown trout	Rock bass
Carp	Shad
Catfish	Silver salmon
Cod	Smallmouth bass
Crappie	Smelt
Flatfish	Striped bass
Gar	Sunfish
Goldfish	Tench
Grayling	Warmouth bass
Haddock	White bass
Herring	Whitefish
Lake trout	American lobster
Largemouth bass	American oyster
Pike	Clam

Railroad Miles Provided Gratis in Typical Years

1892	62,761
1893	97,361
1894	65,093
1895	81,578
1989	63,167
1900	42,746

---

*The terminology follows that used at the time. Actually, in many cases, particularly of marine species, larvae were stocked.

Brown trout had been obtained from Europe during Baird's administration, and by 1884 they were being hatched in Michigan. During that same year, the Alpena and Northville hatcheries distributed 9 million whitefish to Lake Ontario, 7 million to Lake Erie, 16 million to Lake Huron, 4 million to Lake Superior, and 11 million to Lake Michigan.[124]

The free mileage provided by railroads in their own cars does not include the free mileage provided to the messengers, who traveled with the fish on most deliveries (most of their trips were paid for), or the miles covered by the fish delivery cars operated by the Commission. When those miles are added in, the figures become truly prodigious. In 1892, the Commission was operating three railroad cars, and the total mileage covered for all activities was 184,730 miles.[106] By 1895 a fourth fish delivery car had been added to the Commission's inventory. Those four cars, along with the messengers, covered a total of 189,458 miles.[123] In 1900, when the Commission had five fish delivery cars in operation, the following numbers were generated: "The five cars of the Commission traveled 101,796 miles in distributing fish, and detached messengers and employees of the stations traveled 157,297 miles."[121]

The Central station in Washington, D.C., in addition to hatching and distributing fish, was the site of a public aquarium. Freshwater and marine species were on display there for the public. The diversity of fish displayed at the Central station is demonstrated from information complied for the 1897 fiscal year:[125]*

Freshwater species:

Largemouth bass	Smallmouth bass	Rock bass
Crappie	White bass	Yellow bass
Rainbow trout	Brook trout	Lake trout
Steelhead trout	Quinnat salmon	Atlantic salmon
Yellow perch	White perch	English tench
Paradise fish	Golden tench	Goldfish
Golden ide	Pike	Pickerel
Channel catfish	Yellow catfish	Chub sucker
Red sucker	Leather carp	Shad
Sunfish	Mill roach	Common eel
Lacefin dace	Freshwater shrimp	Bullfrog
Crawfish	Freshwater terrapin	Snapping turtle

Marine:

Sea trout	Sea bass	Sea robin
Spot	Pinfish	Scup
Croaker	Pigfish	Striped mullet
Flounder	Mummichog	Tautog
Toadfish	Hogchoker	Moonfish
Swellfish	Striped bass	Black drum
Sheepshead	Skate	Stingray
Lobster	Hermit crab	Blue crab
Medusa		

---

*The names in the list are as presented by Ravenel. Many of the common names have changed in the interim.

What Was Livingston Stone Up To?

We last left Livingston Stone fighting floods and looking over his shoulder in the event of insurrections by some of the local tribes. He continued to produce salmon and trout at the two McCloud River facilities, known in combination as the Baird station. Stone was involved in other activities as well. For example, in 1879 he was responsible for another east-to-west delivery.[126] The California fish commissioners had required a delivery of lobsters, striped bass, eels, and black bass.

Maintaining the saltwater species during the cross-country rail trip was difficult. Of particular concern to Stone was that the saltwater became foul when confined in tanks:[126]

> It is well known that ocean water contains an infinite number of microscopic insects commonly called by the general term animalculæ. It is equally true, though not so well known, that these animalculæ are the cause of the fouling of ocean water when confined in tanks.

Boiling the water killed the "animalculæ" but also led to formation of a precipitate that he thought was another source of salinity reduction. Another way to treat the water was to cover the containers and let them stand for two or three weeks. The organisms of concern would die and fall to the bottom: "The water above will be perfectly sweet and clear, and will remain so indefinitely."[126]

Stone recognized that using ice in the water to keep the animals cool reduced salinity. To handle the cooling problem he came up with three methods:

- Place an ice-and-salt mixture in large stone jars, and hang the jars in the tanks.
- Put the freezing mixture in a vessel surrounded by another vessel containing the water to be cooled.
- Fill a large drain-tile with the ice-salt mixture, and keep it in a reserve tank of water until the water cools enough to be exchanged with the warmer water in the lobster tanks.

All three methods were used during the trip.

Seth Green had obtained 3,000 eels from the Hudson River and had them delivered to Albany, New York, to meet the train. Another 600 eels were obtained from a Mr. Mason, who delivered them from New Jersey. There were also 22 female lobsters with over 1 million eggs (one lobster's eggs had hatched en route to Albany, where the train was loaded), 132 striped bass 3–4 inches long, 30 striped bass 6–8 inches, and 22 black bass.

Contemporary thinking with respect to fish stocking is to take a very conservative approach, particularly with the introduction of species or even strains or races of animals that are not native to a particular region or even to a particular watershed. The assumption in the 1800s was that if more fish were stocked, production would increase. If new species were added, additional production could be obtained. Even so, stocking programs were usually fairly well planned. As in the 1879 delivery by Livingston Stone, the species being delivered to California were requested by the California fish commissioners. Those individuals undoubtedly discussed what they wanted to stock and didn't just say something like, "we'll take whatever you've got."

Not all fish stocking involved a great deal of forethought, however. There was even a bit of superstition involved, though after Stone's dunking in the Elkhorn River after the train wreck, his actions are perfectly understandable. "We crossed the Mississippi River at seven o'clock [on a June] Saturday, throwing into the river, as I always have done before, a few fish for luck."[126] Some losses occurred, but most of the animals reached California, where they were distributed into what were thought to be suitable environments.

In response to a request for 2 million salmon eggs by the California Fish Commission in 1883, an additional hatchery building was added to the Baird salmon facility.[108] Livingston Stone arrived on the first of August expecting several hundred salmon to arrive within a week. However, only one fish made its appearance by August 7.

Salmon were scarce not only on the McCloud River in 1883. The salmon run on the Spokane River at Spokane Falls was also in shambles. The Indian fishery at the falls had taken tens of thousands of fish in 1882, but only a few dozen fish arrived there in 1883.[108]

Stone had expected to see up to 8,000 fish jumping every hour by the time he arrived on station, but he observed none. He approached the Indians to verify his observations, which they confirmed:[127]

> This fact was now not only evident, but it implied that some very unusual agencies were at work on a large scale somewhere below us to prevent the salmon ascending the river. On making inquires, we were told that there were several thousand Chinamen, variously estimated at from 3,000 to 6,000 at work on the California and Oregon Railroad, on the Sacramento River, eight or ten miles below us, and that these Chinamen were doing a very large business in capturing fish by exploding giant powder in the water.
>
> We also heard that the railroad company were putting in very heavy blasts of powder near the river, and it was possible that this heavy blasting kept the salmon back. . . . I sent a man to the scene of the blasting operations to make an examination. . . . This man, on his return, stated that the Chinamen did kill what fish they could with giant powder, but he also gave such a description of the blasting operations of the railroad company as led me to think that while the Chinamen were doing some mischief killing salmon . . . , the heavy blasting on the railroad was the chief agency in keeping the salmon from ascending the river.

Stone requested the California Fish Commission to intervene and remedy the situation. Before the man sent out to make the inquiry could report back, Stone learned from an Indian that some unidentified men had constructed a rack across the river several miles below the Baird station, which allowed them to prevent upstream migration of the fish. Those men were apparently selling salmon to the Chinese railroad workers.

Stone sent one of his own men, Mr. Radcliff, downstream to check out the story:[127]

> Mr. Radcliff found some white men preparing a ground for drawing a seine, but did not find any rack or obstruction in the river. He reported also that in his opinion it was the Chinamen that were keeping the salmon back by exploding giant powder in the river

and not the blasting operations of the railroad. I mention this in order to have the case fairly stated, though I am quite confident myself that it was the blasting on the railroad that is mainly responsible for the disturbance of the salmon in the river. . . .

The result of the poor run was that only 1 million eggs were collected during the 1883 season.

The following year, Stone's report of activities at Baird laid the blame for destruction of the salmon run in 1883 squarely on the back of the Central Pacific Railroad Company. He recounted the 1883 debacle as follows:[128]

> The consequence was that it became impossible to take the usual number of salmon eggs at the McCloud Station. Indeed, it was with great difficulty, and only by preserving efforts under great discouragements, that enough parent fish could be caught to yield the extremely small number of a million salmon eggs.

Thus, in 1884, salmon spawning activity at the Baird station was curtailed. A caretaker was left in charge of the station and a few repairs were made.

Problems also plagued the Baird trout spawning facility during 1884.[129] Loren W. Green, who was in charge of that facility, had indicated that the female trout were slow to mature and that the numbers of eggs produced by individual fish was reduced. The trout problem was attributed to a scarcity of food in the river, and that scarcity was associated with the lack of spawned-out salmon carcasses, which would serve as a nutrient source for the organisms on which the trout depended for food. A storm caused considerable egg mortality in the hatchery because of turbid water, but the station was still able to incubate over 300,000 eggs. Some of the fry that were hatched from those eggs were sent East. The bulk of the remainder were given to the California Fish Commission and retuned to the McCloud River. A few were maintained at the station for rearing.

A small salmon run occurred in 1885.[110] As a consequence, no eggs were taken at the Baird salmon station that year.[130] The facility was left in the caretakership of Robert Radcliff. The trout facility was operated as usual. In 1886 the total number of eggs collected exceeded 220,000. The following year the total was 268,000. Trout numbers were down, in part, owing to what Stone referred to as a "mysterious disease."

Salmon production at the Baird station between 1883 and 1888 was curtailed. From what we have seen in the reports of Stone, the primary reason for halting operations was the lack of fish as a result of the dynamiting associated with railroad construction. Commissioner McDonald put a different spin on the story in his report of 1892.[122] McDonald indicated that attempts over several years to establish populations of Pacific salmon in the rivers of the northeastern United States had failed. For that reason, according to McDonald, salmon egg collection was curtailed. One can only speculate on whether behind McDonald's arguments lay the feeling that to criticize a railroad might lead to a diminution of enthusiasm for providing the Commission with free transportation. The failure of the Commission to establish Pacific salmon on the East Coast was certainly a fact, and perhaps a convenient happenstance. Had Pacific salmon stocking in the East appeared more promising, the decision might have been somewhat different.

In any case, salmon egg taking at the Baird station resumed in 1888 in order to provide fish to restore the Sacramento River runs.[122] Mr. G. B. Williams was put in charge of the station that year.

With the resumption of salmon spawning at the Baird facility, trout production seemed to be deemphasized. For example, in 1892, only salmon were produced.[105] During that year, over 3 million salmon eggs were taken. The railroad had apparently completed its construction activities on the McCloud River, and the salmon runs had been reestablished. In 1894, the take was slightly over 2 million eggs, 50,000 of which were sent to the Chicago World's Fair for hatching.[112]

Capt. W. E. Dougherty (in the U.S. Army) and Livingston Stone were the persons in charge of the Fort Gaston, California, salmonid hatchery. It was the first such establishment to be located on a military reservation in the United States. In 1895, Mr. Williams, the superintendent at Baird, resigned, and Livingston Stone was placed back in charge of that facility.[123] He remained at Baird until 1897, when he requested reassignment. That was the year he took charge of the Cape Vincent, New York, hatchery, where he remained for the remainder of his years of service to the Commission. The American Fisheries Society honored Stone in 1910 by publishing his photograph and remarking that he was the only living founding member of the organization. Stone was unable to attend the meeting held in New York City, though the problem was not physical this time. A letter written to the Society had the following to say about Stone's condition:*

> . . . from [Livingston Stone's son] I learned that his father's condition is such that he is able to be about each day, and is in fairly good physical condition. His mental condition is somewhat peculiar, and, as I understand it, he lives almost entirely in the present—that is, he knows those about him, converses and plays games with them, but does not remember anything in the past. He does not remember anything about his fish-culture work at Cape Vincent or elsewhere, or his associates in the work.

Stone was living in Pittsburgh, Pennsylvania. He was attended by his wife and son as well as a nurse. He died in 1912.[9]

Activities with Atlantic Salmon

Atlantic salmon hatching was already routine by 1880. Charles G. Atkins was the authority at the time. In the *Report of the Commissioner* for 1878, Atkins described how to set up an Atlantic salmon hatchery. He stressed that the first thing needed is a good water supply.[131] Some contemporary aquaculturists could benefit from that sage advice.

Atkins also admonished that once the facility is in operation, constant vigilance is important. He stressed the need for the maintenance of complete records of not only the number and condition of eggs but also temperature (air and water), water flow rates, and general water quality. Finally, he advised that once the eggs have hatched,

---

*Letter from Chester K. Green to Seymour Bower, president of the American Fisheries Society dated September 19, 1910, and published in the 1910 *Transactions of the American Fisheries Society*, vol. 10: 74.

the health and behavior of the fish should be recorded. That advice is as valid today as it was over a century ago.

Atkins was a pioneer in fish tagging. His first attempts to develop tags that would not be lost and that would provide information on individual fish were outlined in Chapter 1. By the mid-1880s Atkins had received enough tag returns to determine, at least to his satisfaction, that Atlantic salmon which survived after spawning would subsequently spawn again two years later.[132]

Other than some changes in personnel, variations in the size of the spawners that were obtained, and the date at which adult salmon became available for purchase, there was little to report on what became the routine procedure associated with the spawning of both landlocked (Schoodic) and anadromous Atlantic salmon. A few modifications were made in the hatcheries, and these and the low fertilization rate of Schoodic salmon reported in 1885 were some of the more significant events in an otherwise routine series of years during the 1880s.[133–138]

So What about Other Species?

If you'll leaf back a few pages, you'll see that the list of species being produced by the Commission expanded significantly during the 1880s and 1890s. Discussing all of them would be tedious, but it might be of some interest to look at what was happening to some of the primary species of interest to the Commission during its early years of activity. I refer, of course, to trout, shad, cod, oysters, and lobsters. (And we shouldn't forget about carp; however, we'll defer that discussion for awhile.)

Trout. Since we've been talking about salmonids, we'll start out with some information on trout production in the late 1800s. Trout of various species were being hatched and distributed throughout the United States by the Commission during the closing decades of the nineteenth century. The year 1885 was probably typical. One of the facilities reported on by Fred Mather was the Cold Spring Harbor, New York, hatchery.[139] It was owned by the New York Fish Commission but operated in large part by the U.S. Fish and Fisheries Commission. The station was responsible for spawning and hatching several species. Some of the activity involved the receipt, hatching, and subsequent distribution of fry obtained from eggs collected elsewhere. In 1885, for example, hatchery personnel handled brook trout and rainbow trout eggs collected at the Northville, Michigan, station and brown trout eggs shipped in from Germany.

The Wytheville, Virginia, station was located on state land, and control of activities was granted to the Commission by the Commonwealth in return for a $500 annual rental fee.[140] Maintenance expenses were the responsibility of the Commission, but permanent improvements were provided by the Commonwealth. The purpose of the station was to produce trout for stocking in the Appalachian region. Commissioner McDonald felt that brook trout and rainbow trout were good species for stocking in cold-water streams located in several Appalachian states.

The Northville and Alpena stations in Michigan were responsible for the hatching of whitefish and trout. Trout species included lake trout, brook trout, brown trout, and rainbow trout. Frank Clark reported that less-than-satisfactory results had been obtained with respect to rainbow trout in fiscal 1886:[141]

> On account of the continued partial failure in the returns from this species, I would recommend that the stock on hand be distributed, and their propagation at the Northville Station be discontinued. At present about one-half of the water supply and pond facilities is devoted to rainbow trout, the returns from which are meager and unsatisfactory. By supplanting them with brook trout and German [brown] trout, or by concentrating the divided forces on the line, the aggregate results would be greatly increased.

Frank Clark's name has come up several times in this narrative, so perhaps it's time to take a closer look at the man. Frank Nelson Clark was born in Clarkston, Michigan, a town named after his ancestors. Nelson W. Clark, Frank's father, was one of the early American fish culturists. His name has appeared previously in conjunction with whitefish spawning. Nelson Clark was the man Spencer F. Baird picked to embark on that particular activity. The two Clarks worked together in commercial fish culture during the period when the vocation was not yet in vogue:[142]

> Father and son alike were born fighters, with splendid physical equipment and mental resourcefulness. Both were throbbing and pulsating dynamos of human energy, and so they swept opposition aside and beat down and conquered adversity.

The Clarks moved to Northville, Michigan, in 1874, having found a better set of conditions for the rearing of brook trout. Nelson Clark died in 1876, leaving Frank to continue the business alone, which he did until the Commission took over the facility and hired him in 1880.

Beginning in 1874, Frank worked on a seasonal basis for the Commission. He would travel to the Atlantic coast at the appropriate times to spawn shad and other marine species. He also made frequent trips distributing fish to various locations, including the West Coast.

In 1880, having proved his ability to Commissioner Baird, Frank Clark was employed by the Commission on a full-time basis to supervise propagation activities aimed at enhancing the fish stocks in the Great Lakes. He operated from the Northville station, which was upgraded by the Commission beginning in 1881.[28]

A few years after assuming his permanent position with the Commission, Clark embarked on a new enterprise at the suggestion of Marshall McDonald. Clark's charge was to attempt producing fingerling or larger trout rather than fry.[143] By the 1890–1891 season, Clark had distributed over 150,000 of the larger fish to several states. There was feeling expressed that fish released from hatcheries after extended stays and fed such things as beef liver would not be able to identify natural foods— and would therefore not survive. That issue was addressed by Clark:[143]

> . . . is it not a fact, . . . that in nine cases out of ten that the stream adjacent to any fish-culture establishment where trout are bred, and where trout partially grown are invariably escaping, does not become, in a few years, better stocked than streams where perhaps 100,000 have been planted[?]

Clark ran into problems convincing people who expected to receive 3,000–25,000 fry for stocking that equivalent results could be obtained by stocking 100–500 larger trout.

Frank Clark died on December 19, 1910.[142] Some of his personal reflections on his association with Spencer F. Baird were published by the American Fisheries Society during the year of his death. He also described his early activities as a fish culturist:[144]

> The first fish egg that I hatched . . . [was] in the winter of 1865–66, [when] I succeeded in hatching and rearing one trout to be three months old. That was the total output for the first season.
>
> Since that date I have been connected with hatcheries that have turned out as many as 600,000,000 eggs in a season—no, I will not say turned out, but collected— probably 450,000,000 fry or thereabouts were turned out.

Shad. Shad propagation continued from the early days of the Commission until well into the twentieth century. Marshall McDonald laid out the costs of the activity with respect to shad production on the Potomac River in 1886 as follows:[145]

> For the conduct of the work, . . . the Commissioner authorized an expenditure not to exceed $5,000. At Fort Washington Station the actual cost of collecting, developing, and transporting the eggs was $2,879.90; at Central Station, for hatching and distribution, $916.55; total, $3,796.45. The total number of eggs obtained was 36,362,000, and the losses during incubation were 6,625,000, leaving the aggregate number furnished for distribution from the Potomac River station 29,737,000. The cost of production was $127.66 per million, or 78 shad for each cent of expenditure.

In addition to the two stations mentioned in the above quotation, shad were collected at Battery Island station on the Susquehanna River and by the steamers *Fish Hawk* (operating on the Delaware River) and *Lookout*.[146] The collecting crew at the Battery Island station in 1886 was described as follows:[147]

> The collection of spawn for the station was done by men and boys hired temporarily for the purpose. As many as 40 men and boys in addition to the station's ordinary force were employed. These were paid monthly wages, each being allowed $10 a month for subsistence.

Collection of shad began on April 19 and continued until June 10:[146]

> . . . the total number of eggs collected [was] 60,766,000. Of this number there were received from the steamer Fish Hawk 2,099,000, and from the steamer Lookout, 2,433,000. . . .

Shad collected at the various stations and from the ships were stocked not only on the East Coast in such places as Narragansett Bay, Long Island Sound, Chesapeake Bay, and the Hudson River, but also in the Mississippi River, the Colorado River, and the Columbia River Basin.[146]

Cod. While the desire to produce cod was in place from the Commission's beginning, the technology to do so was somewhat slow in developing. Some cod eggs were hatched by the Commission at the Gloucester facility in 1878,[8,76] but it was clear that research would be needed to perfect the technique. Baird transferred cod operations to Woods Hole in 1879,[46] where things proceeded slowly. In December 1885, James Carswell proceeded from Washington, D.C., to Woods Hole to work on the problem. He was equipped with Marshall McDonald's hatching apparatus that simulated tidal action.[148] Capt. H. C. Chester, who was in charge of the Woods Hole station, felt that constant water motion was required, but the first experiment with flowing water resulted in complete mortality. McDonald himself showed up a few days later and concluded that there had been too much water motion. Various techniques were tried, but everything failed.

Carswell finally set up the tidal-motion hatching apparatus and, after losing the first batch of eggs to a salinity problem, managed to hatch out about 50% of the batch used on his second try. He estimated that a total of 15 million eggs had initially been taken, most of which were lost during the various experiments. About 2 million fry were ultimately distributed. Conclusions with respect to cod hatching included using as little motion as possible, filtering the water before use, and avoiding all metal in the hatching apparatus.[148]

Both the Gloucester and Woods Hole facilities continued to be focal points for cod hatching once the techniques had been worked out. During the 1886–1887 spawning season 43,575,000 eggs were handled at Woods Hole. The hatching rate was about 50%.[149] Charles Atkins indicated that more eggs would have been collected had the broodfish been available. As it was, most of the spawners were collected by the schooner *Grampus* or delivered to the ship by fishermen in the Gulf of Maine and transshipped to Woods Hole. While there was interest in rearing shad beyond the hatching stage, efforts to do so were not productive.

The Charles Atkins mentioned in the preceding paragraph was the same person who pioneered salmon and trout culture. Born in 1841 and a contemporary of Seth Green, Livingston Stone, and other notable nineteenth-century aquaculturists, Atkins continued to be active in the profession until his retirement at the age of 79 in August 1920.[150] He died in September of the following year.

By 1897 cod were being produced in larger numbers than any other marine fish. Up to that time, 449,764,000 cod fry had been released along the East Coast. The output in the 1896–1897 season alone was 98,000,000.[76]

Mackerel. Activities at Woods Hole were not restricted, by any means, to the hatching of cod. Spanish mackerel and sea bass continued to be of some interest. In 1893, for example, three adult mackerel yielded sufficient numbers of eggs that 368,000 fry were produced. Incidentally, over 1 million sea bass fry were obtained from nine adults.

The East Coast mackerel population was apparently in serious decline during the 1890s, though the cause of that decline had not been determined. Efforts to produce large numbers of fry for stocking had been disappointing because of low broodstock availability and poor hatching success.[151]

Writing in 1899, J. Percy Moore concluded that the Commission would not be able to produce enough mackerel to circumvent the problem. "The few millions of eggs annually secured are so insignificant in comparison with the vast numbers which must be produced naturally that even if all were hatched the fry resulting would be a mere drop in the ocean."[151] Moore came up with a novel alternative approach, though there doesn't seem to be any evidence that the idea was ever implemented.[151]

> It is well known that the purse-seine fishermen operating some miles offshore frequently secure whole schools of spawning mackerel, from which the eggs run so freely that decks of fishing vessels become literally covered with those which have accidentally escaped. This circumstance, coupled with the . . . readiness with which fertilization can be accomplished, leads to the following suggestion, which I recommend as a guide toward a tentative policy of the Fish Commission during the progress of further investigations:
>
> The captains of the fishing schooners should be asked to cooperate upon such terms as may be agreed upon, to the end that when such spawning schools of mackerel are met with the fish should be immediately stripped, the ova and sperm mixed, and, after permitting a few minutes to insure [sic] fertilization, turned overboard to undergo their development amid the natural surroundings from which they were taken. In this way . . . vast numbers of fertilized and healthy eggs could be liberated under conditions . . . favorable to growth.
>
> The work of stripping and impregnation can be accomplished so simply and quickly that it would not materially interfere with the regular duties of fishing, and I have no doubt that the more intelligent fishermen, particularly in view of some small consideration, could accomplish it successfully. Each schooner could be provided with a circular of instructions and such vessels (large, shallow pans) as would be needed. . . . It may even be deemed advisable . . . to place experienced spawners on vessels . . . to conduct the operations.

Lobsters. Invertebrate culture was also of interest to the Woods Hole aquaculturists. In 1893, over 10 million American lobster eggs were obtained. The larvae were released within 48 hours of hatching.[123]

Lobsters had been a species of interest from the inception of the Commission. You'll recall that Livingston Stone was transporting lobsters to California when his aquarium car went into the Elkhorn River. The fish culture manual published by the Commission in 1898 indicated the following:[76]

> If egg-bearing lobsters were not liable to destruction by man, artificial propagation would hardly be necessary. Notwithstanding the enactment of stringent laws prohibiting the sale of "berried" lobsters, the frequent sacrifice of such lobsters, with their eggs, and of many immature lobsters, has seriously reduced the lobster output and rendered active and stringent measures imperative. By the present methods millions of lobster eggs are annually taken and hatched that would be lost, and the females producing them, amounting to several thousands, are liberated.

Egg hatching averaged 54% up until 1889, when some experiments were conducted to determine if better methods could be developed. Among the types of apparatus used, the McDonald jar proved best, leading to an 81% hatch rate of the 4,353,000 eggs incubated therein. In 1894, some 97 million eggs were incubated at Woods Hole. The Gloucester facility also hatched lobsters that year. That facility incubated 10 million eggs.[76]

By 1890, five attempts had been made to transport American lobsters to the West
Coast of the United States. Those attempts were summarized by Rathbun,[152] who
had been involved in the latter efforts. The first trips by Livingston Stone have been
described earlier. The two last stocking efforts in 1888 and 1889 involved somewhat
larger shipments than had been attempted earlier. Confirmation that lobsters were
becoming successfully established had yet to be obtained, but that certainly didn't
dampen the enthusiasm for the initiative. The five stocking efforts and numbers of
animals involved were as follows:[152]

Date of Shipment	No. Shipped	No. Stocked	Stocking Location
June 1873	162	0	(lost in transit)
June 1874	150	4	California
June 1879	22	21	California
June 1888	614	332	California
June 1889	710	233	Washington

The 1888 shipment was stocked, for the most part, off Monterey, California. Lobsters shipped in 1889 were stocked in several locations on the Washington coast and
in Puget Sound.

Other Developments of Interest

Many Commission activities in the 1880s and 1990s, in addition to those already
described, could be outlined, but I'll limit the discussion to only a few. Among them
are an attempt to introduce the American oyster to the state of Washington, a survey
of three Gulf Coast states to determine the status of their fisheries and the possible
need for supplemental stocking or the stocking of additional species, stocking programs in the Great Lakes, and the participation of the Commission in expositions.

Oysters in Willapa Bay. In late 1883, Commissioner McDonald visited Willapa Bay,
Washington, and became convinced by local residents that the introduction of
American oysters might be of considerable help to the existing industry.[153] The local
oysters supported a significant industry—about 350 people were involved and over
$66,000 worth of oysters were produced in 1885—but the animals were small
compared with their East Coast cousins.

Oysters from the East Coast had been stocked into San Francisco Bay since the
mid-1870s, and while the beds were augmented by shipped-in stock at intervals,
there was also some evidence that successful natural spawning was occurring. It was
felt that the San Francisco Bay population would be an excellent source of stock for
Willapa Bay, Washington, because the California oysters had acclimated to the cold
waters of the West Coast. Alternatively, oysters could be imported directly from the
East Coast, but it was thought desirable to obtain them from the northern end of the
range where the water was colder. In order to maintain as large a population as
possible in San Francisco Bay, the decision was made to stock Willapa Bay with
oysters imported from the East Coast.

A carload of oysters left New York for South Bend, Washington, on October 26,
1894. The trip, which was expected to have taken 18 days, was accomplished in 13.

The shipment involved 80 barrels that had been collected from a number of areas along the East Coast. The oysters were stocked within a limited area to increase the chance that they would be able to reproduce naturally. The cost of shipping the oysters was $784.80. Inspections of the stocking locale over the first year after stocking demonstrated that the oysters were surviving in their new home.[153]

American oysters never did become established in the waters of Washington, as successful spawning did not occur. Several decades after the experimental stocking described in the above paragraphs, importation of the Pacific oyster from Japan began. Some reproduction of Pacific (or Japanese) oysters occurred in the waters of Washington State, but annual importation of new stock was required. That continued until the 1980s when local hatcheries were established, thereby eliminating the need for imported oysters.

The Situation in Some Gulf Coast States. In 1897, Barton Warren Evermann, one of America's most famous ichthyologists, evaluated the fisheries of Mississippi, Louisiana, and Texas. He was charge with determining what fishes were present in southwestern Mississippi and deciding whether additional species should be introduced. In Louisiana, the catfish catch had been declining, and Evermann was asked to determine the cause and recommend steps that should be taken to reverse the trend. He was also asked to evaluate the suitability of the Sabine and Neches Rivers in Texas for shad stocking.[154]

Evermann did not recommend development of a massive enhancement program to assist in the recovery of the catfish population in the Atchafalaya River, contrary to the traditional approach of the Commission. Instead, he suggested that the state legislature institute a closed season on catfish from March 15 to May 15 to allow fish to spawn, and to place a lower legal limit of 4 pounds. His visit to Texas was too brief to allow him to come up with definitive recommendations.

Great Lakes Fish Culture. In a paper presented to the American Fisheries Society in 1900, Frank Clark summarized fish stocking in the Great Lakes by the U.S. Fish and Fisheries Commission and some of the state commissions.[155] Whitefish was the primary species of interest. That fish was stocked by the Commission, the government of Canada, and the fish commissions of Minnesota, Michigan, Wisconsin, Ohio, Pennsylvania, and New York. More than half of the whitefish planted in the Great Lakes up to 1900 were placed in Lake Erie. Between 1895 and 1899, 443,677,000 whitefish were stocked in Lake Erie. During the same period, the U.S. Fish and Fisheries Commission stocked 185,938,000 whitefish fry in Lakes Superior, Michigan, Huron, and Ontario.

Clark indicated that Lake Erie's whitefish fishery was in much better shape than that of the other Great Lakes. He attributed that situation to the heavy plantings in Lake Erie and the fact that many of the most productive hatcheries were located in close proximity to that lake. Rather than send the fish several hundred miles for stocking, they were placed in the most convenient location—Lake Erie. Clark blamed the railroads for not providing free transportation for whitefish and lake trout from hatcheries to the other Great Lakes. He recommended the establishment of additional hatcheries to overcome the problems.

Clark was interested in determining the best place for release of whitefish and lake trout fry. While their release in traditional spawning areas might be appropriate, there was no scientific evidence to support the conclusion that the spawning grounds provided the best environment (including, very importantly, the proper food resources) for the young fish.[155]

> . . . we cannot with certainty know that better planting grounds exist until practical investigations are consummated and the proper localities determined by scientific study into the plankton life with the food question in view. It is very natural to suppose that the predatory fishes are familiar with the location of the spawning areas of whitefish and lake trout, etc., and that there are decidedly better localities for the welfare and safety of the fry.

Clark was convinced that the stocking program had been responsible for the maintenance of the Lake Erie fishery. He felt that there had been an ". . . enormous waste of fry somewhere."[155] It was clear to him that most of the stocked fish never entered the fishery. He calculated that at least 80% of the fry planted were lost.

In the discussion that followed the presentation of Clark's paper before the America Fisheries Society, the question of rearing the fry to larger sizes was raised.[155] Clark indicated that it was not possible to retain the fish beyond the fry stage because millions of dollars would be required.

The Commission Shows Off. Annual expositions were held in various states around the nation during the late nineteenth and early twentieth centuries. The expositions were used to show off new technology to the public and were not restricted to American technology, but were international in scope. The U.S. Fish and Fisheries Commission developed exhibits at a number of such expositions. Some of those exhibits were quite elaborate, often involving displays of live fish, operational hatchery equipment, and fishing gear.

One of the major events in which the Commission was involved was the Columbian Exposition. That was a world event, a precursor to the modern world's fairs, so planning began well in advance. A special series of U.S. commemorative stamps was issued, and the Commission was heavily involved. The Commission's exhibit of practical fish culture equipment included trout and salmon hatching troughs, shad and whitefish hatching tables fitted with McDonald jars, and cod boxes. Eggs were actually hatched when available, though artificial eggs were also used in demonstrations of the gear.[156] Mr. S. G. Worth, superintendent of the Central station in Washington, D.C., developed artificial eggs out of resin for use at times when the exposition was open but real eggs were not available.

Beginning in May 1891, the Commission rented a building in Washington, D.C., to put together the exhibit. All of the displays were constructed in Washington over many months, after which everything was shipped to Chicago, Illinois, in early 1893 and set up at the Columbian Exposition. When the show opened on May 1, there were 800,000 shad eggs, 3 million walleye (pike-perch) eggs, and 84,000 yellow perch eggs in the hatchery display. Shad and common sucker eggs were received later in the year, along with additional walleye and yellow perch eggs. Trout eggs arrived in late June. Of 20,000 eggs shipped to Chicago from the Leadville, Colorado, hatchery, about 12,000 hatched and were fed in troughs until an equipment failure led to their demise. Chinook (Quinnat) salmon eggs arrived from California in September, and nearly half of them were successfully hatched. About 54,000 lake trout eggs were shipped from the Alpena hatchery to Chicago and placed on display in hatching troughs until the close of the exposition, after which they were shipped to Frank Clark at the Northville station.[156]

In 1895, an international exposition was held in Atlanta, Georgia. It was officially called the Cotton States and International Exposition. Governmental participation included the Executive Branch, the Smithsonian Institution, and the Fish and Fisheries Commission. The scientific investigation and methods and statistics branches of the Commission displayed models of vessels, fishing gear, fish models and examples of invertebrates collected during dredging and trawling operations. In addition, charts on the various commercial fisheries and their value to the nation were displayed. The fish culture display was similar to that prepared for the Columbian Exposition.[157]

> The fish-cultural operations [were] shown by models and photographs of hatching stations; models and full-size specimens of apparatus used in the collection, transportation, and hatching of eggs; apparatus used in the transportation of fish; charts showing a summary of work done since the organization of the Commission; results obtained with reference to special fisheries and results at the different stations of the Commission during the fiscal year 1894–95; also by the practical hatching of eggs of salmon, whitefish, and trouts.

In addition, aquaria were constructed in which both marine and freshwater fish were displayed. The marine fish were collected from Morehead City, North Carolina and Pensacola, Florida. Coldwater species, including sea anemones, lobsters, and starfish were shipped in from Woods Hole and Gloucester, Massachusetts. Freshwater fishes were obtained from the Commission's hatcheries in Quincy, Illinois; Wytheville, Virginia; and the fish ponds in Washington, D.C. Collections were also made in the Chattahoochee River near Atlanta and a lake near Luluton, Georgia. A total of 1,573 freshwater fishes (23 species) and 1,085 saltwater fishes (40 species) were displayed. In addition, there were reptiles (including an alligator) and various invertebrates on display.[157]

Another major exposition was held in 1901. The Pan-American Exposition, like the Columbian Exposition, was represented in a series of U.S. commemorative stamps and featured a major display by the U.S. Fish and Fisheries Commission. The exposition was held in Buffalo, New York, and featured displays similar to the others that have already been described. Hatching troughs, jars, and other devices were provided for the hatching of Pacific salmon (various species), grayling, various species of trout, walleye, cod, and shad. Over 6,000 square feet of space were devoted to aquaria in which were displayed all the species being cultured by the Commission plus a number of other commercially important species. A large ice machine was installed to provide cold water for the salmonids. Salt water was shipped in by rail. In keeping with the name of the exposition, a number of species were imported from Puerto Rico (spelled *Porto Rico* in those days) and Bermuda.[158]

Attempts to Stock the West. Following a quarter century of transporting fish, molluscs, and crustaceans from the East Coast to the Pacific ocean, Hugh M. Smith summarized the results that had been obtained. He acknowledged the problems associated with transporting aquatic organisms such long distances, but also indicated that failures to acclimate animals that were released in good condition could be caused by a number of factors associated with the release sites:[159]

- Unsuitable water temperature
- Unsuitable food
- Unfavorable topographical condition of the bottom
- Absence of suitable rivers of anadromous fish
- Enemies and fatalities acting on a relatively small number of individuals

Smith attributed the successes that had been achieved, in part, to one of the pioneers of American fish culture:[159]

> Mention should be made of the efficient services rendered to fish-culture by Mr. Livingston Stone in successfully taking fishes across the continent at a time when fish transportation was an undeveloped art and when the difficulties encountered would have discouraged one less enthusiastically interested and less competently informed on the general subject. To Mr. Stone more than to any other person is the direct credit due for the introduction of most of those fishes which have since attained economic prominence.

Not all of the introductions were successful, but a considerable number were. Among them were various species of trout, members of the catfish and sunfish families, goldfish, carp, oysters, and clams. Among introduced species that did not survive was the Atlantic salmon, the species currently being reared by commercial salmon farmers in marine net-pens in both New England and Washington State. Some persons have expressed concern that Atlantic salmon escapees from Puget Sound, Washington, fish farms could become established and compete with Pacific salmon. The failure of Atlantic salmon to become established a century ago should help allay those fears, though there is no way that the possibility of successful establishment can be entirely ruled out.

THE FIRST DECADE OF A NEW CENTURY

According to E. M. Wood, who was writing in 1953, the development of fisheries science after 1900 was influenced by changing social and economic conditions in the United States that provided the public with a great deal of mobility and a concomitant increase in leisure time that was often used for recreational fishing. Both conditions led to increased pressure on sport fishes and a change in the emphasis of management agencies:[92]

> Previous to 1900, the basic goal was to increase the food potential of American waters. The factors of recreation and sport fishing were recognized but were considered of insignificant importance. Today, in contrast, the primary consideration is sport fishing and every effort is being made to minimize the food aspect.

With respect to how the Commission had changed, Wood said the following:[92]

> At the inception of the original Fish Commission, propagation of commercial species was [of] the highest importance. Since 1900, however, there has been an increasing amount of controversy between public opinion and the logic based upon a growing number of biological investigations on the effects of artificial propagation.

While the state fish commissions were developing major programs dedicated to the stocking of sport fishes during the post-1900 period, the federal government program continued to target commercial fisheries enhancement for at least the first quarter of the twentieth century. The Commission was becoming increasingly involved in the production of inland fishes, though in terms of shear numbers, the traditional marine species and inland fishes that had been produced to augment commercial fisheries continued to dominate.

George M. Bowers remained in the job of Commissioner throughout the first decade of the twentieth century. A significant change occurred in 1904 when, after over 30 years in existence as an independent commission, the U.S. Commission on Fish and Fisheries was renamed the Bureau of Fisheries and placed within the Department of Commerce.[160] Activities did not change under the new arrangement. (The major change insofar as we are concerned is that, at least for awhile, the terminology will change. Whereas I have referred to the Commission in earlier pages, I shall now adopt the term Bureau in reference to the federal fisheries agency.) Things became more complicated later. The Bureau of Fisheries was split into the Bureau of Commercial Fisheries and the U.S. Fish and Wildlife Service (Department of the Interior) in 1950. In 1971 the National Marine Fisheries Service (NMFS) was formed and placed in the Department of Commerce. NMFS accepted responsibility for marine fisheries and most anadromous salmonids. The U.S. Fish and Wildlife Service (USFWS) continued its involvement with inland fisheries and some anadromous salmonids. In general, the NMFS was involved with commercially important species, and the USFWS concentrated on sport fisheries.

With the passage of time, not only did the Bureau grow with respect to its annual expenditures, but the numbers of animals and numbers of species produced, as well as the numbers of facilities in operation, were also growing. That growth is summarized as follows:[160-169]

Numbers of Fish and Eggs Distributed

1901	1,173,833,400
1902	1,495,543,374
1903	1,226,057,457
1904	1,267,334,385
1905	1,759,475,039
1906	1,931,834,609
1907	2,511,597,377
1909	3,107,131,911
1910	3,233,392,572

Numbers of Selected Species Distributed in 1901 and 1910
(Includes Eggs, Fry, and Larger Sizes)

Species	1901	1910
Atlantic salmon*	19,839,084	2,864,982
Black Bass†	228,105	1,386,987
Brook trout	19,725,714	12,150,006
Buffalo	0	201,475
Carp	0	22,710
Catfish	2,374	544,350
Cod	202,871,000	220,208,000
Crappie	33,042	414,477
Flatfish	44,230,000	930,755,000
Grayling	1,735,182	106,018
Haddock	0	712,000
Lake herring	51,020,000	71,740,000
Lake trout	19,725,714	48,145,772
Lobster	60,879,000	163,287,052
Mackerel	0	764,000
Pickerel	300	500
Pike perch [walleye]	240,887,200	476,484,760
Pollock	0	38,140,000
Quinnat [chinook] salmon	19,441,784	53,941,498
Rainbow trout	1,037,303	2,860,338
Rock bass	27,131	69,985
Shad	193,287,000	92,236,000
Silver [coho] salmon	300,041	11,293,025
Smelt	0	4,509,000
Steelhead	370,758	3,900,005
Striped bass	0	7,350,000
Sturgeon	20,000	0
Sunfish	1,268	345,635
Warmouth bass	1,031	792
White bass	0	6,050
Whitefish	326,106,295	251,390,000
Yellow perch	621	332,194,245

Shipments of fish overseas continued as well. For example, chinook and Atlantic salmon, along with lake and rainbow trout, were shipped to Argentina in 1904. Whitefish, chinook salmon, and Atlantic salmon were shipped to New Zealand the same year.[166]

In his report for 1903, Bowers provided a list of all the adults, yearlings, fry, and eggs produced by the Commission since 1871.[163] The number was an incredible 12,023,491,533.

*Includes both anadromous and landlocked fish.
†Includes both largemouth and smallmouth bass.

Note from the above list of species that several that were not produced in 1901 or were produced in low numbers were being actively cultured in large numbers by the end of the decade. Examples are haddock, mackerel, pollock, striped bass, and yellow perch. Carp were, for some reason, not produced in 1901, but there was some production in 1910, though nothing like the levels seen in the late nineteenth century. Sturgeon culture, limited in 1901, was not undertaken in 1910. The large numbers of whitefish being produced were not entirely indicative of the quantity being introduced into the Great Lakes. For example, in 1901 the combined efforts of the United States and Canada involved the taking of 800 million eggs.[170]

Catfish culture by the Bureau expanded greatly during the 1900–1910 period. The three catfish that we now recognize as the channel catfish, blue catfish, and white catfish were considered to be a single species early in the century. The black bullhead was known as the black catfish, and the yellow bullhead was called the horned pout or yellow cat.[171]

In a paper presented to the American Fisheries Society by R. S. Johnson in 1910, the continued interest in fish stocking throughout the country was discussed. Applications for fish to stock both public and private waters were continuously increasing. That year, the number of applications was 10,635.[172]

The Bureau continued to deliver fish throughout the country by rail. Also, as one might expect, the budget of the agency generally increased as its sphere of activity broadened, though there were some ups and downs during the decade:[159,162,163,166–169,172]

Examples of Miles Traveled in Delivery

1901   304,445 by railroad
1902   295,203 by railroad
1910   527,245 by railroad (96,263 miles by Commission cars)

Budget of the Bureau of Fisheries

1902	$543,120
1903	530,640
1904	674,940
1905	768,869
1907	676,170
1909	840,100
1910	823,490

In 1899 a fifth railroad car had been purchased to carry marine fish to the Chicago Exposition. A sixth car was constructed for use in conjunction with the Louisiana Purchase Exposition in 1904. After that exposition closed, the sixth car was put into general service by the Bureau of Fisheries.[173] The increase of $100,000 in the budget between 1904 and 1905 is attributable to survey activity and the establishment of a new facility in Alaska.[168]

W. C. Ravenel, who had been the Assistant-in-Charge of the Division of Fish-Culture beginning in 1895, left the Commission in 1902 to assume an administrative position with the U.S. National Museum.[162] By 1903, Barton W. Evermann became Assistant-in-Charge of the Division of Statistics and Methods. He was involved primarily with gathering information on commercial fisheries.[174]

In 1910, Bowers asked Congress to appropriate $50,000 to fund development of a fish disease laboratory.[168] There was no action was immediately taken, though that type of laboratory was put into operation some years later.

By the middle of the first decade of the twentieth century the Bureau was operating 32 production stations, an office in Illinois that served as a headquarters only, and a number of substations. The main stations and an indication of the species being cultured were, in alphabetical order, as follows:[175]

Baird, California	Salmon and trout
Baker Lake, Washington	Salmon
Battery, Havre de Grace, Maryland	Perch and shad
Boothbay Harbor, Maine	Cod and lobster
Bozeman, Montana	Trout, grayling, whitefish
Bryan Point, Maryland	Perch and shad
Cape Vincent, New York	Whitefish, trout, salmon, walleye
Central Station and aquaria	Whitefish, trout, salmon, perch, walleye, catfish, shad
Clackamas, Oregon	Salmon, trout
Cold Springs, Georgia	Black bass, warmouth, crappie, catfish, sunfish
Craig Brook, Maine	Salmon, trout
Delaware River, Pennsylvania	Shad
Duluth, Minnesota	Trout, salmon, whitefish, walleye
Edenton, North Carolina	Shad
Erwin, Tennessee	Trout, black bass, rock bass, sunfish, catfish
Fish Lakes, Washington, D.C.	Black bass, crappie
Gloucester, Massachusetts	Cod, flatfish, pollock, lobster
Green Lake, Maine	Salmon, trout
Leadville, Colorado	Trout
Mammoth Spring, Arkansas	Largemouth bass
Manchester, Iowa	Trout, salmon, rock bass, perch
Nashua, New Hampshire	Trout, salmon
Neosho, Missouri	Trout, salmon, largemouth bass, rock bass, crappie
Northville, Michigan	Trout, smallmouth bass
Put-in-Bay, Ohio	Whitefish, lake trout, herring, walleye
Quincy, Illinois	Office and headquarters
St. Johnsbury, Vermont	Trout, salmon, smallmouth bass
San Marcos, Texas	Largemouth bass, rock bass, sunfish, crappie, catfish
Spearfish, South Dakota	Trout
Tupelo, Mississippi	Largemouth bass, crappie, sunfish

White Sulphur Springs, West Virginia	Trout, smallmouth bass
Woods Hole, Massachusetts	Cod, flatfish, lobster, tautog
Wytheville, Virginia	Trout, black bass, rock bass

THE FIRST 40 YEARS OF THE AMERICAN FISHERIES SOCIETY

The American Fish Culturists' Association was formed on December 20, 1870, in New York City, and William Clift became its first president.[176] In 1878 the name was changed to the American Fish Cultural Association. Six years later, in 1884, the name American Fisheries Society (AFS) was adopted. Among its charter members were a few folks whose names have appeared in these pages; for example, William Clift, Livingston Stone, Fred Mather, T. J. Whitcomb, and Seth Green. Annual membership dues were initially $3.00, increased to $5.00 in 1872. The dues were reduced to $3.00 once again in 1875 and maintained at the lower level until 1898. At that time, the treasury was so flush ($400.00), that the dues were reduced to $1.00. Life membership at a whopping $15.00 was made an option in 1898 and became a better deal for those who had signed up when the dues were raised to $2.00 in 1904 (with a raise in life membership to $25.00). The dues increase was prompted by the fact that the treasury was actually in deficit to the tune of $46.50 by 1904 and drastic measures were necessary. Dues apparently remained at $2.00 through 1910, when a summary of the first 40 years of the society was written.[176]

The treasury of the AFS involved handling a total of $207.80 in 1887.[177] By 1904, the treasurer had more that twice that amount of cash to handle—that was the year the Society went into deficit. The balance sheet for 1904, which detailed the costs that resulted in the deficit, was published in C. W. Willard's Treasurer's report:[178]

Receipts		$485.75
Initial balance	$62.65	
Baird Memorial fund	98.85	
Dues and fees	316.00	
Annual report sales	12.25	
Expenditures		532.25
Travel expenses	39.25	
Stamped envelopes (500)	10.70	
Stenography	158.20	
Livery	1.00	
Other	7.00	
Publishing	203.65	
More envelopes (500)	10.70	
Secretary's stamps & envelopes	70.17	
Blank receipts	0.75	
Circulars & stamps	30.83	
Balance		<46.50>

Dues included a free copy of the *Transactions of the American Fisheries Society*. The Society also published *Forest and Stream* for a number of years, beginning in 1874.[176]

Examination of papers published by the AFS over its first 40 years of existence is revealing in terms of demonstrating how the fish culture art began to become the science of aquaculture. Some of the earliest contributions of general interest have already been mentioned. Included were papers on general fish culture;[69,93] salmon,[22,132] trout,[143] whitefish,[71,169] eels,[95,96] oysters,[85,87] and lobsters.[83] In the next few pages we'll look at some of the other subjects considered by AFS members through 1910, and expand a bit on some of the subjects that have previously been discussed. Much of the work published in the *Transactions of the American Fisheries Society* through 1910 established the foundation on which research continues today.*

Reinforcement of Early Ideas and Some Changes in Philosophy

The perceived need to augment dwindling fish stocks was the impetus behind the early years of U.S. aquaculture. The environment was considered capable of supporting far more fishes and invertebrates (both in terms of individuals and species) than nature was able to produce on its own in the face of heavy commercial fishing pressure. As we have seen, the U.S. Fish and Fisheries Commission launched a massive stocking program, augmented with respect to some species by activities within the various state fishery commissions, to help the fisheries recover. Conventional wisdom of the time also dictated that it was not only appropriate but highly desirable to introduce new species into places where they had never occurred previously. Thus, the introduction of Pacific salmon to the East Coast, rainbow trout throughout the country, brown trout and carp from Europe into various parts of the United States, and so forth. Experts of the day figured that if an introduction was not successful, it was because the environment was not a suitable one. There was no consideration given with regard to whether or not an introduction might have a negative impact on other organisms in the receiving location. It must be remembered that the impetus, at least initially, was to enhance the economy by improving commercial fishing. By the turn of the century there was also a realization that sportfish stocking was appropriate since the number of recreational fishermen was rapidly increasing.

Persons educated in modern ecological theory are often shocked when they learn about the indiscriminate stocking programs that not only occurred in the nineteenth century but continue with respect to a number of species today. The idea that small populations or stocks of fish might be genetically distinct and should not be tam-

---

*Many of the states produced publications in conjunction with their state fish commissions (and later the state fish and game agencies that now go by a variety of names). The *Transactions of the American Fisheries Society* and the *Report of the Commissioner* were primary sources of information. In addition, the U.S. Fish and Fisheries Commission began publishing what has become known as the *Fishery Bulletin* in 1881. It was the *Bulletin of the U.S. Fish Commission* through 1903 when it became the *Bulletin of the U.S. Bureau of Fisheries*. In 1951 the *Bulletin* was renamed the *Fishery Bulletin* of the U.S. Fish and Wildlife Service. *Fishery Bulletin* was continued as the title when the National Marine Fisheries Service was created in 1971 and assumed responsibility for the publication.

pered with by having fish of the same species from other locations stocked on them is relatively new, though it is currently a hot issue. Yet, the idea that stocking programs might not always be appropriate is not new. We can see some evolution in the thinking surrounding that issue by examining what happened during the first 40 years of the AFS.

The prevailing philosophy from the days of Seth Green to the waning years of the nineteenth century was summarized eloquently by John H. Bissell in 1886. He had the following to say about his predecessors:[179]

> You [the members of the AFS] have been honored by papers and addresses from men of your own number who have won distinction by knightly deeds . . . in conflict for the secrets of nature, wresting from nature's willing hands the knowledge that practical men have been gathering and storing up against the day when the millions that are peopling and are to people this continent, shall cry out for more and better and cheaper food.

Bissell also described his view and how the emphasis had changed since the first American fish culturists began their work:[179]

> A younger generation is coming upon the field to take its part in carrying forward fish-culture, to apply the precious stores of knowledge, which have been laid up by the practical observation and scientific research of the past twenty-five years, to the practical solution of some very important economic questions that are beginning to clamor loudly for solution.
>
> The question most urgent just now is not, can fishes be artificially hatched and reared, and acclimated to alien waters, but can the fisheries of this country now be saved?

The question as to whether stocking programs had been effective was addressed by Bissell. His example came from is own experience with whitefish stocking in Michigan. After describing the fact that whitefish stocks continued to decline in the face of a fairly massive stocking program, he said the following:[179]

> Has artificial propagation then been a failure? No, for it has not had a fair chance. . . .
> First—It has not been conducted upon a scale adequate to accomplish the results.
> Where we are hatching about fifty millions of whitefish we need from six to eight times that number every year. . . .
> Second—Artificial propagation has not had a chance in point of time.
> It is only within the first few years of the second decade of its existence—say from 1882 or 1883 that we ever hatched and planted over 15,000,000 of whitefish in any one year. The same period will cover the most extensive operations of the U.S. Commission in this direction.

Two years later, Bissell once again addressed the AFS, this time to discuss the idea of enhancing cooperative activities between the U.S. Fish and Fisheries Commission and the commissions in the various states as well as among the state fish commissions. Following up on the theme of his 1886 title, "Fish Culture—A Practical Art," he said:[180]

I think it is generally agreed, that fish-culture has passed its purely experimental stage. It is in fact fast becoming recognized as a practical art. . . .

Fish-culture, when appreciated and invoked in both its branches, artificial propagation and legal regulation, has demonstrated its ability to restore exhausted fisheries. [The next step is] the working out of a just and comprehensive system of regulation of fishing as an industry, and as a recreation.

Bissell recognized that the Commission had been working closely with various states for some time. Each annual *Report of the Commissioner* documented those relationships. However, he felt that the collaboration could be expanded. He used examples of the need to extend activities like those being conducted to stock coastal waters on the Atlantic and Pacific coasts and the Great Lakes. Surveys of the lakes and their fisheries were something that he felt was needed. Another example was what he called inland lakes:[180]

> Another direction in which co-operation can, I believe, be advantageously employed is in a thorough examination of interior lakes. By interior, or inland, lakes the dwellers along the Great Lakes are wont to distinguish the smaller bodies of water wholly within the boundaries of the several States. . . . these lakes were planted [in the past] with various kinds of fishes, not with any special reference to their adaptability to the fish planted, but because the Commission had fish for that purpose, and in a general way the people in the vicinity of the lakes wanted fish. I do not say this with the design of casting any reflection upon the authorities in those days. The promiscuous planting of fish was then perfectly natural; and our experience is based largely upon their mistakes as it is still more largely upon the notable success of so many of their experiments.

Bissell was not concerned that fish had been stocked in areas where they would disrupt endemic species but knew that in some cases the fish stocked in previous years were inappropriate as they could not survive in the water bodies into which they were placed. He pointed out that some information had been obtained on the inland lakes of Michigan but was of the opinion that the states did not have the human or fiscal resources required for the job, whereas the U.S. Fish and Fisheries Commission did have those resources.

Finally, Bissell indicated that surveys for appropriate species and locations for stocking could be extended to rivers and streams: "A systematic examination of all streams would in this State [Michigan] within a few years secure the planting of trout only in waters entirely adapted in temperature and food supply to trout."[180]

Bissell felt that state cooperation should be developed among states with shared waters, such as occurs among various states bordering on the Great Lakes. He felt a management system that applied to all those waters would be more effective than individual state management programs. He also felt that cooperation should be developed among states bordering the same river, though he recognized that there would be significant difficulties involved.

Collection or maintenance of broodfish, spawning, and distribution of eggs and fry represented the primary management process used by both the states and federal government during the nineteenth century. Transplanting, that is, collecting fish in one location and moving them to another, was also practiced to a limited extent until

1888, when the Fish and Fisheries Commission launched a campaign in cooperation with the Illinois Fish Commission to rescue fish stranded each year after the spring flooding of the Mississippi River.[5] The early transplantation activities might not be even worthy of note except for a couple of telling comments in 1902 made by J. J. Stranahan in a short article in the *Transactions of the American Fisheries Society*: "Transplanting for the purpose of crossing is now acknowledged to be highly beneficial as fishes, the same as other animals, become dwarfed by inbreeding in confined waters."[181] While the concept of inbreeding depression has not been abandoned, current thinking by many fish geneticists stressed what they feel is the importance of maintaining genetic integrity of isolated populations, even small ones.

Stranahan's observations on largemouth bass and catfish that had been moved from one place to another led him to the following conclusion: ". . . fish taken from large lakes and planted in small waters with outlets will not remain, but will return to the larger water or try to; while if taken from waters similar to those into which they are to be placed, good results will follow."[181]

Spencer F. Baird's view that the nation's commercial fisheries were in trouble prompted him to successfully lobby for the creation of the U.S. Commission of Fish and Fisheries in 1871. Nearly a quarter century later, and after billions of eggs, larvae, and fry had been stocked, the situation was still grim. Bushrod W. James, a member and vice-president of the Pennsylvania Fish Protective Association, presented his view to the AFS in 1895. He cited examples of fisheries that were in decline on both coasts, beginning with Pacific salmon. He recommended stricter regulation of the fishery, then said something quite prophetic: "But the fishes, or other animal life or plants on which the salmon feed, must also be guarded from destructive depredation."[182] He then turned his attention to the East Coast fisheries, beginning with Atlantic salmon:[182]

> Salmon is rare in all our rivers; the great fishing banks of Maine and Massachusetts are failing; the lobsters are growing scarce and small; mackerel is almost gone from some quarters. . . . Herring catches in some localities are growing less and less. . . . Some fishermen say that shad is getting scarce in some of our rivers; others assert that they, once so rarely flavored, are now at times tainted with coal oil and sewage or foul mud, and are consequently almost unsalable.

James reflected the earlier call of Bissell for cooperation among the states in the regulation of fisheries:[182]

> A very apparent defect is instituted by conflicting laws made for the control of streams which run through two or more States, whereas, if each State would consult with its neighboring ones before maturing its laws regarding rivers and streams, and fishing therein, conjoint measures might be taken which would improve the local fisheries without injury to any one locality.

Pollution was seen by James as a major factor in the continued decline of the nation's fisheries. He stated that chemicals and sewage should be kept out of streams used for fishing, at least to the extent possible. While recognizing the fact that danger to the public of eating fishes caught in polluted waters had not been ascertained, he asserted that such fish were more heavily parasitized than those from clean waters. He indicated that some of those parasites were harmless to people, while others were "poisonous."[182] James felt that scientific research would resolve that issue and might lead to the banning of some fish for sale and the closure of some streams to fishing.

To resolve the problem, James advocated the imposition of regulations:[182]

> . . . each State should have laws compelling the clearing and lowering of the mouths of all rivers or creeks in which the waters lie stagnant and restricted by rubbish; that each State Commission should have a biologist, who could make known the presence of dangerous parasites, and all who are interested in fish culture and protection should join in trying to discover whether there could not be some plan adopted to destroy them without endangering the life of the fish; that the food animalculæ should be as carefully protected as the fish themselves, and that all deleterious matter should be kept from them as far as possible. . . . States, and especially those that have coast lines and bays, should so regulate the fishing seasons that the strong, mature and fertile fish may be allowed to reach the spawning places unmolested, or else that certain streams in every State shall be closed against fishermen every second year, thus giving them a whole season in which to spawn and multiply.

James was also a realist, so he came up with an alternative:[182]

> If these plans are not practicable then others must be adopted. Perhaps good results would follow if fish culture were made so universal that at the time of the running of the schools to the spawning grounds men were stationed at the mouth of or along every important river to catch the fish, obtain the eggs, and hatch them artificially; then they could be deposited in fitting places, after the season was over, and thus the danger of extinction would be over.

The bottom line appeared to be that if regulation failed, increasing the level of fish culture activity would succeed.

Exotic introductions of aquatic species has been a hot topic in recent years. One of the accidental introductions that has been in the news is the zebra mussel, the range of which is rapidly expanding from the Great Lakes, where it was first reported. The intentional introduction of walking catfish in the 1960s caused an uproar in Florida, a state where the culture of tilapia has been an on-again, off-again activity for a number of years. Increasingly restrictive federal guidelines are being developed with respect to intentional introductions of exotic species. In addition, how to deal with the issue of the accidental introduction of aquatic organisms in discarded bilgewater from transoceanic ships is being hotly debated. In 1886, a paper appeared in the *Transactions of the American Fisheries Society* entitled "Intentional and Unintentional Distribution of Species." The paper, written by R. E. C. Stearns, described how various undesirable species of both animals and plants had been scattered around the world through unintentional introductions. Included in that list were the Norway rat, the cockroach and other insect pests, and weeds. Having laid the groundwork for an understanding of how accidental introductions can be highly detrimental, Stearns turned his attention to aquatic species. He presented examples of undesirable snail introductions and then turned his attention to the incidental stocking of clams from the East Coast into California in conjunction with the introduction of the American oyster in the early 1870s. As a result of that introduction, a significant clam fishery developed. The fact that two formerly abundant and economically valuable molluscs, a cockle and a clam, were almost entirely displaced by the introduced species was mentioned in passing. The desirability of the East Coast species was sufficiently great that it had been intentionally introduced into Washington State.[183]

Another late-nineteenth paper which, at least from its title, dealt with a topic currently being debated, was presented to the AFS by Herschel Whitaker in 1895. Entitled "Some Observations on the Moral Phases of Modern Fishculture," the paper dealt not with whether stocking programs were working, whether aquaculture was an appropriate activity, or what the potential ramifications of exotic introductions are. Instead, it concerned his view that fish stocking programs in the Great Lakes should be curtailed until the commercial fishermen were brought under control.[184]

Whitaker presented information that showed that the whitefish catch in the Great Lakes was declining and that increasing numbers of nets were capturing fewer and smaller fish each year. His description of the fisheries was not at all optimistic:[184]

[Lake] Ontario, with its former wealth of fish of the finest edible character, has long since been robbed of its treasure, and the nets of the fishermen have rotted on the shore. Erie, even richer than Ontario in fine food fish is nearing its last stage, as was demonstrated during the season of 1894, by the exodus of the commercial fishermen from that Lake to the Lake of the Woods. The Lake Michigan fisheries have been in a large measure ruined, and fishing in many localities on that Lake has ceased to be an industry to be followed with profit. Huron and Superior have suffered seriously from the same causes. . . .

He cited the capture of immature fish and collection of gravid fish during the spawning season as primary causes of the declines. Canada, which had imposed a ban on fishing during November to protect spawners, revoked that ban in 1895 in an attempt to force the United States into taking action.

Whitaker was anything but diplomatic in laying blame for the problem:[184]

A more selfish or senseless prosecution of an industry has never been witnessed in any age or country. With the exception of here and there an individual, the fishermen, never extended the hand of co-operation to the State in its attempts to restock the waters. We are met on every hand and at every step by their selfish greed.

Legislators did not escape Whitaker's criticism. He indicated that the lawmakers were aware of the decline and recognized that the imposition of regulations was required:[184]

They [legislators in Michigan] have been interviewed privately and addressed publicly on the need of legislation, which should arrest the practices now fast destroying the fisheries, and while now and then they will admit privately the force of the argument, protective legislation has so far failed. Legislators of fair intelligence admit privately that these practices are wrong and vicious, but in the same breath assure you that their constituents insist that they *must not be interfered with in their vocation,* and as the average legislator has his personal ambitions and future, he weakly succumbs to the influence of a handful of fishermen in his district, and subordinates the public interest to his personal ambition.

Those comments have a good deal of validity in the present.

Whitaker argued that the waters and their fishes were the property of the public, not the commercial fishermen:[184]

> The waters belong to the public, and their rights in the fisheries are paramount to that of the individual fishermen. The fisheries are theirs and whoever exercises the privilege of fishing therein does so by the sufferance of the public, and under implied understanding that he shall not prejudice the public rights therein. The fisherman in the prosecution of his business is enjoying a *privilege* and not a *right,* and he is entitled to enjoy that privilege so long as he exercises it with a due regard to the paramount right of the public to have them preserved for the future, and no longer.

The comments were visionary and, once again, could have been made yesterday rather than a century ago.

The solution proposed was draconian:[184]

> In our opinion every state and government engaged in the artificial propagation of commercial fish on the great lakes should agree to discontinue the work until the fisheries are given such protection as will insure [sic] results of benefit. Fish culture has its uses, but if the object for which commissions are created, viz.: to restore and maintain the fisheries, is to be met with methods which give it no chances of success, further planting [stocking] should cease until a more enlightened public sentiment shall demand the correction of existing abuses, or until the public pulse has been sufficiently quickened to the necessities of proper regulation to demand the passage of just restrictive laws.

Whitaker commented on the original intent of fish culture and how the philosophy associated with the practice should change to reflect the continued decline in the Great Lakes fishery:[184]

> There was a time in American fish culture when it was honestly believed that restocking by artificial propagation, without any other intervention, would restore depleted waters. But that time has passed. . . . We, of the great lake region, have had forced upon us the fact that while to-day we are planting millions of fish in good condition in the lakes, we are hopelessly handicapped as to the results by the war of wanton destruction waged upon the fisheries by the netters, who say we will take fish in season and out of season, we will take them by any and all kind of devices, and nobody must say us nay, it is a matter of no concern to us whether there are fish for those who come after us; after us comes the judgment.

During the discussion that followed Whitaker's paper presentation, a Mr. Peabody from Wisconsin commented that he was not in complete agreement with Whitaker's pessimistic view of the Great Lakes fisheries, and went on to mention conditions in his state:[184]

> . . . last October, we interviewed the fishermen . . . to get their views regarding the restocking of the waters, and we found that these net fishermen (I do not look upon

fishermen with a great degree of fondness; they are a sort of pirates as a rule), but these men are all of them anxious to see that proper legislation is obtained to protect the white fish on these shores, and they said to us that on account of the planting by the Commissioners in that year, they lost $20,000, and one man lost $60,000, and because of that they hope to get legislation which will protect the white fish in their own waters, and one of the points to be considered is that it is illegal to catch white fish weighing less than one pound or one pound and a quarter. . . .

As we shall see, stocking did not cease. Also, while the fisheries scientists of the 1890s did not realize it, problems in the Great Lakes were just beginning.

Following up on the need for regulation, Seymour Bower took a much different approach than the one we have just encountered. Bower, whose paper was presented in 1897, described the problem of maintaining the nation's capture fisheries as being one of permitting the maximum removal of mature fish without depletion of the stocks. His approach was not to eliminate stocking until appropriate regulations were in place, but to protect juveniles from harvest and utilize artificial propagation to rebuild and sustain the fisheries. His view was that natural propagation was inefficient. He also felt that some fishes, having reached maturity, were not useful except for harvest and consumption by man or as prey for other aquatic organisms:[185]

> When fish have matured, it is time, so to speak, to realize on the investment. They should then be converted into food, either for some other fish, or for man. If allowed to remain, they defeat the very object for which they were created, namely, to be caught and utilized. The food which they consume . . . [would be better utilized] by going to the young and growing fish, instead of being wasted on the adults merely to prolong their lives. When a female fish has matured and yielded a crop of ova to the saving process of artificial propagation, she has accomplished more in the way of reproduction tha[n] she could in hundreds of seasons under natural environment. . . .
>
> It is evident that restrictive measures need not apply to the adult fish, provided a sufficient number are available for artificial propagation, but as affecting the young and immature fish such measures should be of the most stringent character. The killing of young fish of the more valuable species is little short of criminal, and should be penalized in every possible way.

Bower did not mean his comments on taking adult fish indiscriminately to apply to all species. He recognized that many species, such as those that guard their eggs and provide some type of parental care, should be allowed to perform those functions, at least once, before being captured. He was apparently not concerned about protecting such species until they had reached an age where they had spawned more than once. Artificial propagation was, he believed, the means by which sufficient numbers of recruits could be produced to sustain a fishery. Natural productivity was, in his opinion, incapable of maintaining fish populations in the face of fishing pressure:[185]

> Natural propagation will never force a water to its highest productive limit, unless fishing is absolutely prohibited for an indefinite period. Fortunately, this course is not necessary, for while we cannot prevent [activities] more or less destructive of one kind or size of fish by another after they leave our hatcheries, we *can* and *do* save the

enormous waste that occurs under natural conditions during the *ova* stage, and thus bring into existence immensely increased numbers of young fish.

Fish culturists of the time were convinced that the majority of the mortality experienced during the life cycle of fish occurred prior to egg hatching:[185]

> Fish culturists and all who have carefully investigated the subject are unanimously agreed that the treatment and protection we extend to the ova multiplies hatching results five hundred to one thousand times, and some place the ratio much higher.

Following that view to the next step, the aquaculturists of the time were of the opinion that ultimate survival of the hatched fish until they entered the fishery would also be several hundred times higher than the rate occurring in nature. The obvious assumption was that the survival of released fish larvae or fry from a hatchery would mirror the survival of those stages produced in nature.

Another paper on fish protection in the Great Lakes that laid blame for much of the failure of enhancement to improve fishing on the states was presented to the AFS in 1906 by Samuel F. Fullerton.[186] His view was that fish were disappearing in the Great Lakes even though the government had stocked large numbers of fry. Since the states lacked coordinated programs for protecting those fish during spawning season and from fishing until they reached certain sizes, Fullerton felt that regulation should be entirely under federal government control. Fullerton also felt that many of the fish being distributed, particularly in inland waters other than the Great Lakes, were stocked in inappropriate areas where the fry had little chance of survival.

Before looking at his proposed remedy for the situation, it might be educational to look at the exorbitant price of fish in the early part of the twentieth century and the way the price had been inflated because the catch was down:[186]

> Twenty years ago in the city of Duluth, Minnesota, you could get all the whitefish you could carry home with you for five to seven cents per pound. Today you are a lucky purchaser if you get any, but if you do, you will pay from fourteen to seventeen cents per pound and then you will have to be careful that you do not get Winnipeg whitefish instead of the Lake Superior.

Fullerton did not expand on why he preferred Lake Superior whitefish to those from Winnipeg.

Recommendations to remedy the situation included the following:[186]

- Distribute the fry properly by placing them in areas as near as possible to the spawning grounds of the species.
- Negotiate a treaty with Canada establishing a uniform closed season in the Great Lakes to protect the fish during the spawning season. (Recall that Canada had such a regulation and retracted it because the United States failed to institute a similar policy.[184])
- Develop a license system covering all commercial fishermen. Included would be a rigid inspection requirement for meeting the recommended size regulations (minimum of two pounds dressed weight for lake trout and whitefish).

- Turn all stocking programs in inland waters over to the federal government, which would require boats and trained individuals to undertake planting operations.

Not everyone was in agreement on the recommendations. In the discussion that followed, various estimates were made as to the cost of turning all stocking of inland waters over to the federal government. A least two individuals expressed the view that the cost would be as much as 20 times what was then being spent, which would be prohibitive. Fullerton had indicated that people were not properly trained in stocking fish delivered to them by the government, but it was pointed out that written instructions accompanied each shipment. The problem seemed to be that people didn't bother to read the instructions, so a public education program was needed. Finally, one participant in the discussion took exception with Fullerton's view that the Great Lakes fisheries were in decline.[186]

Serious questions about the actual contribution to the fisheries of at least some of the fish that had been stocked by the Bureau were raised at the 1901 AFS meeting:[187]

> When the fact was demonstrated that the ova of certain kinds of fish could be artificially fecundated* and hatched, it was then assumed, that if the fry were planted in the waters almost anywhere then the problem of the future supply of the fish was practically settled. So we proceeded to hatch them out by the hundred[s] of millions and dump them into the waters, usually those from which the parent fish had been taken, it being assumed that a large proportion ought to survive to maturity. But somehow there seems to be a hitch either in Nature's plans or our own for after a score of years' trial and a large expenditure of money we have no assurance that there has been any marked success. In fact we are not *sure* that any one of those hundred[s] of millions has reached maturity.

J. C. Parker, the source of the above comments, did not condemn all fish stocking by any means. He acknowledged that in some instances (giving the brook trout as an example) there had been verification that stocking programs had been successful. His criticism was aimed at those species for which no verification had been obtained. Parker used whitefish in the Great Lakes as an example for which there was no verification. He suggested how verification might be obtained in Lake Ontario:[187]

> Now if all who have an interest in the successful propagation of this fish could pool their interest, and first make an exhaustive examination of this lake to ascertain the exact conditions now existing, as to its aquatic life.† Then all the states interested together with Canada and the general government hatch as many millions of whitefish as possible, for say three years, planting them all in Lake Ontario. Then if possible prohibit all commercial fishing for five years it seems to me that it would settle the question for this lake at least, and indirectly for all the others, and possibly demonstrate whether man could be a controlling factor in the larger schemes of constructive aquatic life.

---

*The term used now is *incubated* but *fecundated* was popular for a number of years.
†This is not a complete sentence, but that's what was published. The writing doesn't get any better in the rest of the paragraph, but it is faithfully reported as written.

During the discussion there was opinion expressed that whitefish stocking in Lake Erie had been responsible for increased catches and that few whitefish, "now and then a carload," had been stocked in Lake Ontario. The discussion got a bit lively between Parker and those who felt that stocking programs, including those with whitefish, had been highly successful.

Parker's paper did nothing to change the stocking program of the Bureau. For the next three decades the annual reports of the Commissioner of Fisheries documented the planting of billions upon billions of fry. It does not appear that any major attempt to verify the efficacy of the stocking programs was made either, though as we shall see, the emphasis gradually shifted increasingly toward stocking sport fishes. Many of the sport-fish species were released as fingerlings, and their survival and recruitment into the fisheries of inland waters, at least, was rather easy to document. The contribution of billions of newly hatched whitefish fry and the fry of various species of marine fishes and invertebrates has never been adequately documented, though the consensus is that there was little, if any survival.

The often heard statement that the nation's supply of fish was continuing to decline was repeated once again in 1907 by John L. Leary, an employee of the Bureau of Fisheries working at the San Marcos, Texas, hatchery. In his presentation to the AFS, he pointed specifically to the East Coast and Great Lakes regions as having continuing declines. His recommendation, based on his work in Texas with bass, was to construct rearing ponds in conjunction with the hatcheries and retain the fish to fingerling size rather than releasing them as fry:[188]

> I will say right here that I would rather have for stocking waters 100 fingerlings 3 to 4 inches long than 10,000 fry. Where fingerlings are planted I feel sure of favorable results, and very doubtful as to results where fry are planted.

John W. Titcomb, an assistant commissioner stationed in Washington, D.C., responded to Leary's papers as follows:[188]

> I believe that this is a tabooed subject. The question seems to be partly of finance. When you consider, for instance, the work of the Bureau of Fisheries, taking the output last spring of white perch, something over twelve hundred million fry, it is practically impossible to consider the idea of rearing them to fingerlings. The whitefish comes in the same category. It seems to be a question of when you can do it, and when not. Some species we know, produce good results with fry. I might give specific instances where brook trout fry and lake trout fry have been planted in ponds, where there were no such fish, and in three or four years good fishing prevailed. The policy of the bureau is to plant fingerlings of trout and bass where possible to do so. But the production is so large and the facilities for rearing to fingerlings so limited in proportion, that it is necessary to plant a large proportion of fry.

Titcomb then went on to challenge Leary on bass fingerling production, indicating that if 100,000 fry were stocked in a pond, only 5,000–15,000 fingerlings would be produced because of the fact that they could not be fed and were carnivorous. He felt that stocking 50,000 fry into "suitable waters" would ultimately produce more fish. After expressing his views, Titcomb did toss Leary a small bone:[188]

. . . Mr. Leary's work in Texas has the hearty approval of the commissioner of fisheries, namely about planting fingerlings. It is due to local conditions that he is able to do it, with the bass and sunfish which he propagates, as he has a tremendous amount of food there. The black bass have been known to grow to about eleven pounds in three years, illustrating the great amount of food that they find, and Mr. Leary can gather any great amounts of natural food in the vicinity of his pond.

Titcomb was essentially telling Leary to stick to what he knew and not try to get involved in national policy. Others involved in the discussion supported Titcomb's position.

Leary never did get back into the discussion, which went on for some time. A number of others, including Hugh M. Smith, who was to become Commissioner of Fisheries, waded in, however. The discussion turned away from bass and back to fish of the kinds Leary was originally trying to target. Smith related some of the Bureau's experience in Washington, D.C., ponds in conjunction with rearing shad. He indicated that there had been some successes, but that more recent attempts to produce fingerlings had been something less than spectacular:[188]

We planted in one large pond . . . approximately two million . . . fry . . . and in the fall we drew down this pond, and as near as I can recollect, turned into the Potomac river thirteen hundred small sunfish, and several hundred Warmouth bass, a few black bass, a few hundred yellow perch, and two or three hundred river herring, and one shad. . . . the planting of a few million shad fry in the Potomac river would produce equally good results.

Surely Smith was not concluding that putting millions of fry directly in the river would yield only one fingerling!

Frank Clark, after holding his tongue throughout most of the discussion, tried to put the matter both into perspective and to rest, at least for the time being:[188]

For twenty-five years this question has come up more or less. I think, if I am right, that I was the first one that ever brought it up before the meeting of the American Fisheries Society in New York City. At that time I was firmly convinced and positive that there was nothing right about planting fry. . . . I am frankly willing to listen to the fry man,—and, gentlemen, you that advocate planting fry are right; and gentlemen, you that advocate planting fingerlings are right. It is only a question of the fish and the place. Now, this fingerling vs. fry question is discussed from time to time, but we all are of the same opinion.

As the twentieth century approached, the governmental agencies responsible for producing fish began to get more and more requests for fish with which to stock farm ponds. Constructed primarily to provide livestock with water, farm ponds were quickly becoming recognized as places where fish could be reared to provide food for the farm family. Stranahan, writing in 1902, reiterated the old theme about the status of U.S. fisheries:[189]

With our public waters rapidly becoming depleted through excessive fishing, in spite of the good work being done by the hatcheries, where are we to look for fish to fill the vary rapidly growing demand, if not through water farming? . . . with the rapid increase of our population and the further growth of consumption through improved transportation facilities, the limit has doubtless already been reached and any permanent increase of per capita supply must come through covering what is now unproductive land with water, thus adding to the output of fish beyond what natural waters would make it, and making many fins grow where none grew before.

There are vast areas in all of the states, probably equal in the aggregate to that of the Great Lakes, which now produce virtually nothing and much of which might be made to furnish an abundance of fish, with comparatively little expense.

Those comments reflected the approach of the time. Nature was there to provide for humankind. Unproductive land was considered useless and to be turned into something that would benefit society, or at least the landowner. Stranahan's vision, which stressed fish culture on the family farm, did not lead to the flooding of vast areas of unproductive farmland, though the construction of reservoirs (often in highly productive floodplains) did create large areas of water that were subsequently stocked with fish. Farmers did continue to build farm ponds, and there are now millions of those bodies of water located throughout the nation. Not many are more than a few acres in size, with the typical farm pond occupying perhaps 1 acre or less. Today such ponds are managed mostly for recreation (in addition to still providing livestock with water), but Stranahan envisioned them to function primarily as a food source. He indicated that there were three typical reasons that fish culture on farms tended to fail:[189]

- Ponds were not being properly constructed. He explained how levees should be built and stressed the need for sluiceways and drain structures.
- The selection of what Stranahan called "too high-toned fish" was inappropriate. He felt that bullheads and sunfish were suitable in most farm ponds, though he mentioned other species, including bass.
- Overpopulation of fish was another major reason for failure. Farmers should begin removing fish from their ponds as soon as the fish become large enough to eat. His stocking recommendation was to not exceed 100 black bass and 200 each of sunfish and bullheads per acre of water.

Stranahan's paper seems to represent one of the earliest that dealt with pond management. The debate over which species to use and how many to stock, perhaps initiated by Stranahan, is ongoing, though it is not raging as loudly as it was several years ago.

Miscellaneous Stuff

Feeding. The vast majority of the fish stocked during the late nineteenth and early twentieth centuries were released as fry prior to the time they first required feed. Whether one believed that fry stocking was an appropriate way to increase fish populations or not, as interest in stocking farm ponds and other inland waters increased, attention began to turn toward feeding young fish. Interest also began to develop in evaluating how the food consumed by broodfish might influence egg viability.[190] Early examinations of this nature, and the development of interest in determining the food habits of fish, were studies that can be thought of as the beginnings of fish nutrition.

The use of prepared feeds in conjunction with fish rearing was not new even in the early twentieth century. Various types of feed had been used to augment natural food items in ponds. Included were such things as liver, coagulated blood, and vegetables. Not everyone favored the use of prepared feeds. Reasons why they should not be used were presented in a paper by M. E. O'Brien in 1888:[191]

- It is unnatural.
- It has a tendency to render the water putrid, and consequently injurious to fish.
- It favors the introduction of disease.
- . . . it entails a great deal of expense.

O'Brien examined stomach contents of trout and found that insects, molluscs, and crustaceans dominated. He advocated producing those natural food organisms in preference to providing prepared feeds. While today we would fertilize ponds to encourage the growth of natural foods, his approach can only be described as novel:[191]

> Trenches, or basins, should be dug in close apposition with the ponds, and, if necessary, communicating with them. These basins to be supplied with spring water by means of a pipe connected with the main spring. One should be devoted to Shell culture [molluscs], another to Insect culture, and a third to Crustacea, and so on.
>
> These various foods could be transferred by means of a fine net, or better still, by a running stream of water communicating with the fish pond, or means could be established whereby these forms could creep from the basin into the pond.

When fish culturists developed an interest in providing some type of prepared feed to young fish, salmon and trout represented logical fishes with which to begin. Concentration on those fishes was logical since their first-feeding fry are quite large compared with cod, shad, bass, carp, and various other species that were being cultured. Thus, they could logically be considered able to accept food particles of sizes that were practical to manufacture.

James Henshall described his approach to feeding grayling and trout in papers published by AFS in 1901 and 1904.[192,193] He pressed finely ground liver through a screen and fed the bloody solution that resulted. He recognized that the liquid contained small particles that could be consumed by fry. By initiating feeding before the yolk sacs were completely absorbed, he reasoned the fish would have developed an interest in the prepared food source when they had to begin looking for food in their environment. He also commented on the fact that spring water, which was often used for early hatchery rearing, was devoid of natural food organisms; thus fry would require prepared food if maintained in that type of water beyond the first-feeding stage. In the second paper, he described his observation that fry in nature ". . .are constantly picking at the leaves and fronds of water plants, evidently feeding on small organisms that have found lodgment there."[193] To mimic natural conditions, he dipped bunches of watercress in his liver slurry and suspended the plants in hatching troughs containing fry. That technique worked reasonably well, but he thought the material washed off the plants too quickly and planned additional studies to find a means to help the liver particles stick to the plants. He proposed mixing the liver slurry with gelatin as one possible option.

Charles Atkins was not convinced that the provision of food prior to or immediately after yolk-sac absorption was necessary.[194] He conducted a series of experiments with brook trout, lake trout, and Atlantic salmon in which he began feeding one group immediately upon yolk-sac absorption and other groups after periods of fasting. In general, beginning feeding a few days after the yolk sacs were fully absorbed did not lead to increased mortality. He fed all groups hog liver four times a day.

These early examples of interest in feeding fish did not resolve any major issue, but they did represent the state of knowledge with respect to fish nutritional requirements at the time. The studies were conducted before much was known about nutrition in the broad sense. Analytical techniques to measure the various constituents in tissues had yet to be developed, and there was little or no information available on the nutritional requirements of humans, let alone fish.

A Pennsylvania Story. Fish commissions had been established at the state level throughout the country by the start of the twentieth century. Many states published annual reports similar in content to the *Report of the Commissioner* that was produced by the U.S. Fish and Fisheries Commission and later by the Bureau of Fisheries. All of that material is available and could be covered in this review, but I have elected to restrict the information presented on state activities to information published in the more widely available literature and to be rather picky in selecting information to present in this book. Two papers in the early 1900s by W. E. Meehan, commissioner of fisheries for the state of Pennsylvania, caught my attention.

The first of those papers[195] described the dissolution of the state fish commission and the formation of the Pennsylvania Department of Fisheries, a move that was subsequently adopted by a number of states. With the creation of a state department, the commissioner of fisheries became a member of the governor's cabinet, thereby increasing the power of the office.

The second paper by Meehan[196] was merely a compilation of the fish distributed by the state of Pennsylvania during the period from January to July 1906. That compilation was of interest in that it showed the federal government was not alone in producing prodigious numbers of fishes. The total number of aquatic organisms that had been distributed in Pennsylvania during the period mentioned was 394,646,546, with several hundred thousand additional fish still in the hatcheries awaiting dispersal. Among the species and species groups mentioned were bass, trout, shad, smelt, whitefish, lake herring, yellow perch, and walleye, and frogs.

Transporting Fish. Transporting fish, not only from the hatchery to a nearby waterway, but virtually across the nation, was being undertaken virtually from the time the U.S. Fish and Fisheries Commission was created. Eventually, nine railroad cars were constructed for the purpose of hauling fish for stocking. Railroad transportation was eventually replaced by trucking, and over the decades a considerable amount of attention was paid to improving hauling conditions to help ensure survival. The need for maintaining good water quality was recognized from the beginning, but means by which to accomplish that task had to be developed.

A Bureau of Fisheries fish distribution car.

For freshwater species, it was possible to obtain water along the route to replace that in the hauling containers. Marine fishes that were transported across the country typically did not share that option unless large amounts of excess salt water were carried, which was often not possible or was inordinately expensive if the haulers were paying the shipping costs. Aeration was the only method used to maintain saltwater quality, at least through the nineteenth century. One novel idea that sprouted up as that century drew to a close was to carry fish in hermetically sealed containers.[197] A patent was apparently granted for the idea, which involved charging the water with compressed air before sealing the container. William Seal's comments on the potential of such a system included several reservations and precautions. Included was the need to determine how long various species and sizes of fish within a species could survive while being essentially canned. He as aware that temperature would influence the rate at which oxygen was consumed in the can. While the oxygen introduced into the water through aeration prior to sealing the can would be retained to be utilized only by the fish, there was also the problem of the carbon dioxide increase from respiration. That gas could not escape and would produce carbonic acid, eventually leading to mortality. So far as I can determine, the movement of fish in sealed containers never became a reality at that time in our history. It was certainly an interesting idea, however, and is not much different from the current practice of shipping fish in sealed plastic bags that have been charged with oxygen.

Finding suitable shipping containers has long been a subject of interest. Today, we have fiberglass, various types of plastics, and aluminum containers that can be used to hold and transport fish safely. During the early years of fish culture, wooden barrels, sealed with materials that could certainly be toxic in many instances, were typically used. Metal containers were also employed, though the danger of transporting fish in galvanized metal containers was recognized early in the twentieth century.[198]

We will look at other observations and innovations associated with fish transportation later. In the meantime, it is obvious that while hundreds of millions of eggs, larvae, and larger animals were being transported during the late nineteenth and

Water towers along the railroads not only supplied steam engines, they were also sources of makeup water for fish tanks.

If problems were experienced in keeping fish alive, they were sometimes stocked from the nearest convenient tressle.

early twentieth centuries, there was precious little control over water quality and not much understanding of water chemistry. The amount of success that was achieved is quite amazing given the level of knowledge and available technology.

Disease. The study of fish and aquatic invertebrate diseases* has become an important part of aquaculture. Typically, as research and development provides culturists with the information required to successfully rear an aquatic animal in captivity, diseases begin to proliferate. That doesn't mean that the species was immune from disease prior to having been placed in culture. It just means that as we learn more about a species we begin to discover the diseases to which it is susceptible.

The identification and study of fish and aquatic invertebrate diseases was in its infancy in the early twentieth century. I will not attempt to chronicle the discovery of the myriad diseases that have been described or the methods that have been developed by those attempting to deal with those diseases. Suffice it to say that one of the early papers on the subject appeared in the *Transactions of the American Fisheries Society* in 1901.[199] Research on the topic progressed relatively slowly but steadily in the various government and university laboratories that were eventually developed.

In Charles Atkins's 1901 paper on the subject, he made the assumption that fishes are subject to diseases, just as are humans. He indicated that the death of fishes in

---

*Included in the term *disease* are such things as pathogenic viruses and bacteria, along with fungi and a variety of parasites.

culture and fluctuations of fish populations in nature could occur as a result of disease outbreaks; however, proof of that option was not available:[199]

> An official report lying before me gives a list of 104 different diseases from which human deaths occurred in the state of Maine during the seven years from 1892 to 1898. Is there any inherent reason why fish should not have as many diseases as men? Observation has already gone far enough to indicate the probable existence of a very considerable number of diseases among the fishes we cultivate.

What can only be considered surprising after some 20 years of fish culture under the auspices of the government was that there were apparently no named diseases of cultured fish. Atkins did list a number of afflications that he had recognized, but they were largely, though not completely, attributed to causes unknown. Among those that he did attribute to specific pathogens were fungal infections of eggs, fry, and adult fish along with parasitic trematodes in young lake trout.

Genetics. Livestock has undergone domestication through selective breeding for many centuries, or even millenia in some instances. The same can be said for a few fishes such as koi. As fish culturists began to think about nutrition and diseases, they also started to look into improving their charges through selective breeding. This was the foundation upon which the discipline of fish genetics was built. An early paper on the subject was published in the *Transactions of the American Fisheries Society* in 1902. In that paper Arthur Sykes approaches the subject as follows:[200]

> Much has been said and written about methods and results of propagation; but little thought, it seems, has been given to the foundation on which we work or the quality of the material of which it is composed, i.e., the potency and vigor of the parent fish and the embryo.
>
> So far as the writer has been able to ascertain in principles underlying the breeding of domestic animals, here exploited, have not been applied in fish culture excepting in a very limited way; and no fish culturist has put principles to a practical and complete test.

Sykes used trout to demonstrate what he was talking about. He first described how broodstock replacement was often undertaken:[200]

> The eggs [of trout] . . . are laid down in hatching troughs [following fertilization] and incubated. The first of the season's crop of fry is usually saved for the use of the hatchery to increase and replenish the breeding stock in the ponds. The remainder of the crop is sold by the private hatchery or planted in public streams by the state hatchery as the case may be.

The rationale for keeping the first portion of the season's production for broodstock replacement was based on the observation that those fish tended to be stronger, grew to larger size, and appeared to be more hardy than fish hatched later in the season. In addition,[200]

> The early hatched fry usually comes from the older fish, hence has not been inbred as much as that hatched later. Fry is purchased from other hatcheries for the infusion of new blood in the breeding stock.

The fact that fish culturists went outside of their own facilities and introduced new fish into their broodstock populations meant that there was some attempt to reduce inbreeding. However, Sykes indicated that there was no selection involved in the process:[200]

> The fish culturist does not as a rule make a selection of his stock with a view to increasing the size and hence the usefulness of the individuals. His matings are haphazard and the results of a corresponding nature. It is true that he introduces new blood into his ponds, but of what avail is such new blood if it is of the same quality as the old?

Today, among fish geneticists involved with fish released into nature there is a considerable amount of interest in maintaining genetic diversity within fish populations; that means random mating rather than selective breeding. Commercial fish farmers, on the other hand, have followed the agricultural approach, which involves selecting for traits that allow the fish to grow rapidly to large size and perform well in captivity. In 1902, and for many years thereafter, culturists involved in enhancement seemed to be following the model advocated today of random mating (with the exception that bringing in broodfish from other hatcheries would now be considered inappropriate by some geneticists). Sykes felt that wild stocks could be improved through selective breeding and that the practices of the time were causing deterioration in ". . . color, size, vigor and productiveness."[200]

CHAPTER 3

Expanding the Science

THE BUREAU OF FISHERIES FROM 1911 THROUGH 1941

Activities of the Bureau of Fisheries from 1911 until the outbreak of World War II included more of what we have already seen in terms of fish propagation and distribution along with a considerable expansion of activities. There were changes in personnel and in focus. Many new hatcheries and satellite stations were constructed, and a few were actually closed. Fisheries management developed as a discipline distinct from fish culture, and the many scientific disciplines that underpin aquaculture began to mature.

Relative to its propagation and stocking activities, the Bureau began interesting new programs involving the culture of some unusual species such as sponges and freshwater mussels. It also became more actively involved in annual efforts to capture fish stranded after flooding in the Mississippi River valley and stocking them back in their waters of origin or into new locations. Reservoir construction led to expansion of the sport-fish production program, and the need to maintain fish populations damaged by dam construction and other human activities led to the development of mitigation hatcheries for salmonids. Stranahan's appeal to inundate unproductive land with water and produce aquatic animals[189] didn't result in the creation of vast areas of submerged private lands, but tens of thousands of farm ponds were being constructed by 1914,[92] and demand for fish to stock them with was very high. The Bureau became involved with that activity as well.

Billions of fish were released by the U.S. Fish and Fisheries Commission and its successor the U.S. Bureau of Fisheries. Rather than get bogged down in details, we shall examine the period of years in question in a summary fashion and then move on to some of the new activities and advancements in aquaculture that were occurring.

Before we look at the truly prodigious numbers of fish being produced, moved about, and salvaged from sure death after floods, let's take a look at the folks who

Hugh McCormick Smith (1866–1941)

were at the Bureau's helm during the period of interest. Annual reports of the Bureau showed that George Bowers was Commissioner of Fisheries through 1912,[201] after which Hugh McCormick Smith assumed the position.[202] Smith remained in the job until 1922 when he was replaced by Henry O'Malley.[203] O'Malley was in the position for the next decade, and was followed by Frank Bell,[204] who held the job through 1938. Charles E. Jackson assumed the job and was responsible for producing the last report of the Bureau that was produced before World War II began.[205]

Of the Commissioners mentioned in the above paragraph, Hugh M. Smith was arguably the best known. Smith, who was born in 1866, had been raised in Washington, D.C., where he completed his education at the Georgetown University medical school.[202] He began working for the U.S. Fish and Fisheries Commission under Spencer F. Baird in 1886 and became Assistant-in-Charge of Statistics and Methods of the Fisheries in 1893. Smith was Commissioner during World War I, was a moving force behind the establishment of the first fisheries school in the United States, and was appointed to represent the United States in international fisheries venues including being appointed by President Howard Taft to represent the nation on the Permanent Council for the Exploration of the Sea.

Following his retirement from the Bureau, Smith served as adviser to the King of Siam (now Thailand) on fisheries matters and was appointed director of the Siamese Department of Fisheries in 1926, a position he held until 1934. He returned to the United States with specimens from Siam and worked until his death in 1941 as an Associate in Zoology in the Smithsonian Institution.

Let's take a look at the numbers of aquatic eggs, fry and fingerlings distributed by the Bureau during most years from 1911 to 1941:[201,206–231]

1911	3,646,294,535	1926	5,232,373,000
1913	3,863,595,282	1927	6,481,073,000
1914	4,047,643,417	1928	7,036,317,200
1915	4,288,757,800	1929	7,060,369,500
1916	4,847,262,566	1930	7,570,482,300
1917	5,158,963,295	1931	7,121,805,700
1918	4,098,105,159	1932	7,073,935,200
1919	5,876,985,350	1933	7,202,155,625
1920	4,770,355,720	1935	5,071,725,000
1921	4,962,489,405	1936	8,171,200,579
1922	5,125,101,320	1937	7,919,100,100
1923	4,314,859,029	1938	8,121,131,985
1924	5,361,810,654	1939	8,024,540,685
1925	5,301,862,583		

The data for 1939 were published in 1941. The invasion of Pearl Harbor on December 7 of that year and the entry of the United States into World War II undoubtedly led to the discontinuation of annual reports from the Bureau of Fisheries.

The Bureau and its predecessor had considered the stocking of commercial fishes to be of primary importance. However, as the federal fisheries program expanded, a new policy developed that broadened the Bureau's role into the stocking of fishes of primarily recreational interest into inland waters. Both individuals and groups could apply for fish. The Bureau accepted responsibility for delivery but depended upon the recipients to properly stock the animals and ensure that the fish were legally harvested. The Bureau justified distribution of fish primarily to areas where they would enter the recreational fisheries as follows:[215]

> The benefits accruing from this phase of fish-cultural work are considered invaluable. Not only is there an economic gain in the increase of the food supply by the

utilization of otherwise unproductive waters, but there is an educational effect that develops and fosters a sentiment favorable to the protection and growth of fish life. Moreover, innumerable persons derive direct and important benefits from a day's fishing in the open places.

In 1922, over 23 million fish were distributed as a part of the program. The fish, which came both from hatcheries and as a result of rescue efforts, included catfish, Atlantic salmon, various species of trout, grayling, crappie, largemouth and small-mouth bass, rock bass, warmouth, and sunfish.[215]

Various foreign countries obtained fish from the Bureau during the early twentieth century. A few examples include:[201,211]

1911	Brazil	Smallmouth bass
	Canada	Walleye
	Cuba	Rainbow trout
	Germany	Rainbow trout
	Portugal	Rainbow trout
1918	Canada	Sockeye salmon
	Japan	Chinook salmon, rainbow trout
	Mexico	Largemouth bass, yellow perch
	Canal Zone	Largemouth bass, carp, catfish, sunfish

The budget of the Bureau increased throughout the period, as might be expected. Examples of annual Bureau budgets are:[201,203–205,208–211,213,214,216–218,222,232–239]

1911	$1,003,470.00
1914	1,047,180.00
1915	1,118,471.66
1916	1,075,340.00
1917	1,144,850.00
1918	1,263,560.00
1920	1,206,190.00
1921	1,216,310.00
1922	1,240,430.00
1923	1,262,090.00
1924	1,223,490.00
1926	1,589,140.00
1927	1,814,253.00
1928	1,948,568.00
1929	2,092,108.00
1931	2,631,885.00
1932	2,905,540.00
1933	1,976,020.00
1935	1,325,327.00
1936	1,565,920.00
1938	1,967,000.00
1939	2,220,200.00

No reason was provided for why a whole-dollar amount was not budgeted in 1915. In any case, there was a general increase in the budget over time into the 1930s, when the Depression undoubtedly dictated the budget reduction. By the end of that decade the budget once again exceeded $2 million.

The variety of species being reared by the Bureau was considerable. In 1913, it included the following:[206]

Catfish	Carp	Buffalo	Shad
Whitefish	Lake herring	Coho salmon	Chinook salmon
Pink salmon	Chum salmon	Sockeye salmon	Steelhead trout
Rainbow trout	Atlantic salmon	Lake trout	Brook trout
Grayling	Smelt	Pike	Crappie
Rock bass	Warmouth	Smallmouth bass	Largemouth bass
Sunfish	Walleye	Yellow perch	White perch
Striped bass	Yellow bass	Cod	Pollock
Haddock	Flatfish	Lobster	

Carp, pink salmon, smelt, and yellow bass were not being produced by 1939, but Rio Grande perch and mackerel had been added to the list of species cultured by the Bureau.[231]

The proliferation of fish culture stations operated by the Bureau of Fisheries can be seen in the following list from 1913, which does not include satellite facilities:[206]

Stations in Place and Primary Species Produced

Afognak, Alaska	Pacific salmon
Baird, California	Pacific salmon
Baker Lake, Washington	Pacific salmon, steelhead
Battery, Maryland	Yellow perch, shad, white perch
Boothbay Harbor, Maine	Cod, flatfish, haddock, lobster
Bozeman, Montana	Trout, steelhead,* grayling
Bryan Point, Maryland	Yellow perch, shad
Cape Vincent, New York	Whitefish, cisco, trout, walleye, yellow perch, Atlantic salmon
Clackamas, Oregon	Salmon, trout, steelhead
Cold Springs, Georgia	Salmon, trout, steelhead
Craig Brook, Maine	Atlantic salmon, Pacific salmon, trout
Duluth, Minnesota	Trout, steelhead, whitefish, walleye
Edenton, North Carolina	Shad, bass
Erwin, Tennessee	Trout, bass, sunfish, yellow perch, rock bass, catfish, carp

---

*Notice that trout and steelhead (which are trout) are listed separately in this list. I elected to separate them out to distinguish the sea-run rainbows from their nonanadromous cousins. The "trout" designation includes lake trout, brook trout, rainbows, and others.

Gloucester, Massachusetts	Cod, pollock, haddock, flatfish, lobster
Green Lake, Maine	Atlantic salmon, trout, smelt
Homer, Minnesota	Trout, buffalo, catfish, yellow perch, bass, crappie, sunfish, rock bass
Leadville, Colorado	Trout
Mammoth Spring, Arkansas	Bass, crappie, catfish, sunfish, rock bass
Manchester, Iowa	Trout, bass, rock bass
Nashua, New Hampshire	Trout, steelhead, Pacific salmon, bass
Neosho, Missouri	Trout, bass, rock bass, crappie, sunfish, carp
Northville, Michigan	Trout
Put-in-Bay, Ohio	Whitefish, walleye, yellow perch
Quincy, Illinois	Bass, rock bass, walleye, yellow perch, sunfish, catfish, crappie, buffalo, carp
St. Johnsbury, Vermont	Trout, steelhead, Pacific salmon, Atlantic salmon, bass
San Marcos, Texas	Bass, rock bass, crappie, sunfish
Spearfish, South Dakota	Trout, steelhead
Tupelo, Mississippi	Bass, rock bass, sunfish, warmouth, crappie, catfish, yellow perch
White Sulphur Springs, West Virginia	Trout, bass
Woods Hole, Massachusetts	Cod, flatfish
Wytheville, Virginia	Trout, bass, rock bass, sunfish
Yes Bay, Alaska	Salmon

By 1930, the following stations had disappeared from the list:[223]

Battery, Maryland
Bryan Point, Maryland
Green Lake, Maine

Additional stations, such as Afognak, Alaska, Yes Bay, Alaska, and Cold Springs, Georgia, were off the list by 1935.[227]

The Homer, Minnesota, station became a substation of one of the new stations, LaCrosse, Wisconsin, that were listed in 1930.[223] Other new stations were:

Berkshire, Massachusetts
Fairport, Iowa
Louisville, Kentucky
Orangeburg, South Carolina
Quinault, Washington
Saratoga, Wyoming
Springville, Utah

Eleven new stations had been added by 1935:[227]

Crawford, Nebraska	Leetown, West Virginia
Dexter, New Mexico	National Forest of New Hampshire
Flintville, Tennessee	Pittsford, Vermont
Fort Belvoir, Virginia	Marrion, Alabama
Hagerman, Idaho	Yellowstone Park, Wyoming
Lamar, Pennsylvania	

In the 1939 *Report of the Commissioner* of the U.S. Bureau of Fisheries, the following stations were listed, many of which were responsible for one or more substations (shown in parentheses):[231]

Berlin, New Hampshire (St. Johnsbury, Vermont)
Birdsview, Washington (Mount Rainier, Washington; Spokane, Washington)
Boothbay Harbor, Maine
Bozeman, Montana (Glacier Park, Montana; Ennis, Montana, Miles City, Montana)
Cape Vincent, New York (Barneveld, New York; Cortland, New York; Watertown, New York)
Carson, Washington (Little White Salmon, Washington; Big White Salmon, Washington)
Clackamas, Oregon (Battle Creek, California; Butte Falls, Oregon; Mill Creek, California)
Clark Fork, Idaho
Craig Brook, Maine
Crawford, Nebraska
Dexter, New Mexico (Santa Rosa, New Mexico)
Duluth, Minnesota
Edenton, North Carolina (Weldon, North Carolina)
Elephant Butte, New Mexico
Erwin, Tennessee
Fairport, Iowa
Flintville, Tennessee
Fort Belvoir, Virginia
Gloucester, Massachusetts
Hagerman, Idaho (Salmon, Idaho)
Hartsville, Massachusetts
LaCrosse, Wisconsin (Bellevue, Iowa: Genoa, Wisconsin; Guttenberg, Iowa; Lake Mills, Wisconsin; Marquette, Iowa)
Lake Park, Georgia
Lamar, Pennsylvania (Ogletown, Pennsylvania)
Las Vegas, Nevada
Leadville, Colorado (Creede, Colorado; Eagle Nest, New Mexico)
Leetown, West Virginia

Louisville, Kentucky
Mammoth Spring, Arkansas
Manchester, Iowa
Marion, Alabama (Cohutta, Georgia; Lyman, Mississippi; Marianna, Florida; Tupelo, Mississippi; Warm Springs, Georgia)
Nashua, New Hampshire
Neosho, Missouri (Forest Park, Missouri; Natchitoches, Louisiana; Tishomingo, Oklahoma)
Northville, Michigan
Orangeburg, South Carolina (Jacksonboro, South Carolina; Hoffman, North Carolina)
Palestine, West Virginia
Pisgah Forest, North Carolina
Pittsfort, Vermont
Put-in-Bay, Ohio
Quinault, Washington (Duckabush, Washington; Quilcene, Washington)
Rochester, Indiana
Rochester, New York (Carpenter's Brook, New York)
Saratoga, Wyoming
Spearfish, South Dakota
Springville, Utah (Bear Lake, Utah)
Uvalde, Texas (Fort Worth, Texas; San Angelo, Texas; San Marcos, Texas)
Walhalla, South Carolina
Welaka, Florida
White Sulphur Springs, West Virginia
Woods Hole, Massachusetts
Wytheville, Virginia (Harrison Lake, Virginia; Norris, Tennessee; Smokemont, North Carolina)
Yellowstone Park, Wyoming (Jackson, Wyoming)

In 1888 the Bureau of Fisheries began a program to rescue fish that became landlocked as a result of annual flooding by the Mississippi and Illinois Rivers, an activity that continued until 1940.[5] The annual floods left many fish stranded in temporary water bodies left when the water receded. Those water bodies eventually dried up, leading to the loss of millions of fish trapped therein. Rescued fish were either returned to the rivers of origin or were delivered to the various applicants who applied to the Bureau for fish. The numbers of fish involved in the program were prodigious, though they fluctuated widely from one year to the next depending on the severity of flooding and subsequent rainfall. For example, in 1929, flooding was significant but the rains that followed led to the slow withdrawal of floodwaters, thereby allowing the fish to return to their home streams rather than becoming stranded. The number of fish rescued that year clearly shows the impact on the Bureau's program. The following is a summary of the numbers of fish rescued during representative years after the program was well developed and in full operation[204,208,212,215,221,223,225,227–229,231,235,239–242]

1915	8,357,000	1931	182,534,861
1919	55,783,075	1932	51,611,367
1921	120,656,420	1933	72,180,000
1922	179,475,069	1935	47,162,505
1925	7,625,740	1936	43,519,371
1928	145,176,900	1937	50,572,399
1929	5,202,253	1938	42,202,000
1930	161,354,609	1939	1,864,820

Channelization of the upper Mississippi River down to the twin cities of Minneapolis and St. Paul, Minnesota, was progressing rapidly by 1938, and had been nearly completed by 1939.[231] The impact that increasing the channel depth to 9 feet had on the need for fish rescue is reflected in the dramatic reduction in 1939 as compared with previous years. With completion of the channelization process, the need for fish rescue was eliminated.[5,239] Species involved in the rescue operation included the following:[212]

Bass (largemouth and smallmouth)	Rock bass
Buffalo	Sunfish
Carp	Walleye
Crappie	White bass
Drum	Yellow bass
Pike	Yellow perch

The use of railroads for fish distribution by the Bureau continued through the first four decades of the twentieth century. Shipment on trucks increased progressively, reaching 245,000 miles by 1938,[230] and there was even some experimentation with airplane shipments as early as the 1920s.[221]

Nine car numbers were ultimately assigned by the Bureau of Fisheries. By 1923, Car Nos. 3,4,7,8, and 9 were in operation, Nos. 1,2,5, and 6 having been taken out of service.[173] The Bureau developed plans to replace two of the aging cars, Nos. 3 and 4, by the mid-1920s. By 1928, Car No. 4 had deteriorated to the point where it was considered unsafe. That car was placed on a siding at Lakeland, Maryland, where it was used as a home for Bureau hatchery employees. A new steel car was being constructed in 1928 to replace at least one of the old cars.[221] The new car, constructed at a cost of $59,000, was placed in service late in 1929.[240]

The railroad cars continued to pile up the miles through the 1930s. In 1937, the Bureau paid for transport over a total of 50,342 miles and had an additional 19,331 miles donated. Fish were delivered to applicants and transferred between hatcheries.[229] A year later paid miles amounted to 60,255 miles, and there were an additional 12,307 free miles.[230]

Moving Fish Around

In previous pages the numbers of fish and invertebrates distributed by the federal fisheries agency and locations of primary federal hatcheries have been described. An indication of the scope of the distribution network has been provided through a few examples of the number of miles covered by aquatic animals on the nation's railroads. Today we often hear warnings about the stocking of fish outside of their native waters. There have been many "concerns"* raised about loss of genetic diversity by diluting native populations with stocked fishes. Nowhere is the intensity level higher than in the Pacific Northwest, where salmon stocks are now being listed under the Endangered Species Act. Some geneticists argue that the same approach should be applied to various other species. The old cliché about closing the door after the horse has left the barn comes to mind again with respect to most species, since hatchery fish were being distributed virtually throughout the nation almost as soon as the U.S. Fish and Fisheries Commission was formed. Many examples of the scope of fish distribution activities could be presented, but a couple examples from 1917 are representative. The first involves states and the numbers of locations into which largemouth bass were stocked that year. Fish were distributed into creeks, rivers, ponds, and lakes.[243]

State	Locations Stocked	State	Locations Stocked
Alabama	95	Montana	3
Arizona	7	New Hampshire	2
Arkansas	70	New Jersey	10
Colorado	21	New Mexico	22
Connecticut	10	New York	29
Delaware	2	North Carolina	123
Florida	27	North Dakota	21
Georgia	97	Ohio	67
Illinois	53	Oklahoma	112
Indiana	65	Pennsylvania	128
Kansas	16	South Carolina	106
Kentucky	74	South Dakota	15
Louisiana	12	Tennessee	71
Maine	2	Texas	79
Maryland	25	Vermont	2
Massachusetts	7	Virginia	139
Michigan	69	West Virginia	22
Minnesota	58	Wisconsin	218
Mississippi	146	Wyoming	3
Missouri	60		

The original distribution of largemouth bass was from Canada in the north to the Gulf of Mexico in the south, the Atlantic coast in the East, and the Rocky Mountains in the West, though by 1883 they had been stocked into every state then in existence within the Union.[244] Most of the bass stocked in 1917 were planted within the original distribution area, but there was no attempt made to stock hatchery fish within the watersheds from which they originated.

Salmon stocking in 1917 included the following:[243]

*There's that word again.

Coho salmon	1 California location
	5 Oregon locations
	12 Washington locations
Chinook salmon	6 California locations
	1 Kentucky location
	8 Oregon locations
	8 Washington locations

Chum, pink, and sockeye salmon were also being stocked in 1917. Alaska, Maine, Oregon, and Washington were recipients of one or more of those species.[243] It is of some interest that salmon were still being stocked in the East, though no sustaining populations were established.

Freshwater Mussels

The Bureau of Fisheries devoted a considerable amount of attention for a number of years to the culture and distribution of freshwater mussels. Studies on mussel life history and culture methodology development began soon after 1900,[245] and the first plantings were undertaken in 1913.[232] Mussel harvesting became intense after 1891, when the pearl button industry developed. Mussels were thereafter in demand for the mother-of-pearl buttons cut from their shells. Some 98 tons of mussels were harvested in 1894, but the fishery grew rapidly, reaching 24,000 tons in 1899 and 38,000 tons in 1908.[246] There was also a demand for the irregular pearls produced by the mussels. They were used for jewelry. It was estimated that 10,000 fishermen became engaged in the fishery, with the annual value of the mussels being $800,000 for shells and $400,000 or more for pearls. The value of the buttons produced was estimated between $5 million and $9 million.[245] Mussel meat had little value, though some research on its use as fish food was conducted beginning in 1918.[247]

With the demand for mussels exceeding the supply, the Bureau's studies led to the establishment of spawning facilities beginning in 1913. During the second year of planting activity, 227,536,814 mussels were produced, more than double the production of the inaugural year. Larval mussels were produced at the Fairport, Iowa, biological station.[209] They were distributed in the following rivers and lakes: Mississippi near Fairport, Iowa; Black, Arkansas; Wabash, Indiana; Grand, Michigan; Maumee, Indiana; Lake Pepin, Minnesota; and Lake Pokegama, Minnesota.[232] Bureau scientists knew that the larval mussels were parasitic on fish, so they infected 167,819 fish with the larvae and liberated the infested fish into the stocking locations. The fish, all adults, were among those that had been rescued from flooded lands.

Though the activity was only in its second year in 1914, commercial mussel fishermen were already reporting that the numbers of juveniles on the mussel beds was increasing. Bureau personnel doubted that any increase could be attributed to their plantings and estimated that several years of planting would be required before the results could be ascertained with certainty.

Larval mussel distribution via infected fish continued for several years. The numbers of mussels planted varied in relation to the numbers of fish that were rescued from year to year and the availability of ripe mussels for breeding.[209,210] Numbers of mussel larvae distributed during representative years were as follows:[208-211,213,214,248]

1915	344,655,260	1919	136,907,365
1916	331,451,490	1920	183,021,720
1917	252,486,200	1921	169,740,050
1918	209,132,800	1926	2,803,625,100

In 1922 the mussel yield was some 52 million pounds with a value of $1 million. The products manufactured from the mussels were worth some $8 million.[218] It was estimated that between 1913 and 1923 the value of mussels increased from $10.00–15.00 per ton to $60.00–90.00 per ton.[249] Second-generation cultured mussels were first produced in 1917, only three years after the methods required for culturing the animals were developed.[250]

MORE DISCUSSION ABOUT AQUACULTURE AND ITS RAMIFICATIONS

The twentieth century brought with it increased interest in recreational fishing and, as we have seen, increased activity on the part of the Bureau of Fisheries to accommodate recreational anglers by producing and stocking sport fishes. At the state level, fishing licenses were being adopted in many states, the first being Nebraska, which put its licensing system in place during 1901. By 1911, nearly a dozen states had begun issuing licenses. Those states, and the year they began requiring recreational fishing licenses, are as follows:[251]

Arkansas	1911	Oregon	1909
Colorado	1909	South Dakota	1911
Idaho	1905	Utah	1907
Minnesota	1911	Wisconsin	1909
Montana	1905	Wyoming	1911
Nebraska	1901		

Licensing was sufficiently unpopular in Montana that it was repealed a year after enactment, but it was reinstated in 1909. Women and children were often exempted from license requirements, though regulations varied widely among the states. Marine fish licenses are a phenomenon of the late twentieth century.

Stocking the Great Lakes

The issue with respect to whether fish stocking was having any positive influence on the Great Lakes commercial fisheries continued to rage into the twentieth century. Downing, writing in 1911, felt that there had been a positive impact from the introduction of hatchery fish. His contention was based on the following argument:[252]

. . . the supply of eggs for hatching are all obtained from the fish caught for market by the commercial fishermen, the eggs being secured either by having men go out in the boats with the fishermen to strip the ripe fish as they are taken from the nets, or by purchasing the fertilized eggs from the fishermen at a certain price per quart. The number of eggs so secured necessarily depends upon the number of fish taken by the fishermen, and as the number of eggs collected from year to year has steadily increased, it is safe to say that the number of fish caught by the commercial fishermen has increased in like proportion.

There was no indication as to whether the increased egg take was also associated with increased numbers of fishermen and/or increased intensity of effort. The discussion that followed the presentation of Downing's paper before the American Fisheries Society (AFS) supported his claims and extended his hypothesis to other species and regions. Regardless, the commercial fishery declined by 50 million pounds between 1918 and 1925.[253]

The Great Lakes issue provided an early U.S. example of the type of conflict that can develop between commercial fisheries and aquaculture. The issues and approaches to their resolution with respect to the Great Lakes were discussed in some detail by Lester Smith, a commercial fisherman, in 1936.[254]

The feelings of commercial fishermen was summarized as follows:[254]

Hatcheries . . . have failed in results. Since the time hatcheries have been established there has been a steady decline in the production of the species artificially propagated, especially lake trout and whitefish, while at the same time other fish, such as herring and smelt, have maintained themselves by natural propagation.

Objections of commercial fishermen opposed to hatcheries were attributed to both controllable and uncontrollable factors. Controllable factors included inefficient and disinterested help involved with the taking and hatching of eggs, the killing of too many males and too many unripe female fish, and problems with water supplies and the use of chemicals in hatcheries. Uncontrollable factors were associated with weather; the killing of adult fish for their eggs and sperm when, if left alone those fish could spawn in successive years; and depression of market price when the fish taken for spawning were sold. Smith acknowledged that the controllable factors were being addressed to a large extent. The view of the fishermen opposed to hatcheries was that only natural propagation would bring back their industry to its historical levels, though evidence in support of that claim was lacking.[254]

On the other side, there were commercial fishermen who favored artificial fish propagation. That group argued that taking eggs and incubating them in hatcheries ensured a higher level of fertilization and hatching than could be obtained under natural conditions and that taking adults impacted only a small percentage of the spawning populations.

Smith proposed a series of actions that he felt would improve the situation:[254]

1. Change gill netting practices to reduce the chance of taking dead fish.
2. Impose mesh size regulation on gill nets.

3. Require that each state involved in taking Great Lakes fish conduct research to determine ways in which fish could be returned to the water in good condition following spawning.
4. Teach commercial fishermen that artificial propagation is beneficial even when fish are plentiful, and that if they artificially fertilize eggs on their boats and return those eggs to the lake, they may help augment the fishery.
5. Develop uniform hatchery programs for all hatcheries located on the same lake, thereby making it possible to compare results.
6. Obtain more information on life histories of the fishes being cultured.
7. Stock fish in the waters from which the eggs were obtained.
8. Protect fish, through legislation, from destruction in nets until they reach maturity.

Ultimately, the controversy surrounding the value of hatchery activities to the Great Lakes fisheries was greatly overshadowed by other problems, beginning with the invasion of the lakes by the sea lamprey. That controversy continues to rage today. Currently it is perhaps most visible in the Pacific Northwest, particularly in the Columbia River Basin with its salmon problems.

Stocking Marine Waters

Documentation of contributions of marine fish stocking was extremely difficult to obtain, as pointed out by a leading ichthyologist, C. M. Breder, in 1922.[255]

> The actual value of the cultivation of marine food fishes has long been open to question, and in consequence has become the target for both just and unjust criticism. That this condition has existed for such a long period of time is largely due to great difficulty to be encountered in any attempt to measure the effectiveness of fish cultural work on marine fishes. Among the prime reasons for this difficulty is the fact that many of the little understood factors contributing to the production of ocean conditions, cause annual fluctuations of considerable size in the abundance of fish life which tend to invalidate any deductions based on the statistics of catches made by commercial fishermen, not to mention such other factors as have been introduced by man himself.

Breder argued that an increase in landings might as easily be the result of natural fluctuations caused by environmental factors as of the effects of stocking. Using cod as an example and making some simple assumptions (which gave the cultured fish a clear advantage in terms of their contribution to the overall fishery), he calculated that among large cod, one fish in every 13,375 landed could be attributed to the hatchery. Based on landings for 1921, he calculated that hatchery fish amounted to 0.08% of the cod catch. Breder advocated the conduct of studies aimed at determining how oceanographic conditions impacted fish production and was quick to point out that his comments on the probable low impact of hatchery fish on marine fish landings were not meant to be offensive:[255]

> The writer . . . wishes to emphasize that the foregoing remarks are not intended to be mere destructive criticism of marine fish culture in America and wishes to reiterate the hope that they may stimulate constructive thought in the minds of fish culturists in

such a direction that marine fish culture may some day, if possible, be raised from its position of questionable value to that of its companion operating in the fresh waters which has its worth absolutely established.

Another point of view was expressed by D. C. Booth, to whom you have already been introduced. Booth didn't address the issue of whether marine fish stocking was a worthwhile activity. Instead, he discounted the value of commercial fisheries, in general.[256]

> The so-called commercial fishes are important solely for their financial and food value. It is possible to obtain an equivalent in wealth and food in other ways so that the commercial fishes are not indispensable. Too much stress, however, cannot be laid on the esthetic and recreational value of the game species, and whoever heard of a man planning a summer vacation catching the commercial fishes? A popular vote in America, excluding certain elements, would probably show a large majority in favor of the game as compared with the commercial fishes.

The Bureau of Fisheries was not dissuaded from their activities as a result of the comments of such critics. In his report of 1924, Commissioner of Fisheries Henry O'Malley called for increased emphasis on hatchery production, including the production of marine species.[217]

> . . . at no stage in the bureau's history has the demand for larger scale operations been so keenly felt. . . . Deforestation, pollution of waterways, construction of power dams especially affecting the runs of anadromous fishes, reduction of fish-nursery areas through the reclamation of bottom lands, destruction of the young in improperly screened irrigation ditches, land improvement, and many other factors have contributed to the difficulty of keeping our waters adequately stocked with fish by natural means. The construction of good roads and mountain trails and the increased used of the automobile have made it possible for a much larger number of persons to seek recreation in the open country and more remote places.
> . . . [the effect of the above-mentioned impacts] on the runs of anadromous fishes, such as shad, salmon, and sturgeon, and on the fishes of interior waters has been especially serious. Forms having a fixed abode or restricted movement, such as oysters, mussels, clams, crabs, and lobsters, also are suffering depletion wherever they are not afforded adequate conservation and conditions suitable for their existence. Commercial fishing for salmon on the Atlantic coast is practically nonexistent, and the future of the salmon fishery on the Pacific coast is in jeopardy. The catch of shad has decreased by 70 per cent, and the lobster catch is less than one-third that of 30 years ago.
> Under these conditions the need for supplementing the natural supply of food fish by cultural methods is felt as never before, and the bureau has striven to keep pace with the increasing demands made upon this branch of its service. Production in 1924 was four times as great as in 1904, and was accomplished at considerably less than double the cost of operations then.

The Need for State and Federal Cooperation in Fish Culture

Glen C. Leach, who was a long-time employee of the Bureau of Fisheries, pointed to the automobile as a factor in the decline of inland fisheries. In 1925 and 1926, Leach presented his views to the AFS.[257,258] His was an appeal for the Bureau and the state fisheries agencies to cooperate more fully in order to meet the continuing increase in demand for more fish for stocking:[257]

> It is clearly evident that the demands for fish for stocking purposes are growing at a rapid rate. Among the many causes for this condition may be mentioned the rapidly extending use of the automobile by fishing parties and the facility by which distant and heretofore inaccessible waters may be reached by means of it.

By 1927, there were some 22,000,000 automobiles in the United States. That amounted to one car for every five people. Over 500,000 miles of surfaced roads had been constructed.[259] That, coupled with the fact that working hours had been reduced meant that people had more time for angling and an improved means of getting to good fishing spots.

Leach laid some of the blame for declines in recreational fisheries on habitat degradation: "Streams and lakes which at one time were surrounded by forests and comprised an ideal environment for trout have, since the denudation of the land, passed through various transitory stages until at present they are hardly fit for carp."[258] Leach obviously did not share Spencer F. Baird's view of a few decades earlier that carp were a desirable fish. As previously promised, the subject of carp will be revisited.

Among the things that Leach pointed out in the 1925 paper was that the federal program had not added any new hatchery facilities in several years. He felt that the pressure on the states to produce fish for stocking was much higher than that being applied to the federal government. Also, the need for fingerling fish had been recognized and was being addressed much better by the state agencies than by the federal government. The hatchery and distribution system of the federal government had not been designed to produce fingerlings, and Leach didn't seem to think that it could be modified very conveniently, particularly in the face of a budget that was not growing appreciably. His idea was to deliver young fish to fishing clubs and other cooperating citizens groups that would watch over the fish for several months, after which stocking of the resulting fingerlings would occur. In his presentation of 1926, Leach provided more details of the procedure and discussed instances where it was already working: "Our plan has been to furnish to an organized club of sportsmen a certain number of fish along in May after they had been feeding for about six weeks, thus relieving the crowded condition of our hatcheries."[258] The venture was limited to trout which were eventually stocked in streams near the locations where the supplemental rearing occurred. The fish were initially stocked in troughs but were later moved to ponds.[258]

> In October the Bureau sends a man to the co-operative hatchery to get the fish out of the ponds, place them in cans, and give them to applicants. In many instances the applicants are members of the club but we are not interested in that. We only want to be sure that the fish are distributed in that immediate region.

The need for cooperation between state and federal hatchery programs was enthusiastically supported by Seth Gordon, president of the American Game Association, in a paper he presented to the AFS in 1932. Gordon also made a plea for a more scientific approach to the management of fisheries so that the fish produced would be wisely used. He felt that much of the effort being exerted by public hatcheries benefited only private landowners and not the general public and that much of the hatchery work aimed at augmenting commercial fisheries (such as in the Great Lakes) had actually been detrimental to those fisheries.[260]

Swenson Earle, Liaison Officer with the Bureau of Fisheries, reported in 1936 that cooperative efforts between the federal and state agencies were working extremely well and that the resource was being enhanced as a result.[261] Earle had spent 14 months visiting federal and state hatcheries before reporting on his findings. The desires of Leach and Gordon to develop cooperative programs appeared to have been heard, and action had been taken.

Since we have had some insight into the philosophy of Glen Leach (1872–1942), this might be a convenient place to take a brief look at his career. Leach was born in Ackley, Iowa, and received his high school education in Oswego, Kansas. He attended Washburn University in Topeka, Kansas,* for two years and later took courses in business and engineering in colleges in St. Louis.[262] Leach joined the federal government in 1901. He spent his first year with the post office in St. Louis and then became a fish-car laborer. During the next dozen years he worked his way up the ladder in the U.S. Fish and Fisheries Commission, assuming positions at the Put-in-Bay (Ohio) station and the Yes Bay (Alaska) salmon hatchery; and ultimately he became superintendent at the Manchester (Iowa) station in 1914. Two years after that Leach was assigned to a position in Washington, D.C. Hugh M. Smith, then Commissioner of Fisheries, appointed Leach to succeed Henry O'Malley as the person in charge of the Division of Fish Culture in 1918.

As we have seen, budgets were tight during and in the years immediately following World War I, but funding levels ultimately increased, providing Leach with the funds required to greatly expand the fish culture facilities and activities of the Bureau. Funding for hatchery construction was augmented in 1930 with passage of the White Act. That legislation provided Leach with an additional decade of construction activity even in the face of severe budget constraints brought on by the Great Depression. During his tenure directing the Division of Fish Culture, Leach oversaw an increase in the number of federal hatchery facilities from 75 to 133. State hatcheries were also proliferating, of course. A compilation of the state and federal hatcheries in existence in 1937 is as follows:[263]

State	Federal Hatcheries	State Hatcheries
Alabama	1	2
Arizona	0	5
Arkansas	1	1
California	3	26
Colorado	3	17
Connecticut	0	8

---

*Your humble chronicler also took courses at Washburn, and his wife, like Glen Leach, attended Washburn for two years. That is the only known association we have had with Mr. Leach.

State	Federal Hatcheries	State Hatcheries
Florida	0	2
Georgia	2	5
Idaho	2	8
Illinois	0	7
Indiana	2	5
Iowa	4	7
Kansas	0	1
Kentucky	1	4
Louisiana	1	5
Maine	2	14
Maryland	0	5
Massachusetts	3	6
Michigan	1	15
Minnesota	2	7
Missouri	2	9
Mississippi	1	0
Montana	4	12
Nebraska	1	4
Nevada	0	2
New Hampshire	2	11
New Jersey	0	2
New Mexico	3	6
New York	4	19
North Carolina	4	6
North Dakota	0	1
Ohio	1	12
Oklahoma	1	5
Oregon	2	16
Pennsylvania	1	10
Rhode Island	0	2
South Carolina	1	8
South Dakota	1	7
Tennessee	3	2
Texas	4	9
Utah	2	10
Vermont	3	4
Virginia	3	2
Washington	9	32
West Virginia	2	3
Wisconsin	3	32
Wyoming	3	7

Leach served as president of the AFS for 1922–1923 and made many contributions to the *Transactions of the American Fisheries Society*. He also wrote several compilations on fish culture that were published by the Bureau of Fisheries. Included were documents on the culture of whitefish, grayling, and lake trout; shad; walleye, yellow perch, and pike; and rainbow and brook trout.[264-267]

Activities within the States

By 1937, the number of hatcheries being operated by the states, as can be seen from the information in the preceding section, had significantly surpassed the number operated by the federal government. The state hatchery systems were filling local demands and making decisions as to how the waters within their jurisdictions were to be managed. State salmon hatcheries were in place in the Pacific Northwest, and various inland states as well as those on the Atlantic and Gulf coasts had developed either distributed or centralized hatchery systems. The following paragraphs provide an admittedly superficial glimpse of the types of activity that were under way early in the twentieth century.

In Oregon, the Board of Fish Commissioners developed an artificial propagation program aimed at salmon enhancement, with particular emphasis on chinook salmon. By about 1910 facilities were being designed to retain young salmon for several months in rearing ponds prior to release, the idea being that fish released at five to eight months of age would be less subject to predation and other forms of mortality than would younger fish. Ten salmon hatcheries were being operated in Oregon during the 1911–1912 season, all of which produced chinook salmon. Sockeye were reared at one of the facilities, steelhead trout at four hatcheries, and silver salmon at eight. The total number of fish produced was 31,924,251. A few million trout were also produced.[268]

Some individual Pacific Northwest hatcheries had high capacities for hatching eggs. The Green River hatchery in Washington, for example, which began operation in 1899, had grown from a capacity of 2 million eggs to nearly 30 million by 1926. By 1928 the need to provide rearing facilities for fry was also acknowledged.[269]

Kansas took the approach that one large hatchery could, if properly designed, produce the fish required for stocking the entire state.* In 1919, the state legislature appropriated $80,000 to expand the small hatchery at Pratt. The facility was designed to contain 100 acres of ponds plus various support buildings. The source of water was from a reservoir created by damming a stream. Water was fed from the reservoir by gravity to the ponds.[270,271] The facility was opened in 1912 and experienced bank erosion problems along with gopher burrowing during its first year of operation. Bass production was good that year in spite of the problems, but distribution of the fish had to be curtailed because drought conditions had severely reduced water levels in receiving waters.[272]

The Pennsylvania Fish Commission, formed in 1903, was responsible for operation of the state's hatchery system. In 1916, Commissioner Buller discussed the formation and activities of the commission and made the following comments about the plight of hatchery workers:[273]

---

*That's probably true. Having lived in Kansas, I can attest to the fact that there isn't a whole lot of water to stock.

The work of the fish culturist is hard and the hours long, and it is only after years of training that a man attains that efficiency which is so essential in the propagation of fish. The men are now not only overworked, but are unable from the smallness of their number to get all out of the hatcheries that these would do if properly manned.

Lack of proper financial support was responsible for the problem:[273]

> It has been difficult to keep men in the employ of the Department owing to the inadequate salaries that the Department is able to pay under the appropriations made by the Legislature. As the men are trained . . . and become efficient they are sought for and bought up by offers of a much higher salary than the Department is able to pay. The result is that the Department makes the man and some one else gets the benefit of the training.

Buller had closed two hatcheries by 1913 and concentrated efforts on updating the six that remained. Hatcheries often produced a diversity of species. The Wayne County hatchery, for example, hatched bass and trout. It was also a source of bullfrog tadpoles that were distributed to applicants around the state. Shad were produced at the Torresdale hatchery. In addition, walleye, whitefish, and yellow perch eggs were hatched, and plans to produce sunfish and catfish were being developed. A hatchery on Lake Erie produced fish for stocking that water body. Buller was interested in producing not only game fish, but also fish sought as food by average citizens; thus his interest in yellow perch, sunfish, and catfish, among others.[274] Buller's approach was obviously continued beyond his tenure as commissioner. In 1938, for example, over 235 million fish were stocked in Pennsylvania. The list of species stocked included brook, brown, and rainbow trout; black bass; walleye; yellow perch; bluegill; catfish; minnows; whitefish; pickerel; and suckers.[275]

Minnesota was a quarter century ahead of Pennsylvania in the formation of its fish commission. In 1874, the Minnesota Fish Commission was created, and it appropriated the sum of $500 for the operation of a hatchery system. Given the level of the appropriation, the state contracted with two private hatcheries,[276] though a number of state hatcheries were constructed after funding levels increased.

The recreational fishery for Pacific salmon in the Great Lakes and its development in the 1980s is well known, but what is not generally recognized is that both Atlantic and Pacific salmon had been introduced into Minnesota over 100 years earlier. In the late 1870s and early 1880s various lakes and streams in southern Minnesota had been stocked. Salmon were also stocked into the upper Mississippi River, though those efforts failed:[276]

> In the light of our present knowledge it would seem ridiculous to attempt to stock our waters with the Quinnat [chinook] salmon [or] the Atlantic salmon . . . because every condition existing in our waters is against success, but we should not criticise the early commissioners, or the fish culturists whom they employed, for attempts to introduce these desirable species because we can hardly realize at this time the exact conditions existing in those days when there was no drainage, no destruction of our forests and peat bogs by fire, and no cultivation of areas around the headwaters of our streams, all of which necessarily follow an increasing population.

Apparently no one considered putting salmon in the Great Lakes, though that can be understood since the emphasis was all being placed on recovery of the native fishes through massive stocking programs.

Thaddeus Surber, superintendent of fish propagation for the Minnesota Game and Fish Department, placed the appearance of modern fish culture in Minnesota at 1914. He conceded, however, that the siting of state hatcheries was not always appropriate until after 1920 because siting decisions were usually based on politics, not biology. Surber felt that the change from stocking fingerlings rather than trout fry that began in 1923 was a significant advancement:[276]

> The planting of fingerling fish . . . has clearly demonstrated that many waters which we deemed unsuitable can be made excellent trout streams so that we are now enabled to attract anglers over an area of the state fully double the area utilized ten years ago.

Surber commented on various other types of fishes being cultured but indicated that in some cases rearing fish to fingerling size was not possible:[276]

> There has been a great deal of agitation demanding that we rear pike to fingerling size before planting. As most practical fish culturists know this idea is preposterous on account of the well-known cannibalistic propensities of this fish.

Surber might be impressed to learn that northern pike and muskellunge are now successfully being reared to fingerling size in various states.[277]

Among the successes that Surber felt had been achieved in Minnesota was the stocking of walleye in many bodies of water to which that fish was not native. He indicated that stocking that species in the southern part of the state was not successful because carp consumed the food required by the young walleye.[278]

In 1927, the Maryland legislature enacted a license fee, the proceeds from which were used to expand hatchery facilities with the idea of rearing trout to legal size before stocking. By 1930, anglers were able to obtain their limits in half day of fishing.[279]

The importance of maintaining the fisheries of Chesapeake Bay was not ignored by Maryland hatcheries. In 1930, shad, herring, and perch were stocked in waters tributary to the bay. The total number of fish stocked was approximately 700 million.[279]

We'll conclude our tour of state hatchery activities with a Gulf of Mexico state, Louisiana, where all the attention was on the production of freshwater fish, and a good deal of it was associated with largemouth bass.[280,281] Percy Viosca, Jr., director of fisheries for the Louisiana Department of Conservation, described the development of fish culture in Louisiana in his 1929 paper read at the annual AFS meeting. He segmented the history into three periods:[281]

> During the first period, the experimental period or age of migration, cold water species such as the rainbow trout and the northern small-mouthed bass migrated via the fish car far from their native home into the State of Louisiana; but, as we have later learned, only to serve as tempting morsels for our more voracious native species.

During the second period, namely that in which the practical and scientific aspects of fish culture merge, there arose, as a result of the confusion and diffusion, our wonderful Beechwood Fish Hatchery, a project which might easily be considered the last word in warm water fish hatcheries. Fry in large numbers were hatched and planted throughout the State in proportion to political pressure, regardless of whether or not there was a life expectancy for them, and perhaps little did our political ancestors care as long as the public was fooled into believing that due to their efforts the frying pan would not be relegated to the scrap heap.

Viosca called the third period of the age of scientific curiosity, where much was learned about the various factors that impacted fish production, including such things as cannibalism, predation, natural food supplies, pollution, and diseases. He described contemporary fish culture policy as follows:[281]

First, we have gotten away entirely from the planting of fry and we are getting away from the planting of fingerlings as fast as that is practicable. We are raising all of the important native game fishes and no foreigners, and are planting each species only in such waters as are suitable for their development. Owing to the almost prohibitive cost of transporting adult fish, Louisiana is getting away from the plan of a central culture station and has entered a period of decentralization.

Angling and Fish Culture

Charles O. Hayford, then superintendent of the Hackettstown, New Jersey state fish hatchery, presented a paper to the AFS that questioned the value of stocking fingerling trout in the streams within his state.[282] He contended that catchable fish (stocked at 8–12 inches) seemed to support the sport fishery and that the hundreds of thousands of fingerlings that were being stocked did not seem to reach catchable size. During the discussion at the end of his paper, he indicated having killed water snakes that had each consumed a number of trout. The implication seemed to be that fingerling stocking was doing a good job of maintaining the snake population but not much in the way of satisfying anglers.

Views as to whether hatchery fish were superior or inferior to their wild cousins found their way into the literature in the 1920s. Kenneth Lockwood expressed the view that early fish culturists were interested only in production and had little or no concern about what happened to the fish after their release.[283] Anglers, on the other hand, were interested in catching fish and knew little about fish culture or, for that matter, had any interest in the subject. It was Lockwood's view that the situation had changed by the late 1920s and that culturists had developed an appreciation for the desires of the angling public through interacting with them. The anglers, for their part, had developed an appreciation for the efforts of the culturists. That appreciation was demonstrated in an experience related by Lockwood:[283]

On a fishing trip last spring I watched a stranger battle a fourteen-inch brown on a popular New Jersey stream. It was a thrilling fight and there were times when it seemed the fish would be the victor. Eventually, however, the angler won, killed his trout and creeled it.

"Man!" he exclaimed, "Charlie Hayford must have seasoned that trout's meals with red pepper!"

That man's comment certainly reflected an interest not alone in the fish, but in the culturist and in artificial propagation. No man who did not recognize the relation between angler and culturist would have had such a thought at the successful culmination of his battle with a good fish.

"Who is Charlie Hayford?" I asked.

He looked at me in surprise. He was so surprised he stuttered.

"Wh-why, man!" he exclaimed, "he's the superintendent of the state hatchery—the man that raises our fish. Say, you ought to take a day off and go up there and see how it's done."

Lockwood was not only convinced that anglers and culturists understood each other's needs, but he further believed, as had been previously expressed by the one and the same Charles Hayford,[282] that the understanding was best in states where catchable fish were being stocked.

Lockwood recognized that not everyone was a proponent of cultured fish, but dismissed that as a minority view.[283]

> Of course, we all know that now and then a caustic critic makes a big hullabaloo about "liver-fed trout fresh from the hatchery." I would rather not at this time and in this place give any definition of such a person. For every one in my own state I can pick a thousand who commend the present system.

Concluding his remarks on the subject, Lockwood provided his opinion on how the typical hatchery worker viewed the situation: "When things go well with him, the cockles of his heart are warmed by the knowledge that, largely through his efforts, fishermen are finding that unbounded pleasure which is the lure of angling."[283]

Not everyone held the opinion that fish culturists cared about the fate of their charges. An opposing view was presented by Richan in 1936:[284]

> . . . why [is] so much emphasis . . . placed on hatchery production and production records and so little on proper planting and looking after the fish after they are planted and the results of these plantings[?] To be associated with a hatchery one is soon impressed with the idea that all that is necessary is to forever increase production, as if this will solve all the problems. Very little is said and very little interest is shown about what becomes of the fish after planting. The hatchery man is more concerned with increasing production and having a lot of big figures to show at the end of the year than he is with what becomes of the fish that he does produce. The fish have been reared, put out, and forgotten.

A sportswriter, Roy Wall, published a short paper in 1940 that also called the performance of fish culturists into question. He accused them of dealing with "paper fish," in that what was planted was often never seen again.* He was less than complimentary about the success of fish culture:[285]

> Have we not let the fish-culturist headline the show long enough? Certainly it is not enough just to hatch fish fry unless we know something of what becomes of this

---

*Interestingly enough, marine fish managers of today often receive the same criticism. Modern fisheries managers employ computers to model fish populations, so they might better be described as dealing with electronic fish.

planted fry. In short, what is the survival percentage? Civilization is so speeding up the evolution of our fishable waters and at the same time demanding larger fish crops each year that it would appear to me, if my observations have been reasonably accurate, that food and water surveys, stream and lake improvement and ecological balance are now imperative.

Wall seemed to be ahead of his time in desiring to take an ecosystem approach to fisheries culture and management. It was many years before the scientific community began to follow his suggestion.

The quality of wild versus hatchery fish has long been argued, and that debate continues today. Back in the 1930s, Russel F. Lord, a worker for the Bureau of Fisheries in Vermont, was a strong defender of the angling quality of hatchery trout. He documented his viewpoint in both data and parody. He evaluated the behavior and quality of hatchery fish in the early 1930s by releasing fish and recapturing them by angling after different time periods had passed. He concluded that the fish began eating natural foods very quickly after release, behaved similar to wild fish with respect to avoiding human contact, fought well, and were sufficiently difficult to catch that they were ". . . well earned."[286] In terms of their appearance, Lord said the following about the brook and rainbow trout he had studied:[286]

. . . both species were very easy to look at. They were not the typical, dull-colored, pot bellied type of trout which used to be found all too often at hatcheries. They were nice heavy-shouldered fish, lithe and trim. A diet containing a liberal percentage of salmon-egg meal had resulted in colors that no wild fish would have been ashamed of.

In 1935, a paper describing a mock trial in which Mr. Average Angler was found guilty of libel for claiming hatchery trout were inferior was published by Lord. Mr. Angler's punishment was as follows:[287]

Mr. Average Angler, on each Sunday from May 1 to September 1, you shall go, RODLESS, into a stream well stocked with hatchery trout. There you shall remain and watch the happy anglers landing fish. . . . And then, Mr. Average Angler, if in the course of the season you do not experience a change of heart, if you do not feel the urge to go forth and obtain some of the worthy fish for yourself, you shall be banished forever from places where alders bend over a crystal current and darting forms— hatchery trout—slant upward from the shadows to the teasing flies.

While acknowledging that heavily fished streams often served anglers best by being stocked with catchable trout, Needham expressed the view that streams that received little fishing pressure could maintain good populations through natural propagation. He used data from national forests to support his contention. One example he presented was from Squaw Creek in the Shasta National Forest:[288]

. . . 10,000 eastern brook, 120,000 rainbow, and 140,000 Loch Leven trout were planted between 1932 and 1936. Records for the 1937 season show a total of 337 catches, amounting to 2,504 trout. Of these 2,497 were rainbow and 7 were Loch Leven trout. No eastern brook trout were reported caught. Even the low survival ratio indicated by these figures may be too high, for at least some of the fish caught must have been naturally propagated.

CHANGING APPROACHES TO AQUACULTURE

The term *fish culture* was almost exclusively used in published information relating to the propagation of aquatic animals until the 1920s when the term *aquiculture* (later replaced by *aquaculture*) started to appear. The scope of the discipline also began to broaden in the minds of at least some practitioners. In the 1980s the U.S. Department of Agriculture finally acknowledged that aquaculture was a form of agriculture, though many aquaculturists had been making that argument for a number of years. In fact, the argument had been advanced by Lewis Radcliffe, Deputy Commissioner with the Bureau in 1926, when he referred to aquaculture (he used the term *aquiculture*) as "water farming." Radcliffe was an advocate of making aquaculture a scientific discipline and recognized that much better understanding of the dynamics existing in the culture environment was required:[289]

> Increasing the productivity of a pond area involves many factors including a satisfactory water-supply, control of the emergent plant growths; the artificial feeding of the fish with the types of food most productive of growth and health; fertilization of the pond itself, to make it produce the highest maximum yield of natural food organisms, etc.

H. P. K. Agersborg reinforced Radcliffe's view that aquaculture development should follow the agriculture model.[290]

Understanding and manipulating the culture environment was becoming recognized, as was the need to diagnose and treat diseases,[259,291,292] develop hatchery foods, improve transportation methodology, and even determine the proper sizes of ponds for various activities.[277] There was also a growing recognition that stocked fish would survive better if released into appropriate habitats. The need for habitat improvement was expressed beginning in the 1920s.[258,260,293] H. S. Davis, a pioneer in the study of fish diseases and nutrition, was an employee of the Bureau of Fisheries in Washington, D.C., for many years. He stressed the need for selective breeding and for more experimental work in conjunction with making progress in the fish culture arena.[292]

H. S. (Herbert Spencer) Davis was born in Oneida, New York, in 1875. He received his college education at Wesleyan and Harvard Universities. He took his Ph.D. at Harvard and then taught at the University of Florida for 15 years. He then joined the Bureau of Fisheries as Chief of Aquiculture Investigations. He held the position with the Bureau until he retired in 1944. Retirement was a relative term since Davis, like many fisheries scientists today, continued to be active. He was chief fish pathologist for the Oregon State Game Commission until 1947 and was then a research associate at the University of California, Berkeley. Davis published books on the care and diseases of trout and on the culture and diseases of game fishes and helped develop the field of fish pathology. He was president of the AFS in 1932–1933. He died in 1958.[294]

The compensation received by individuals working in public hatcheries was addressed during the discussion session that followed presentation of a paper to the AFS by Lewis Radcliffe in 1927. Among those who participated in that discussion was John N. Cobb, who was the first director of the College of Fisheries at the

H. S. Davis (1875–1958)

University of Washington.* He had obtained statistics on the pay provided for hatchery employees in the various states:[259]

> If you consider [$110 per month] a proper compensation for the superintendent in a hatchery who is supposed to know something of what he is doing, I don't; I think it is a beggarly compensation. You are never going to get anybody from the outside to come in and help you if you cannot pay him more than that.
>
>   I am in the business of training men along fish cultural and other lines, and naturally we are trying to sell these men to the industry; but we have had practically no desire to sell them to the State Fish Commissions unless the compensation is increased. These men are going out and they are going into the private hatcheries. The private

---

*More on John Cobb and the program with which he was involved is presented later in this chapter.

hatcheries are waking up and are paying compensation that is far in excess of that of the state or even of the federal government; and as a result the best men are being taken away from the federal and state service.

. . . I think that if you were to work it all out you would find that the average compensation for state hatchery superintendents throughout the country is in the neighborhood of $100 to $110. You know that you cannot go out and hire a man to dig a ditch for that money to-day, and it speaks well for the sense of loyalty of many of these men that they have continued to work year after year for this beggarly compensation when they could have quit fish culture and gone out and done better as carpenters or something else. . . . In other words, they have loved the industry and have stayed with it, even when the industry was slipping out from under them and leaving them virtually to carry the burden of the increased cost of living.

CARP REVISITED

Hugh Smith discussed the history of carp introductions and the results of those introductions in 1897. That discussion serves as a good review and backdrop for what followed during the first few decades of the twentieth century.[159]

> The carp . . . has been planted in all the States of the Pacific and Rocky Mountain regions, and is now one of the most widely distributed fishes. At a comparatively early date the local fish commissioners became impressed with the desirability of planting carp in the sloughs, bayous, and shallow waters generally, which were either destitute of fish or, to quote the California commissioners, contained only "the worthless and unpalatable fish of the warm waters of the great valleys in the interior of the State." From the outset a very active interest in the cultivation of the carp sprang up in most of the States, and numerous demands for fish for stocking local waters came from farmers and others.

Smith provided more details on the success of stocking in the western states, and then described the economic importance, food value, and desirable qualities of carp. He felt that opinions of the edible qualities of the fish, which differed widely, were often influenced by prejudice and preconceptions. Disappointment in carp as a table fish came, in part, from the expectations of those who planted them in their ponds after having heard testimonials to their superior flavor.

Beyond the flavor issue, there were a number of other problems associated with carp. One involved feeding activity, which includes rooting around in the sediments, thereby muddying the water. Another problem, or at least one that was perceived in a number of quarters, had to do with disturbance of aquatic plants. In Smith's words, ". . . the carp consumes or uproots the wild celery on which wild ducks feed, and it is reported that these game birds have diminished in numbers of late wherever . . . their feeding-grounds have been invaded by the carp."[159] Smith indicated that carp had reportedly been accused of eating other, more desirable foodfishes, but he dismissed that activity as being rare at best given what was known about the life history of carp. Claims that carp were responsible for eating the eggs of other fishes could not be lightly dismissed, however: "The carp is very unpopular in the upper Columbia, at The Dalles and Celilo, on account of its supposed destructiveness to salmon spawn."[159] Regardless of opinions to the contrary, Smith concluded that carp might not be as destructive as they were thought to be by detractors and that more information needed to be gathered before a definitive conclusion could be drawn. In response to claims that carp were eating the eggs of more-desirable fish, Smith countered by indicating:[159]

> . . . no examinations, by competent persons, of the stomach contents of carp appear to have been made in the Pacific States or elsewhere. Even if it should be demonstrated that the carp consumes large quantities of fish spawn, it would not differ in that respect from a host of native species whose shortcomings in this respect are usually overlooked. If we condemn the carp for this pernicious propensity, without conclusive evidence, what are we to do with the basses, trouts, salmons, sturgeons, and the entire sucker and catfish tribes, with known spawn-eating tendencies?

Smith acknowledged that beyond the feelings of many that carp were highly destructive, they were often perceived as having little or no economic value in the Pacific Northwest:[159]

> It is used to some extent by the German families of [the upper Columbia River] and also in the fishing-camps, but the consumption is light. Along the same river, at Payette and Ontario, some favorable mention of the carp as a food-fish is made, but the sentiment of the people is generally against it and the fish has no economic value.
> At Huntington, on the Snake River, Oregon, some carp are caught which find a sale among the farmers of the neighborhood. Along the same river, at Payette and Ontario, some favorable mention of the carp as a food-fish is made, but the sentiment of the people is generally against it and the fish has no economic value.
> At Spokane, carp are sold in limited quantities to German families at 3 to 3½ cents a pound.
> Mr. Wilcox* reports that carp are found constantly in the Portland market, although the abundance of the fish is such that at times it can not be sold at any price.

That having been said, Smith reported that by 1890 there were 1,006 carp culturists in the states of California, Oregon, Washington, Nevada, Idaho, and Utah. Over 1,200 ponds and other water bodies had been stocked with over 100,000 carp in total. The value of those fish was placed at in excess of $15,000. The weight of wild carp sold in San Francisco markets from 1889 through 1902 was over 234,000 pounds and was valued at $7,915.

In 1901, a discussion on carp, led by S. P. Bartlett, an employee of the Commission, was held in conjunction with the 30th annual meeting of the AFS. In his opening remarks, Dr. Bartlett sang the praises of carp. His first sentence to the group assembled has to be one of the more interesting ones ever published in the *Transactions of the American Fisheries Society*: "From a practical standpoint I want to say to you that the United States Fish Commission builded [sic] a great deal wiser than it knew when it introduced carp in the waters of Illinois."[295] That having been said (whatever it means), Bartlett gave a long introduction to the symposium, excerpts of which follow:[295]

---

*Mr. Wilcox was apparently an employee of the Fish and Fisheries Commission.

There is, perhaps, no one here that has been a stronger advocate in years gone by of protection than myself. I early made up my mind that any law the enforcement of which would kill a fisherman was next to gospel. I have changed my mind as to that considerably and believe now in propagation rather than protection. . .

I want to say to you briefly . . . and without giving you any reasons for it . . . that the carp have produced in the State of Illinois more money than all other fish put together. That seems like a pretty hard statement to make, but it can be verified, and I want to say to you that there are more carp eaten on the hotel tables in the State of Illinois than any other fish. I have been served with "red snapper" which turned out to be carp. This cry against the carp is a great big humbug—it is an outrage—they are a good fish if you know how to cook them, but not so good if you don't. . . .

Illinois produced three quarters of a million dollars in coarse fish last year. It would be as much as your life is worth to take a trip down the Illinois river and tell the men there that carp is not a good thing. They would take you out and duck you gracefully into the river. More than one-half the towns on the Illinois river depend mostly for their existence on the fish industry, and considerably over two-thirds of the fish taken are carp. They grow anywhere and everywhere; they grow with the black bass, and the black bass are as plentiful as ever. . .

I have no patience with the newspaper talk that says that the carp are an enemy of the game fish. I do not believe anybody can prove it. I would like to hear it if it is so.

The carp in this state are accused of eating up all of the water plants,—in fact they have been accused of destroying the duck hunting in the states of Illinois and Indiana; they have been accused of almost every crime that fish can be accused of, but I do not believe any one can prove that the carp has ever been an enemy of the game fish or destroys its young or spawn. That is a pretty bold statement to make, but we have representatives here from all over the country, and I would like to hear what they have got to say on the subject.

If I get a little bit extravagant, please attribute it to old age and forgetfulness. I do not wish to make any mistake or to exaggerate. I came here just to provoke the discussion on the carp.

Bartlett's colorful (and clearly unbiased) remarks were followed by those of various other individuals.* Here are some examples:[295]

[*Mr. Peabody:*] I have run up against a number of very strong statements regarding the injury that carp do the fishing and shooting interests of Wisconsin.

[*Mr. Townsend:*] It may be that the carp has been introduced in some places where it was not needed, where other kinds of fish were more important; it might not be advantageous to introduce the carp into the beautiful little lakes of Mr. Peabody's state; but there are many waters in this country teeming with carp, and people are finding out in many places that carp is a food fish. There is a market for carp in the big eastern cities and carp will sell there. They sometimes sell even for a high price; generally they sell for a low price and are bought by poor people.

[The carp] is here to stay and we can not exterminate it any more than we can exterminate the green grass of the fields. I do not wish to pose as an advocate of the carp—I prefer other fish for myself—but I maintain that the carp has a place in good

---

*Statements are not necessarily presented in the order they occurred. I've taken the liberty of lumping together the comments of each individual.

and regular standing in our big eastern markets, and I do not think that our great republic with its rapidly increasing population can afford to sneer at even so cheap a source of food.

[*Dr. Bartlett:*] So far as [carp] eating up the growth in the water and destroying that is concerned, that it to some extent true, but I do not think it is extensive enough to drive away the black bass from their breeding grounds or in any way interfere with them. . . .

. . . I have worked faithfully for carp all these years. For the first few years, fishermen would take the carp, open them up and dress them for sale the same as buffalo, and I had free access to the stomachs of the carp and failed to find to any considerably extent evidences that the carp has interfered with the spawn of other fishes. That is true at least for the muddy waters of Illinois that abound with plenty of other food for the carp.

At certain seasons of the year [carp] do make the water very roily. But we are to consider that our black bass are taken from waters that frequently have six or seven inches of mud at the bottom, and so it makes no difference.

[*Mr. Bower:*] I think that where bass and carp inhabit the same water it is natural that the bass should increase.

There is no danger of a carp ever looting the spawn from a black bass bed. On the other hand I do not think the carp can retaliate against the bass in any way, shape or form. In other words, in the interchange of hostilities between the two species, the bass gets the better of it at every stage of the proceedings, and I think it is a perfectly natural result that the bass should increase in waters where there is an abundance of carp.

[*Mr. Lydell:*] I have never known but a single instance where the carp has destroyed the spawn of the black bass, and I never knew of their destroying any other spawn.

[*Dr. Parker:*] The carp is the most omnivorous of all fishes. He is a hog and will eat everything. He will eat spawn if he gets it, but I do not think he will search it out. I believe . . . that the black bass will increase as a result of the presence of the carp, but we will see a depletion of the perch.

[*General Bryant:*] The greatest trouble we have in some of our lakes in Wisconsin is that the carp have got in there. I do not know of a fisherman in Wisconsin that would catch one if he could, and I never heard of one being eaten either by anybody in the circle of my acquaintance.

. . . I always believe what the Illinois people say about the carp, and I do not question their veracity or their judgment at all, but the people in our section of the country are not educated up to the idea of appreciating the gospel according to St. Bartlett (applause and laughter) and other disciples and brethren of the faith.

Within a radius of five miles of Madison there are billions of carp. Every fisherman sees them, curses them, and refuses to catch them.

Participants in the discussion who were generally in favor of carp outnumbered those opposed. The discussion uncovered the fact that there remained much to be learned and that the opinions related in large degree to the part of the country each speaker represented.

The discussion in 1901 included comments of the value of carp in the marketplace, and there was a general lack of agreement even on that subject. In 1904, the Bureau published information on the actual value of carp in the New York market in 1901.[296] That year 6,906,950 pounds valued at $197,451 passed through that major fish market. The overall U.S. market was placed at 17,160,873 pounds valued at $407,633. Nearly a third of the total carp marketed came from the Illinois River.

Bartlett attributed part of the problem with the acceptance of carp by consumers as stemming from errors made when the fish were first stocked in Illinois.[297]

Literature on the subject [of carp] showed it up as a great fish, immensely prolific and of rapid growth, all that was required was a mud hole and a dozen carp to insure [sic] a year's supply of fish. The farming community went wild over them, and in a single year 3,000 ponds were constructed or arranged for the rearing of fish in Illinois.

. . . the effort to get something for nothing prevailed, and any old mud hole was utilized to raise the carp in. Hogs, and stock of all kinds, ducks and geese, had access to these ponds, but the carp, true to their nature, lived and grew fast, and as spring approached, began to show on the surface of the waters. From these conditions they were taken and cooked and naturally proved a disappointment, soft, oily and muddy in taste.

Then followed a year in which more intelligent attention was given them. Men made ponds for fish culture and gave their fish the same chance they gave their stock or poultry, good food, good water, and intelligent care. When they were wanted for the table they were properly treated and properly cooked, and they found the carp at least on a par with the former favorite coarse fish, the buffalo, and gradually they came into more general use.

Of course, those comments led to further discussion following presentation of the paper before the AFS.* Mr. Meehan had the following comments:[297]

I think myself I am prepared to modify my former opinion of the carp for cultivation purposes, to this extent: If they can be kept or raised in ponds where they cannot get out into the open streams, and can thus be prevented from diminishing the supply of fish in those streams, I am in favor of so raising them. . . .

They do destroy plant life. There is no question about that with us. It cannot be disputed in Pennsylvania that they destroy the water vegetation, and destroy it in large quantities. It is also undoubtedly true that they keep the water muddy and drive the bass and other fishes therefrom.

. . . I am not a friend of the carp, and . . . have opposed its further introductions in waters, and . . . have heartily approved in Pennsylvania of the enactment of a measure which makes it a misdemeanor to plant the fish in any public waters [of the state].

In short, I consider the German carp undesirable fish.

In 1905, Leon J. Cole told the carp story once again in a Bureau publication.[298] He traced the history of carp introductions into the United States, discussed the life history of the species, and delved into the issues that were polarizing people relative

---

*The paper was actually read by a Dr. Smith, so Bartlett might have been hiding out to escape the "fallout," but that wasn't documented.

to how they viewed the presence of carp in U.S. waters. A major complaint came from duck hunters who felt that their sport was being negatively impacted by the destruction of aquatic vegetation by carp. Some authorities felt that plant destruction was incidental to the habit of carp to root around in the sediments in search of food. There as also evidence that the impacts on even the same type of aquatic vegetation varied from one location to another. Cole's conclusion was that carp were detrimental to certain species of ducks.

Cole also discussed the increased turbidity (roiliness of the water) associated with the activities of carp. There was no question as to whether carp often did cause muddiness of the water; the question was, what could be done about it. One solution was to remove the carp through frequent seining and, if that wasn't effective, to drain the lake or pond and remove the fish. Stocking large predaceous fish to prey on the carp was another recommendation.

The question of whether carp interact to the detriment of other, more desirable, fish species hinged on four charges by sportsmen:

1. Carp eat the spawn of other fishes.
2. Carp eat the young of other fish species.
3. Carp prevent the nesting of such species as bass.
4. Carp drive other species away (principally by muddying the water).

Cole went through an extensive discussion in which he addressed each point. His final summary statement was that[298]

> . . . whereas the carp undoubtedly does considerable damage, from the evidence at hand it seems reasonable to conclude that this is fully offset by its value as a food fish and in other ways; that it can not be exterminated, and that the problem is how to use it to the best advantage. . . .

In 1911, a W. T. Hunt had the following to say about the subject of carp food habits:[299]

> As to the claim that the carp destroys game fish by eating the young, I failed utterly to prove this by making an examination . . . of the stomachs of a hundred carp, all over five pounds. . . . We found roots of many kinds, wild oats, grains of corn, wheat, and almost all kinds of vegetable matter that grows along the stream, but no sign of fish or meat, except earthworms and waterworms, with an occasional crayfish or helgramite.
>
> The most peculiar circumstance noted was the presence of fresh water mussels in at least ninety per cent of the fish, which seems to me to explain the disappearance of these shellfish from many streams.

You might recall that the Bureau initiated a program to culture freshwater mussels because of overfishing at about the same time Hunt was placing the blame for low mussel productivity, at least in the areas with which he was familiar, on carp predation.

Surber was convinced that the high concentrations of carp in southern Minnesota waters made those waters unsuitable for the stocking of pike because the carp outcompeted young pike for food. The competition also extended to bass, crappie, and other fishes. In addition, Surber felt that carp disrupted the nests of many species.[278]

Surber recounted the introduction of carp to Minnesota:[276]

> About 1880 or 1881 the carp craze penetrated the state and the report for the years 1881 and 1882 devotes a good deal of space to a history of the attempt made to obtain a supply through Professor Baird, U.S. Commissioner of Fisheries.

Frank N. Clark of Michigan, of whom we have heard a good deal in earlier pages of this book, shipped 600 carp fry to Minnesota in 1882, though smaller batches had been obtained from Washington, D.C., earlier in the year. As to attempts to correct what was seen as a poor decision, Surber said:[276]

> Little did these commissioners realize at that time that many thousands of dollars would be spent in a frantic effort to eradicate carp from the southern lakes some 50 years later, but they were right in one thing, the lowly carp has produced since its introduction in Minnesota more real revenue to the state than any other aquatic animal.

The controversy continued, of course, and continues to some extent to the present. Not only were carp the target of controversy, but in the early twentieth century there was at least some questioning of the practice of stocking inappropriate exotic fishes. One of those who was asking such questions was James A. Henshall. Among his targets of interest was, of course, the common carp:[300]

> A lamentable instance of inconsiderate stocking with an alien species, and one with which we are too well acquainted, was the introduction of the German carp. At that time the unfortunate results of its introduction to our waters could not very well be seen or anticipated. It was intended, primarily, for the stock ponds of farms, as it was supposed that it would thrive in any kind of water, muddy or stagnant. It was supposed to be herbivorous and harmless to other species. In fact it was all supposition, and it was planted without much thought or consideration. But . . . it proved to be a Goth, a Hun and a Vandal to all our native species, though commercially it is somewhat of a success.

Little discussion of the suitability of carp for culture occurred in the literature, and as Bartlett stated in 1905, early attempts at culture had been disappointing. In 1910, Bartlett once again presented a paper on carp at the AFS meeting in which, after once again praising the lowly carp, he made a pitch for the development of its culture.[301]

> Few, if any, economic experiments looking to increasing the food supply of the people have shown such satisfactory results as has the introduction of the carp into the waters of the State of Illinois by the Bureau of Fisheries.
> It is not necessary to recall the prejudice existing against carp in the early years of its introduction, how they were thrown away by the fishermen when taken in their nets and how they were pronounced worthless by nearly all who undertook their culture. The press throughout the country denounced them, and those responsible for their introduction. . . . it is simply the result of a lack of knowledge of the value of the carp and ignorance of the possibilities connected with its culture.

Bartlett went on to indicate that thousands of acres of water in the country (hundreds of acres in Illinois alone) could be used for carp culture and that problems of flesh quality and off-flavor could be overcome through proper management, provision of good-quality feed, and harvesting at the proper time of year.

The goal of Spencer F. Baird with respect to introducing common carp to the United States was not only met but was perhaps achieved far beyond anything he imagined. The Bureau discontinued hatching carp for a number of years, but began production again in 1919 in response to a strong demand for more carp by fishermen on Lake Erie and the fish dealers who purchased those carp:[212]

> The Bureau's operations consist in taking eggs from the fish caught in seines . . . which would otherwise be lost, and hatching them at the Put-in-Bay station, the fry being returned to the local waters in which the carp spawn naturally. During the fiscal year 1919 the young carp thus produced and liberated numbered 22,800,000.

Demand for carp came not only from commercial fishermen, but also from private landowners who continued to make application for shipments. The Bureau, sensitive to those who were not in favor of the indiscriminate stocking of carp, elected to only act on applications after state endorsement for each request had been obtained.

The controversy surrounding common carp continued, of course, but it paled into insignificance during the 1960s when a related species was introduced to the United States. The infamous grass carp is a subject discussed in the next chapter.

LEARNING HOW TO CULTURE ADDITIONAL SPECIES

Federal and state hatcheries became increasingly involved in the spawning and rearing of freshwater sport fishes in the years prior to World War II. Bass and catfish were of particular interest, but there were advances made in conjunction with other species, including oysters. As culturists developed their art, they began to take increasing interest in such subjects as feed development and disease identification and treatment. Aquaculture was becoming a more scientifically oriented discipline, and some of the people who would shape that science for several decades were beginning to make their presence known. We'll look at some of the technological developments later, but for now let's focus on catfish and bass. Channel catfish culture breakthroughs during the early part of the century set the stage for today's large commercial industry. For a variety of other species in which foodfish producers eventually became interested, some progress was made, but large-scale production was not developed either in the public or private sectors until the post–World War II period. Details on some of those species appears in the following chapters.

Catfish

Within a few years after the formation of the U.S. Fish and Fisheries Commission, fish shipments began to include channel catfish, some of which were hauled to California.[108] Those shipments involved young fish that were captured, not cultured, and the situation did not change for many years because of the inability of culturists to spawn channel catfish and closely related species in captivity. In 1910, for example, John L. Leary, who was superintendent of the Bureau's San Marcos, Texas, station, made the following remarks to the AFS: "I have tried to propagate both the spotted [channel] catfish . . . and the blue catfish . . . for the past four years and have met with no success. . . . "[302] Based on observations of where the fish were found in nature, Leary concluded that they spawned in streams over gravel or sandy areas where the current was swift. He acknowledged that others held the opinion that the fish sought out holes in the stream banks in which to spawn (which was, in fact, one of the preferred nesting sites of those catfishes). Leary had also heard stories from fishermen that catfish would spawn in ponds if eels were present but dismissed that notion as absurd.

In the discussion that followed Leary's paper, John Titcomb indicated that attempts to propagate channel catfish in Mississippi in ponds had failed and agreed that the preferred spawning sites of the fish had not been identified. W. E. Meehan also indicated failure in attempts to spawn channel catfish but indicated that yellow bullheads could be spawned in ponds without difficulty.

Leary and others who participated in the discussion were apparently unaware that aquarium spawning had been achieved with white catfish in 1883[303] and channel catfish in 1890.[304] According to Ryder, who described the spawning of white catfish, about 2,000 eggs were deposited in an aquarium in facilities operated by the Fish and Fisheries Commission in Washington, D.C. A detailed description of larval development was also provided, as was the behavior of the male catfish in guarding and aerating the developing eggs.[303] Details of the reproductive habits of channel catfish in an aquarium at the Commission's Central station in Washington, D.C., were provided in the *Report of the Commissioner* for 1911.[201] Parental care of the developing eggs, provided by the male of the species, was once again described.

In 1917, Austin Shira, director of the U.S. Biological Station at Fairport, Iowa, reported on the first successful pond spawning of channel catfish.[305] In April 1914, he stocked 66 broodfish in a small pond. When the pond was drained in November, nine juvenile catfish were captured. Shira repeated the process in 1915 with 61 adults and collected seven juveniles ranging from 4 to 8 inches in length at the end of the growing season. In 1916, he altered the protocol by stocking 34 broodfish in a larger pond and providing four nail kegs as nests. Shira did not indicate why the nests were provided, but in any event they worked. The pond was drawn down in July. Fry were found in one of the nail kegs and an egg mass in another. The problem associated with spawning channel catfish in ponds had been solved with the simple expedient of providing nesting sites. In 1917, Shira modified the nests by providing lids so inspections could be made without disturbing the broodfish.[306] Spawning pens, equipped with nail-keg nests, were being used in Oklahoma by 1931 to selectively breed catfish.[307]

Kansas has never been known as a hotbed of aquaculture activity, but workers in that state can be credited for contributing to our understanding of the biology of the channel catfish (though not without a slight error in the understanding of heritability of a certain trait), and to the development of hatching technology that is pervasive in the modern catfish farming industry. A Kansas game warden, J. B. Doze, summarized information on the channel catfish in 1925. He discussed how one identifies the species and how to spawn it in captivity, along with methods for feeding the fry. First-feeding fry were provided with a mixture of cooked clam and corn meal. Lights were placed over the pond to attract insects at night. As pointed out by Doze, "Kansas is noted for producing many varieties of things. It has an enormous bug crop."[308] Doze can be credited with providing a good deal of valuable information on the spawning and rearing of channel catfish, but his understanding of genetics was a bit flawed. Noting that a drawback of the species was the presence of very sharp spines in association with the dorsal and pectoral fins, Doze described an experiment designed to alleviate the problem:[308]

> The Kansas Fish and Game Department is undertaking an experiment at the State Hatchery, near Pratt, the purpose of which is to attempt to develop a muley Channel cat. Yearlings were dehorned, so to speak, by carefully clipping the barbed spines from the fish near the body. These dehorned fish will be paired off and their offspring given a dehorning treatment in the hope that finally the young will be hornless. While the treatment may not produce the results desired, it may. It would seem, however, that the experiment would offer more chances of producing muley Channel catfish were the treatment directed organically, rather than surgically.

Doze must be given credit for having some doubts about the efficacy of the approach. insofar as I can determine, the above was the only mention of the experiment designed to poll catfish. Having been personally spined on numerous occasions, I can attest to the desirability of eliminating catfish spines, though if the proper technique is employed, the spines provide a convenient aid in picking up the fish.

Virtually all channel catfish hatcheries employ hatching troughs equipped with slowly rotating paddle wheels that slosh the water, causing it to flow through the egg masses, thus supplying the eggs with oxygen and carrying away waste products. The paddles simulate the actions of the male catfish, which accomplishes the same objective by fanning the eggs with his fins. The first report of employing paddlewheels in conjunction with hatching channel catfish eggs was by another Kansas game warden, Alva Clapp, in 1929:[309]

> We have arranged a hatching house . . . with a series of troughs twelve feet long. The spawn is placed in an eight inch mesh hardware cloth cylinder which fits well down in the trough. In the trough five paddles of galvanized iron are attached to bars across the trough. An electric motor which was first tried gave a uniform impulse to the water and proved unsatisfactory. It was replaced by a twenty-four inch water wheel. A three-inch water line comes in and a two-inch pipe runs out on each side, with an eight-inch nipple on the end. One of these nipples pours the water into the trough and keeps it running through. The other one pours water on to the water wheel. The cups on the water wheel gradually fill and it will make a sudden jerk which will throw the paddle over suddenly. Then the wheel may hang for a minute or two minutes, but when the cups become full of water again it will make another irregular jerk. By this method we can hatch one hundred per cent of catfish eggs.

Paddle wheels are now operated with electric motors and reduction gears that typically provide about 30 revolutions per minute. Clapp's feelings about jerky movement were unfounded, but the reliability of his water-driven system cannot be criticized, and water wheels continued to be used at least through the 1930s.[310] Power failures during storms are a common problem in modern hatcheries, and while most now incorporate emergency backup generators, gravity-fed water wheels could still have a role to play.

Feeding catfish fry was a problem for Clapp. He reportedly tried blood, liver, heart, buttermilk, peas, and menhaden meal but was unable to keep the fry in troughs beyond three weeks of age. Stocking them in ponds prior to the onset of starvation was a convenient solution to the problem.[309]

Bass

The culture of both smallmouth and largemouth bass appears to have begun in the 1870s.[311,312] James Nevin believed that he was the first to confine bass in ponds for the purpose of encouraging natural production and also claimed responsibility for being the first to hatch walleye eggs. Nevin stocked bass broodfish in 1873 and was able to produce fry.[313] The basic techniques for the pond spawning of bass, which were developed in the nineteenth century and described in some detail by Dwight Lydell in 1902,[314] changed little through the 1930s.[311,312]

Lydell attempted to strip eggs from gravid female bass on several occasions, but had success with fertilization only once and abandoned the idea. His comments on feeding adult bass in ponds are colorful:[314]

> At first we were unable to feed the stock fish on liver, but after a time we found that by cutting the liver into strips of about the size and shape of a large angle-worm and by throwing the strips into the water with the motion one uses in skipping stones, they wriggle like a worm in sinking and are then readily taken. The liver must be *fresh!*
>
> In order to bring the fish through the winter in good condition it is necessary to begin feeding minnows in September, and to continue this until the fish go into winter quarters.
>
> The bass eat minnows until they go into winter quarters, after which they take no food until spring.

The importance of zooplankton as a fry food had been established by Lydell's time, and methods for fertilizing ponds to encourage the development of appropriate food organisms had been developed.

Mark Riley of the San Marcos hatchery in Texas was one of the many hatchery men involved in bass culture during the first third of the twentieth century. He placed gravel nests in spawning ponds. Whether that approach was unique to his hatchery, developed by him, or generally employed and adopted from others is not clear. What is clear is that Riley felt more progress in bass culture could have been made if hatchery men had been able to interact more readily.[315]

> Had the privilege of attending a meeting of superintendents each year been possible, I am confident that the results accomplished from applied information there obtained would have well recompensed the Government for the slight expense so contracted, for one who can not acquire valuable ideas from associates engaged in the same line of work and utilize them for his own particular needs, in my opinion, falls far short of getting the best possible results from the station over which he has supervision.

Writing in 1930, H. S. Davis contrasted the rearing of bass and trout, opening his remarks with the statement that bass culture had received much less attention than trout culture:[316]

> This is no doubt largely due to the common impression that bass culture is relatively difficult and expensive and for that reason should only be undertaken as a last resort. However, I believe that this is an entirely mistaken idea and I can see no reason why fingerling bass cannot be produced at a reasonable cost.

High costs were associated with the need for extensive pond facilities. Davis was quick to point out that the feed bills at trout hatcheries did much to offset the facilities' cost differences.

Feeding was a major problem in bass culture, both with respect to providing food for broodfish and to their offspring. Davis indicated that when high production was desired and costs were not a major consideration, culturing food (both zooplankton and minnows) in ponds dedicated for that purpose and transferring the food to the bass ponds was a good method. However, he believed that a more economical approach was to produce the food within the bass ponds and accept the lower production levels that could be achieved using that technique. A. M. Powell, superintendent of fish hatcheries for the state of Maryland, found that canned fish could be used to feed smallmouth bass fingerlings. After presenting his paper at the 1931 AFS meeting, he was asked what kind of fish he was providing. "It was canned salmon," he replied.[317] Further questions from the audience elicited the information that it was salmon flesh and not eggs that was fed. Moreover, the cost of the canned salmon was about 6¾¢ per pound delivered in Maryland from a firm in California. The same food was also used in conjunction with trout rearing.

Removal of bass eggs from spawning ponds to the hatchery for incubation was initiated in the 1930s.[318] The technique, as described by G. E. Sprecher, an assistant superintendent of fisheries stationed in Madison, Wisconsin, involved the use of hatching jars. Following hatching the fry remained in the jars for a few days and were then transferred to rearing troughs where they were provided with natural foods. Sprecher felt that because of such things as temperature fluctuations, no more than about half of the eggs incubated in spawning ponds survived to hatching. Because of that, ". . . it becomes our duty to improve nature's method in the same manner as has been done with . . . many other species of game fish."[319]

O. Lloyd Meehean provided a summary of bass culture techniques in 1939 and stressed the need to better understand pond ecology in order to enhance fish production.[318] This ecosystem approach to fish production, while not attracting much attention by either aquaculturists or others at the time, put Meehean ahead of his time. He stressed the need for understanding the ecological relationships among the various species within the pond (bass, sunfish, and other forage species) and the relationships between fertilization and bass productivity.

The commercial sale of bass as a foodfish was being hotly debated in the United States by the 1920s. Some states had prohibited such sales, while others had not. The issue ultimately led to federal legislation controlling the interstate transportation of both smallmouth and largemouth bass under what is now known as the Black Bass Act. Passed in 1926, the act was flawed by not including a practical method of enforcement. Thus, it was redrafted and became effective in July 1930.[320,321] Simply put, the act was as follows: ". . . the Federal Black Bass law prohibits the transportation of fresh-water largemouth and smallmouth black bass from one state to another that have been taken, sold, transported or possessed in violation of some existing state law."[321] Penalties for violation of the act included fines not to exceed $200, imprisonment for not more than three months, or both. Because of the Black Bass Act, many state laws were subsequently amended to preclude the sale of bass as a foodfish, thereby underwriting its status as a sport fish of increasing importance.

Yellow Perch and Walleye

The culture of yellow perch was becoming established early in the twentieth century. Buller described the maintenance, spawning, and early rearing of yellow perch in 1905.[322] By 1913, yellow perch were being cultured by various hatcheries.[206,273] A considerable amount of the effort expended on yellow perch culture went into stocking programs in the Great Lakes aimed at enhancing the commercial fishery. Increases in yellow perch harvests in the 1930s occurred, but an analysis of causes for those increases led to the conclusion that natural oscillations in productivity, not the stocking of cultured fish, explained the phenomenon.[323] Commercial interest in yellow perch culture developed in the 1970s and early 1980s, largely in such states of Wisconsin, Michigan, and Minnesota, where dairy farmers were facing severe economic problems. Promoters (sometimes known as bioshysters, or snakeoil salesmen) convinced dairy farmers that their grain silos could be converted to yellow perch culture facilities that would be highly profitable. There is no indication that anyone who invested in such ventures ever made a profit.

Walleye or pike-perch culture had long been of interest to fish culturists because of the commercial value of the species. As Americans found increasing amounts of leisure time and sportfishing developed, the value of walleye in recreational fisheries increased. Difficulties with hatching walleye eggs were a problem, with a 50% hatching success apparently being exceptional.[324,325] Fry stocking was apparently the norm well into the 1930s, though techniques to rear walleye to fingerling size were worked out by late in that decade.[326]

Trout and Salmon

The culture of trout and salmon was virtually routine as the twentieth century dawned. Rather than developing techniques for producing those types of fishes, refinements in technology and the development of new subdisciplines were focal points. Relationships between human activities, such as irrigation, and the success of enhancement programs were also being discussed for the first time.[327]

Among the refinements being made in trout culture were an appreciation for determining in advance of stocking whether a particular water body was even suitable. Trout, like other species, had long been stocked as fry, but the advantages of stocking fingerlings was appreciated, which meant that better understanding of the food requirements of young fish was required. By the late 1920s, brook, brown, and rainbow trout were being reared to fingerling size prior to release in at least some localities.[328] In the 1930s an economic evaluation of trout culture was undertaken. Frederick F. Fish* concluded that if trout could be reared to a size of 3 inches for a cost of $30.00 per thousand, the cost of hatchery operations was justified. He indicated that for every 100 fish released at 3 inches, 29 legal-sized fish would be produced, which was 27 fish more than could be expected from natural production.[329]

The hatching of salmon eggs was undertaken in troughs without artificial substrates from the time U.S. culturists began their activities. Apparently, conventional wisdom indicated that while salmon constructed their nests (redds) and spawned over gravel, eggs that fell into the gravel would fail to survive. It was not until 1910 that John Pease Babcock published the results of research showing that proper burial of the eggs in gravel led to strong offspring. Babcock felt that burial of salmon eggs under 6–7 inches of gravel would lead to better production at less cost than placing the eggs in wire baskets in troughs. He predicted that hatchery practices would move toward incubating salmon eggs in the gravel of natural or manmade streams.[330]

Oysters

The term *oyster culture* conjures up for some a vision of hatcheries, larval rearing units, algae culture facilities to provide food, and highly controlled rearing on strings hung from rafts or long lines. However, the reality has been, and continues to be for the most part, that oyster culturists typically lease public intertidal or subtidal beds, though the beds are held in private ownership in some states. In most cases, old oyster shell is placed on the beds to provide substrate (cultch) for larval oysters (spat) that are provided from natural spawning beds. Thus, often, culture is basically management of the beds and does not involve high levels of control throughout the life cycle of the animals. Hatchery production of larval oysters is now routine in the Pacific Northwest and has been used on the East Coast, but spat produced and settled on cultch in hatcheries are usually stocked onto natural bottoms rather than being suspended from rafts or reared in trays.

Management of natural oysters beds, which has long been called oyster culture, began long before fish culture begin in the United States. By 1775, consideration of propagating oysters in the vicinity of Boston, Massachusetts, was already being given. Oyster beds in New York and Connecticut were failing by 1850, and hundreds of thousands of bushels of oysters were being shipped from Chesapeake Bay for transplanting. The movement of large amounts of oysters from the Chesapeake Bay region led to declines in that region by 1880.[331] Leasing programs for the use of oysters bottoms by culturists were established in Connecticut and Rhode Island in the late nineteenth century, but culture was in decline a few decades later. Reasons for the decline included excessive lease fees:[332]

> The same official who in 1908 assessed a piece of ground at between eight and nine thousand dollars, assessed it in 1914 for $128,938, more than fourteen times what he had assessed it in 1908. The rule of assessment, according to law, on both of these dates was the "fair market value," and it was worth much less in 1914 than in 1908. This is

---

*An appropriate name for a fishery scientist, and one who was quite active for a number of years.

only one illustration of many hundreds in which oppressive injustice was inflicted upon the oyster farmers.

An underlying reason for the inability of oyster culturists to make a profit was associated with failures of oysters to reproduce effectively enough to provide the numbers of spat required to repopulate the beds. Reproduction failures were associated with pollution.[332,333]

Successful spawning, egg fertilization, and hatching of oysters had been achieved in 1879, but no means was developed for feeding the spat that were produced until 1920. Scientists realized that the oysters needed microscopic organisms for food, but did not have the technology for changing water without losing the larval oysters. That changed when a cream separator was used to concentrate the spat:[334]

> . . . the tiny oysters could be separated from the water as readily as specks of dirt are separated from milk. At first it might appear that such delicate animals would be injured by passing through a centrifugal machine, but we found that, being enclosed in shells which afforded protection, they could be concentrated in the machine without injury. In this way millions of little oysters were separated from a large volume of water, and transferred into a small bowl to another volume of water containing fresh food and other necessities of life.

The settling and attachment of spat to cultch was first achieved in the laboratory as a result.

A paper in the *Report of the Commissioner of Fisheries* in 1924 argued that producing oysters in hatcheries could be economically viable given the fact that one-year-old oysters collected from nature and sold for restocking sold for as much as $1,000 per million animals. Since a single adult oyster could produce tens of millions of offspring, it was reasoned that artificial propagation could be less expensive than collection and replanting, particularly if the biological requirements of the oysters could be understood and duplicated by culturists.[335]

While most of the oysters produced in the United States came from the East Coast and Gulf of Mexico, the Pacific Northwest was also a source. Initially based on native oysters, such as the Olympia oyster, West Coast harvesting began with Native Americans, who were employed by Europeans to harvest oysters commercially once colonization of the region had occurred. Production reached about 100,000 bushels a year from Willapa Bay, Washington, but dropped precipitously in the 1890s because cultch material was not returned to the beds to provide substrate for new oyster settlement.[336] Some culturists eventually obtained success by culturing Olympia oysters inside diked areas established in intertidal areas.[337]

American oysters from the East Coast and Pacific oysters from Japan were introduced to enhance production in the Pacific Northwest. American oysters normally did not reproduce, and the reproductive success of Pacific oysters was also generally poor, requiring continuous annual introductions.[336] Importation of American oysters to Willapa Bay, Washington, involved 21 carloads in 1903, 95 in 1906, and 80 in 1907. In 1908, only two carloads were delivered, and while the deliveries increased again from 1910 to 1912, it then diminished into insignificance.[337] Poor growth and an inability of the culturists to market the animals led to the decline. American oysters continued to be present in small numbers, and young animals were found in the late 1920s, indicating that at least some spawning was occurring.

The Pacific oyster (also known as the Japanese oyster) grew well in the Pacific Northwest, reaching market size in two years as compared with four or five years for Olympia oysters.[337] However, there was some public hostility in the early twentieth century to items in the marketplace that were affiliated with things Japanese. The Japanese firm that established the species in Washington State faced difficulties in obtaining funding from U.S. banks.[336] In 1922, the Washington state legislature restricted the ownership of land by aliens, and the Japanese were forced to sell their holdings to a domestic corporation.[337] The oyster industry in Washington continued to depend largely on Pacific oysters obtained from Japan until well into the 1980s (with disruption in supply occurring during World War II, of course). In the late 1980s, hatchery technology was finally developed to the extent that the needs of the industry could be satisfied.

THE EARLY EDUCATIONAL PROGRAMS

For over four decades after the establishment of the U.S. Fish and Fisheries Commission, there was no formal program in fisheries at the college level in the United States. Many universities offered courses that included some aspects of fish taxonomy and biology through their departments of zoology or biology, but fisheries science programs were lacking. That lack began to be addressed during the second decade of the twentieth century.

The first school of fisheries was founded at the University of Washington.[338] The idea for the school came from a presentation by Hugh M. Smith, Commissioner of the U.S. Bureau of Fisheries at the AFS meeting in 1913: ". . . in the entire country there is not a single university, college, academy or school where even the rudiments of an education in fishery technique may be obtained. . . . "[339] In February 1914, Smith wrote to the University of Washington and commented on ". . . the great desirability of establishing at the University of Washington a school of fisheries, or at least a comprehensive course in fisheries. . . . "[338] There were signs of encouragement for establishment of a fisheries program from the university administration, but, as is typically the case today, any new program would have to have financial resources in place in advance of gaining approval.

One means of moving programs forward is, of course, the application of outside pressure. In June of 1914, the Pacific Fisheries Society held its meeting in Seattle, with Hugh M. Smith in attendance. The publisher of *Pacific Fisherman* magazine, Miller Freeman, used that magazine to promote both the meeting and the idea of a fisheries school at the University of Washington. Trevor Kincaid, who headed the Zoology Department at the university, was asked by the administration to examine the feasibility of establishing such a school and presented a paper on the subject at the Pacific Fisheries Society meeting.[340] Kincaid indicated that he envisioned a program patterned after an existing one in Japan that included fishery technology and fish culture as areas of concentration. A resolution passed at the meeting read, in part:[338]

> . . . resolved that the Pacific Fisheries Society heartily endorses the plan of establishing such a department and urges upon the authorities of the University [of Washington] the carrying out of this project with the least possible delay, and furthermore, the Society pledges itself to aid and support such department by every means within its power in case it is brought into existence.

Things languished until April 1919, when the College of Fisheries at the University of Washington was formally established.[341] John N. Cobb was appointed director of the college by University President Suzzalo. Cobb was a somewhat interesting choice in that he had never completed college himself, though he professed to having taken a course in a business college at one time. Born in Oxford Furnace, New Jersey, on February 20, 1868, Cobb took his first job with a newspaper when he was 20. He began his formal association with fisheries in 1895, when he became a field agent with the U.S. Commission on Fish and Fisheries (later the Bureau of Fisheries), a position he held until 1912. He then became editor of *Pacific Fisherman* for two years before becoming assistant general superintendent of the Alaska Packers Association. Since there was no fisheries educational program in the United States at the time, Cobb took with him his extensive experience rather than a string of diplomas. Cobb continued to lead the program until 1929, when a heart condition prevented him from working. He suffered a heart attack that year and died on January 13, 1939, in La Jolla, California.[338]

Particularly noteworthy is the fact that during the first quarter that fisheries classes were offered (fall of 1919), a woman (Ruth Studdert) was among the 33 students enrolled. In a profession totally dominated by men, Ruth Studdert was truly a pioneer, though what became of her has not been recorded in any of the literature that I have been able to uncover.

Cobb recruited George C. Embody from Cornell University to teach a course in aquaculture during the 1920–1921 academic year. Embody, who returned to Cornell after less than one year at the University of Washington, had 105 students enrolled in his course, which covered such subjects as the culture of trout, striped bass, perch, smelt, black bass, catfish, carp, goldfish, frogs, lobsters, terrapins, clams, and oysters.[339] Fish culture facilities were as follows:[341]

> Our hatchery is equipped with troughs, batteries and tidal box, and we can handle there about 7,000,000 eggs at a time. The arrangement of the hatchery is somewhat out of the ordinary. Each trough is seven feet in length, set crosswise, and has its own intake and outlet pipes. Each student is responsible for his particular trough, and that alone. Alongside [the hatchery] building there are a number of rearing ponds, and we are adding to them as fast as the need becomes apparent.

Embody, who was born in Auburn, New York, in November of 1876, had attended Colgate College, where he received a bachelor's degree in 1900 and a master's in 1901. He did additional graduate work at Johns Hopkins University but eventually took his doctorate in 1910 at Cornell University, where he taught beginning in

1911. Embody served as president of the American Fisheries Society in 1924. He is credited with developing the first planting table for trout, which showed the number of fish that should be stocked in relation to various stream conditions and food availability. He had nearly completed two books on fish culture when he died in Florida on February 17, 1939.[342]

The University of Washington fisheries program celebrated 75 years of existence in 1994. It has undergone various changes in status—including dissolution at one point. It was changed from a college to a school, then back to a college, and in the early 1980s became a school once again within the College of Ocean and Fisheries Science. The program has been responsible for training thousands of fisheries scientists who hold or have held positions throughout the world. The research conducted by faculty and students in the program has been far-reaching; much of it has focused on salmon fisheries management and aquaculture.

The University of Washington remained one of few universities offering any courses in fisheries within the United States for a number of years. Writing in 1928, Cobb indicated that in about 1922 the Massachusetts Institute of Technology established a fisheries engineering course. He indicated that it was his understanding that only a few students had availed themselves of that course. Furthermore, he mentioned that a number of other institutions had been considering fisheries curricula but had not taken any action.[343]

Cobb indicated that part of the rationale behind establishment of the School of Fisheries was to train superintendents of fish processing facilities: "Of recent years the consuming trade has demanded better and safer food products and this has extended to all kinds of preserved fishery products." Those words could just as easily have been uttered today as in 1928.

Fish culture was stressed by Cobb along with seafood technology. With respect to fish culture, Cobb said:[343]

> In no branch of fisheries is the need of scientifically trained workers more obvious than in fish culture. This branch, dealing as it does with the restocking of our depleted waters; the development of as good, if not better, breeds of fish than previously inhabited our waters; the introduction of new species from other sections of our own country or from abroad; the hatching, care and rearing of young fish in our hatcheries, including the onerous problem of the proper feeding of same; all cry loudly for scientifically trained men.

While the University of Washington fisheries program can boast a number of well-known aquaculturists as having been members of its faculty, the most renowned is Lauren Russell Donaldson, who began his association with the program as an entering graduate student in 1930. His tenure on the faculty began in 1932. Born on May 13, 1903, in Tracy, Minnesota, "Doc" Donaldson attended the Tracy high school and then went off to college at Intermountain Union College in Helena, Montana, an institution from which he received an A.B. degree in 1926. After graduation he was principal, science teacher, and athletic coach at the Shelby, Montana, high school and was also a staff member of the Montana Department of Fisheries until going off to graduate school.[338]

The achievements of Lauren Donaldson are legendary and have been chronicled in not one but two books by Neal Hines.[344,345] More of his achievements are mentioned in Chapter 4.

In 1987, Donaldson was enshrined in the National Fish Culture Hall of Fame in Spearfish, South Dakota. The plaque dedicated in his honor reads, in part:

> His experiments in hybridization and selective breeding of trout and salmon resulted in the famous "Donaldson Rainbow Trout," an especially fast-growing, early-maturing and prolific egg-producing strain. He achieved similar results with salmon.
>
> Salmon and trout strains developed by Dr. Donaldson have been introduced throughout the world. These introductions have had a direct success effect on the economy and diet of both rich and poor nations.
>
> Dr. Donaldson developed a research hatchery at the University of Washington where he also performed extensive nutrition studies. These and other studies have improved growth and survival of salmon and trout in governmental and commercial fish farms.

In addition to the above, Donaldson had a hand in introducing Pacific salmon to the Great Lakes in the 1980s. Those introductions led to the establishment of a very large and economically highly successful sport fishery.

Doc retired in 1973 but has rarely missed spending at least part of his day in his office interacting with students or out in the flower beds that he has tended for decades. Once a week, 12 months a year, he delivers fresh-cut flowers to about 20 different offices, all from plants he has personally tended. He also keeps a close eye on what is going on in the fish hatchery and has not been shy about providing sage advice to the faculty members and students now associated with the facility.

The second major fisheries program to be developed in the United States was associated with Auburn University, Alabama. Initiated by Homer S. Swingle, who was born in 1902 and held a Master of Science in Zoology-Entomology from Ohio State University, this was another program in which the academic credentials were not a good match for what came to be one of the premier fisheries programs in the nation. Swingle, when working as an entomologist with the Alabama Agricultural Experiment Station at Auburn, became interested in developing a management program for recreational fisheries and, while it was a hard sell, he was able to obtain some funding and to begin developing facilities in 1934. Labor associated with Franklin D. Roosevelt's Work Projects Administration (WPA) constructed the first ponds in 1934, and in 1940 and 1943 the Experiment Station purchased a total of over 1,200 acres north of the town of Auburn on which ponds of various sizes were developed.[346]

In a tribute to Swingle published just after his death, the following information was provided:[347]

> While working on pond management for sport fishing, Swingle became interested in producing fish in ponds as a food crop. His contributions to the development of aquaculture as a means of food production gave him worldwide fame. He served as a fisheries consultant to the Governments of Israel and Thailand in 1957 and India in 1962, was the U.S. pond fishculture [sic] representative to the Pacific Science Congress,

and gave leadership to many foreign fisheries meetings and programs. He was a member of President Lyndon B. Johnson's Panel of Specialists of Food for Peace in Vietnam in 1966.

Swingle conducted studies to determine optimum numbers and sizes of fish for stocking and on methods of increasing natural food supplies in ponds. He developed the theory that a given pond can only support a certain poundage of fish, no matter how many fish are stocked. As an example, he concluded from his studies that bluegill production is limited to about 100 pounds per acre in unfertilized ponds.[348]

Additional facilities were subsequently added, bringing the total area of watershed under management to about 2,000 acres.[349] The Department of Fisheries and Allied Aquacultures was created in 1970 from the program that had been housed for 36 years in the Department of Zoology-Entomology. Swingle, who had been involved for all those years, became head of the new department. He also established the International Center for Aquaculture. In 1972, a new fisheries building, Swingle Hall, was dedicated to the program's founder. Swingle died the following year.[347] A portion of the plaque associated with the induction in 1986 of Dr. Swingle to the National Fish Culture Hall of Fame reads as follows:

> During his career, he authored over seventy scientific papers, bulletins and circulars. His contributions to fishery management and aquaculture have had a profound impact on fishery development in the United States and throughout the world. He gained an international reputation for developing methods and techniques to manage and increase the harvest of ponds and lakes. These methods led to vastly increased fish production both for sport and food fish. His fish management methods are used worldwide to increase protein for hungry people.
>
> Homer Swingle became an international traveler, directly assisting and advising more than twenty countries on fish production techniques. Numerous times he was assigned to represent the U.S. Government on fishery matters. Dr. Swingle's leadership led to the expansion of Auburn University into one of the world's largest fishery research and teaching institutions.*

Various universities subsequently offered courses in fish and wildlife, but programs with the types of emphasis being placed on fisheries and aquaculture by the University of Washington and Auburn University did not develop until many years after those programs were established. There was not even a textbook on fish culture through the 1930s.[350] A survey of 80 U.S. and Canadian universities in 1940 found that 43 had some level of training for fisheries biologists, though in most cases support courses were offered.[351] In the United States, fish culture courses were offered at the University of Maine, the University of Maryland, Massachusetts State College, the University of Michigan, Cornell University, Ohio State University, A & M College of Texas,† and the University of Washington.‡

The U.S. Fish and Wildlife Service opened a school for fish culturists at Leetown, West Virginia, in 1941. The rationale behind the program was that government

---

*The plaque also indicated that Swingle had been awarded an Honary Doctor of Science degree in 1958.
†Now Texas A & M University.
‡Auburn University was not a respondant.

hatchery workers had acquired their knowledge from experience rather than education. Students were to attend classes for periods of one to six months at Leetown, where classroom, laboratory, and practical training would be provided. The emphasis of the courses would change seasonally—warm-water fish being stressed in the summer and cold-water fish in the winter. H. S. Davis was placed in charge of the program.[352]

THE PROGRESSIVE FISH-CULTURIST

The *Progressive Fish Culturist* (with no hyphen) was first published as Memorandum I-131 by the Bureau of Fisheries, U.S. Department of Commerce, in 1934.[353] The publication was designed to provide information to hatchery workers on advances in fish culture and was an outgrowth of a previous publication, I-130, by C. M. McCay at the State College of Agriculture in Ithaca, New York. That publication was entitled *Notes for the Progressive Fish Culturist*. From its inception until the outbreak of World War II, each issue of the *Progressive Fish Culturist* carried a new volume number. A significant change was made in 1939, when a hyphen was added to the name, and the publication became the *Progressive Fish-Culturist*.[354]

Publication was suspended after the bombing of Pearl Harbor in 1941,[355] but was resumed in 1947,[356] this time with a new numbering system: the new series began at volume 9, whereas the earlier version had surpassed 50 volumes over a period of eight years. Beginning with the 1947 volume, each volume covered a full year.

The *Progressive Fish-Culturist,* or *PFC,* began as a primarily informative vehicle for the hatchery personnel affiliated with the federal and state governments. Techniques of various kinds were described, new types of apparatus detailed, and approaches to construction and maintenance of facilities outlined. Bibliographies of pertinent literature were published, and professional meeting abstracts were reprinted. Historical tidbits were sometimes presented, a few of which we have already seen. As the *PFC* and the profession matured, the subject matter was expanded to include philosophical papers and papers on fishery management. As aquaculture moved from an art to a multidisciplinary science, the articles in the *PFC* reflected that transformation. The *PFC* has long been, and continues to be, a primary source of information for not only government biologists, but also for private-sector aquaculturists.

It is certainly not feasible to present a complete summary of the contributions to the art and science of aquaculture that have been published in the *PFC* during its early years. However, several interesting and sometimes amusing examples come to mind. Most concern items published prior to World War II, but a notable exception is a series of papers on potato ricers that caught my attention. That series continued into the late 1940s.

In 1941, a paper was published in the *PFC* that celebrated the value of the potato ricer to feed fish. In those days, organ meats were commonly used as food. The potato ricer was an apparatus that could be used to produce particles of a size acceptable by fish. The first paper that described the modification of a potato ricer for use in conjunction with fish feeding recommended a couple of modifications from the off-the-shelf model.[357] A few years later, Roger Burrows and John Manning described an enlarged version of the venerable potato ricer.[358] They indicated that the standard model would only accommodate about 3/4 pound of food and designed a mega-version with a 10-pound capacity.

Maurice Bryant[359] suggested further modifications to the standard potato ricer in 1949, and there was a further modification made by Raymond Wilmot in 1952.[360] Not too long after that, prepared feed pellets were adopted by aquaculturists.*

Diseases and their treatment received a considerable amount of attention in the pages of the PFC. Frederick F. Fish, of whom we have already heard, was one of the active players in terms of government workers involved in fish health. Treatment of aquatic animal diseases has always been a problem because of the relatively few chemicals that are efficacious, and treating large numbers of animals is difficult. Methods must be found that allow the chemical to be placed in the environment or feed, or the cost of treatment can become so high as to be uneconomical. Among the treatment chemicals that have proved useful, particularly with respect to the control of external parasites, are potassium permanganate and formalin. The use of those chemicals was first reported for some of the more important fish parasites by Fred Fish.[361] In 1936, a notorious chemical, malachite green, was being advocated as a fungicide and antiseptic.[362] That chemical, a dye, was widely used to control fungus on fish eggs and as a control for certain parasites, but it is now banned because it is known to be carcinogenic.

Fred Fish published papers in the PFC that described fish bacterial and parasitic diseases.[363–365] Fish also discussed the use of the compound microscope for diagnosing fish diseases.[366] Included in that paper was a primer on how to use a microscope. Fish had been involved with the diagnosis of fish diseases for several years[367] before the official establishment of two fish pathology labs by the government. He was appointed director of the laboratory in Seattle, Washington, and H. S. Davis directed one in Washington, D.C.[368]

Some other interesting subjects covered during the early years of the PFC included the following:

- The need for keeping good hatchery records and to consult them frequently. Mentioned was the hatchery at Marion, Alabama, where the superintendent kept a large map of the facility on the wall and filled in data on each pond for quick reference.[369]
- A floating fish nest suspended from a barrel was described in conjunction with bass spawning.[370] (That notion struck my fancy because of research I was involved with in Illinois strip-mine lakes that involved spawning catfish in nests suspended from floats.[371])
- Bird predation, a problem of particular importance to commercial fish farmers today, was discussed in the PFC beginning in 1936. Cottam and Uhler considered the problem to be minor because game species were not being targeted.

---

*Modern feed pellets are not all that different in form from what was produced by potato ricers. In both cases the feed ingredients are pressed through holes that control particle diameter. To form pellets, the extruded material is cut off to a desired length. Strings of food exiting a potato ricer probably fell off into the water as a result of being overtaken by gravity—or the fish culturist could wipe them off the ricer screen when they reached a suitable length.

Also, they described various methods that could be used to keep birds away from fish culture facilities.[372] Three papers published in the *PFC* later in 1936 took exception with the position of Cottam and Uhler,[373–375] though R. F. Lord said herons were welcome at his hatchery because their bills could be used to fashion letter openers.[376] How he got the heron bills was not revealed.

- The merger of the U.S. Bureau of Fisheries with the Bureau of Biological Survey occurred in 1940,* and the role of the federal government in fish distribution by the U.S. Fish and Wildlife Service under the new system was discussed that same year. The primary role was to stock federal waters, with state water stocking being secondary. If excess fish remained after those demands were met, they could be used to stock private waters.[377]

- A controversy about whether dirt- or concrete-bottomed ponds or pools were best for trout rearing was examined in 1938, and it was concluded that both were suitable culture chambers.[378]

THE NUTRITION AND FEEDING OF FISH

We have already examined some of the developments associated with the rearing of selected species between 1910 and 1941, but have not stressed feeding and nutrition to any extent. Rather than include that information in conjunction with the discussion of individual species of fish, it seems more appropriate to look at the sequential development of information on the subject since it was during that period of years that attempts to develop prepared feeds were initiated and during which nutritional requirements were investigated in a scientific manner for the first time.

As pointed out by H. S. Davis in 1927, a wide variety of foods has been provided to trout, but standard fare had become livers, hearts, and lungs from domestic animals.[379] A number of studies in the 1920s and 1930s involved modifications in fish diets what would allow less expensive ingredients to be used and would provide feed in a form that was easier to store. Fresh liver, which had sold for $0.03 per pound in earlier years, had reached an average of about $0.15 per pound by 1920.[380] The price of beef liver as $0.22 per pound in some locations by the end of the decade. In 1928, at least 6,879,439 pounds of various food items were fed to hatchery fish in the United States. The cost was placed at $270,795.15.[381] A 1934 survey showed that 11,455,000 pounds of food was used in U.S. hatcheries. That food was valued at $608,000.

At least 35 different items were being fed as the 1920s were ending. Included were sheep plucks† and liver; beef heart, spleen, and lungs; pig liver, milts, plucks, hearts, and lights;‡ horse and cow meat; fish meal; salted and fresh salmon; condemned canned salmon; herring and other species of fresh fish; middlings; red dog flour; beans (including pinto and soybeans); and cod liver oil. Less frequently used

---

*The Biological Survey was reinstituted in 1994 as a part of the U.S. Fish and Wildlife Service. The now National Biological Service is charged with cataloging plant and animal resources on both public and private lands within the United States. That activity is related in large part to the Endangered Species Act.
†I don't know anything about sheep plucks.
‡Don't know about those, either.

food items included blood, boiled eggs and potatoes, milk (including buttermilk), shrimp, shrimp heads, and fish processing waste.[381] Another survey of feed ingredients published in 1935 showed that there were 41 products in use by commercial fish culturists in addition to fresh products from packing houses. Federal government hatcheries used 30 products in addition to fresh meats.[382] Variable results from feeding trials undoubtedly created some confusion among culturists. Ingredients that performed well with respect to one species might be a poor food ingredient for another. Sometimes, in fact, two studies on the same species with the same ingredient led to different results.[383] Some of the published conclusions included the following:

- "Raw fish retards growth in chinook salmon, as does herring meal."[380]
- "Salmon meal and eggs lead to good growth in chinook salmon."[380]
- "Cooking is detrimental to liver as a fish food."[384]
- "Trout fed solely on dried horse liver meal grew more rapidly to a size of 2.5–3.0 inches than those fed fresh liver."[385]
- "Clam meal is an excellent food for young trout, with dried buttermilk coming in a close second. Dry skim milk is also suitable but less so than buttermilk."[385]
- "Fingerling trout above 4 or 5 inches perform well on vacuum-dried fish meals and shrimp meals."[386]
- "Fish meal, cottonseed meal, and dried skim milk are all suitable foods for trout if supplemented with raw meat. Cottonseed meal is not toxic to trout, even when fed at high levels."[387]
- "The best food for trout is dried salmon eggs, though fish and meat meals, along with dried buttermilk are also good foods."[388]
- "Cottonseed meal should not be fed at levels above 30% of the diet to trout, and all dry foods should be fed in combination with fresh meat."[388,389]
- "A combination of air-dried salmon meal and beef liver produces rapid growth with low mortality in chinook salmon."[390]
- "Beef liver is inferior to sheep and pig liver. On the other hand, sheep liver alone or a 1:1 mixture of beef liver and sheep liver give the same results."[387]
- "Adult largemouth bass can be fed in ponds on a diet of ground cornmeal, wheat shorts, tankage, and cod liver oil.[391] Young bass can be trained to accept such feeds as beef liver."[392]
- "Salmon viscera [sic] a satisfactory food for chinook salmon. Salmon viscera combined with seal meal is [sic] unsatisfactory when used continuously,[393] though seal meal can be mixed with salmon meal for feeding young salmon."[394]
- "Apple flour can be used to provide bulk in feeds without negatively impacting growth."[393,394]

Increasing costs provided much of impetus for the evaluation of alternative feedstuffs for use in conjunction with fish. Availability was another consideration. Rough fish, which could be obtained in the mid-1930s for $0.02 to $0.03 per pound, was not available everywhere and had to be used fresh. Horse meat could be used but was only available at reasonable prices in the western states. Dry ingredients were recommended for inclusion in trout diets after the fish reached about 2 inches in length. After evaluating the available information, H. S. Davis recommended that hatchery workers select from among the following series of suitable diet formulations:[395]

- "pig or sheep liver, 50%; salmon egg meal, 20%; meat meal, 30%"
- "pig or sheep liver, 50%; salmon egg meal, 20%; dried milk or buttermilk, 30%"
- "pig or sheep liver, 50%; meat meal, 25%; fish meal, 25%"
- "pig or sheep liver, 50%; meat meal, 25%; dried milk or buttermilk, 25%"
- "pig or sheep liver, 30%; salmon egg meal, 20%; meat meal, 25%; fish meal, 25%"
- "pig or sheep liver, 30%; salmon egg meal, 20%; meat meal, 25%; dried skim milk or buttermilk, 25%"
- "pig or sheep liver, 30%; fish meal, 20%; meat meal, 25%; dried skim milk or buttermilk, 25%"

Davis indicated that vegetable meals could be used, but to what extent was unknown. Since trout had been determined unable to digest raw starch, the recommendation was to thoroughly cook vegetable meals. On suggested formulation containing cottonseed meal was:

- "pig liver, 30%; cottonseed meal, 30%; fish meal, 20%; dried skim milk of buttermilk, 20%"

In 1938, Lauren Donaldson and Fred Foster summarized the results of studies that had been conducted using various diets to evaluate the performance of trout and salmon. Nearly 100 diets had been used by that time. Food conversion ratios (based on pounds of food fed on a wet-weight basis to pounds of fish gain) ranged from less than 1.0, for such diets as 100% dried carp meal and a mixture of 30% beef liver and 70% salmon meal, to 10.0 for cottage cheese, and 20.0 for a wet diet containing 30% salmon liver and 70% frozen spawned-out salmon.[396]

While much of the work involved the "let's try this and see what happens" technique, the newly developing science of nutrition was beginning to be applied to fish culture by the late 1920s. Casein (milk protein) was used in experimental diets fed to brook trout in 1927 in a study aimed at examining the effect of varying levels of protein, fat, vitamins, and minerals on that species. Conclusions reached in the study included the determination that trout required a minimum of 10% dietary protein for growth. The study also prompted the conclusion that some factor other than protein, carbohydrate, minerals, and the known vitamins was required. Whatever the missing factor was, it could be provided to at least some extent by feeding dried skimmed milk.[397] The unknown required nutrient was subsequently dubbed factor H and became the object of further study. Factor H was determined to be present in large quantity in fresh raw meats[387,398] and was heat labile.[396] Lauren Donaldson mentioned in 1935 that the addition of 30% liver to a canned-salmon-and-flour diet provided the necessary amount of factor H.[390] Fish that were deprived of factor H became anemic and had pink or white, rather than red, gills. Addition of the vitamin folic acid to dry feeds seemed to alleviate the problem.[399]

Other findings about the nutritional requirements of brook trout were uncovered in a study by Tunison and McCay in 1935.[400] Those investigators demonstrated that the fish were able to absorb calcium from the water as well as from the diet. They also determined that hydrogenated cottonseed oil was not digested as well as liquid oils and that liquid cottonseed oil and salmon oil had equivalent digestibilities. It was not until 1972 that the essential fatty acid requirements of trout were established.[401,402]

CUTTING OFF THE SALMON RETURNS

As the U.S. population grew, concentrations of humans developed along the coasts and in the floodplains of rivers. The sometimes unruly rivers tended to utilize those floodplains periodically, much to the chagrin of the residents. Controlling the rivers became a high-priority task undertaken by the federal government. Not only was it thought desirable to reduce the threat of flooding, but there was also a need to provide water for irrigation and, as the demand for electricity grew, for power generation. Dikes were constructed to move water past regions subjected to flooding, and the great era of dam construction began to control flooding, generate electricity, improve navigation, and supply water for irrigation. Dams throughout much of the nation were constructed by the U.S. Army Corps of Engineers as well as by local and state governmental agencies and utility companies. The U.S. Bureau of Reclamation was established in 1902 to harness the major rivers in the western United States: Colorado, Sacramento, Snake, and Columbia. That was accomplished by constructing a number of often extremely large dams over a period of several decades.[403]

Dam construction led to changes in habitat and prevented or severely inhibited the ability of fishes to migrate. With respect to salmon, the blocking off of significant upriver stretches prevented the fish from returning to their home streams for spawning. Dams had begun having an impact on Atlantic salmon runs as early as 1798.[404]

Impediments to salmon migration in the Columbia and Snake Rivers began with the completion of two dams in 1910. After a hiatus of two decades, additional dams were added. Many of the dams were constructed without the benefit of fish ladders. Some, in fact, were so high that it was not feasible to add such ladders when the need for such passage devices was recognized. The dams in place today are as follows:[405]

Dam	River	Year
Swan Falls	Snake	1910
Lower Salmon Falls	Snake	1910
Rock Island	Columbia	1933
Bonneville	Columbia	1938
Grand Coulee	Columbia	1941
Bliss	Snake	1949
C. J. Strike	Snake	1952
McNary	Columbia	1953
Chief Joseph	Columbia	1955
The Dalles	Columbia	1957
Brownlee	Snake	1958
Priest Rapids	Columbia	1959
Rocky Reach	Columbia	1961
Oxbow	Snake	1961
Ice Harbor	Snake	1961
Wanapum	Columbia	1963
Wells	Columbia	1967
Hells Canyon	Snake	1967

Dam	River	Year
John Day	Columbia	1968
Lower Monumental	Snake	1969
Little Goose	Snake	1969
Lower Granite	Snake	1969

Tributaries of the Columbia River other than the Snake were also heavily dammed:[403]

Dam	River
Libby	Kootanai
Albeni Falls	Pend Oreille
Boudary	Pend Oreille
Cabinet Gorge	Clark Fork
Noxon Rapids	Clark Fork
Kerr	Flathead
Hungry Horse	Flathead
Chandler	Yakima
Roza	Yakima
Dworshak	North Fork of the Clearwater
Anderson Ranch	South Fork of the Boise
Pelton	Deschutes
Round Butte	Deschutes
Big Cliff	Santiam
Foster	Santiam
Green Peter	Santiam
Detroit	Santiam
Cugar	South Fork of the McKenzie
Dexter	Willamette
Lookout Point	Willamette
Hills Creek	Willamette
Merwin	Lewis
Yale	Lewis
Swift	Lewis
Layfield	Cowlitz
Mossyrock	Cowlitz

Not all the dams were sanctioned by a state or federal agency. As reported in 1920, a dam constructed on the Sacramento River had cut off salmon returns to the Baird station on the McCloud:[404]

In California certain state officials have suggested that since the dam was constructed without a permit from the War Department, action to correct the evil should be taken by the United States authorities. But since the Sacramento River at the point in question has not been adjudged a navigable stream, no permit was required and the matter falls wholly under the control of the State of California. It is pertinent to ask whether that state is so lacking in foresight and its officers so devoid of responsibility for public interests that they will continue to permit conditions that menace thus directly the public welfare.

Salmon runs were declining in various parts of the Pacific Northwest during the early decades of the twentieth century, and not all the blame could be placed on dam construction, which was in its early phases:[404]

> The run of Pacific salmon has entirely disappeared in some streams. In others it has been tremendously impaired. In districts like Puget Sound it has sunk to a fraction of its former size and during 1919 only one district in Alaska reported a catch that equaled 100 per cent of the number for the preceding ten years. Furthermore, these results were obtained by the use of more boats, more men, more gear and other destructive appliances than had ever been in service before.
>
> . . . unless the taking of salmon can be subjected to reasonable restrictions, that splendid fish will in a short time be as much a luxury on the Pacific coast as its congener is today on the Atlantic.

Three-quarters of a century later, we are still attempting to resolve the situation with respect to allocating the resource without causing its destruction.

As the era of dam construction began, the salmon were essentially ignored by governmental agencies. That decision was not intended to cause the demise of salmon in the Pacific Northwest. In fairness to those who made the decisions, there was a lack of understanding of the salmon's life history. It was not until the 1950s that the mechanism by which returning adult salmon could find the streams in which they were hatched was developed and verified.[406,407] The loss of spawning habitat and of access to upstream spawning areas was widely enough recognized in the 1930s, however, that federal action was taken. In 1938 the Mitchell Act was passed. That act authorized expenditure of federal funds to mitigate against salmon and steelhead losses in the Columbia River Basin. The method employed to implement mitigation involved the construction of numerous state and federal salmon hatcheries.

Hatcheries had, of course, been in operation in the Pacific Northwest for decades prior to passage of the Mitchell Act. The general superintendent of hatcheries for the state of Washington reported in 1927 that the state hatchery system had grown in 30 years from three facilities capable of hatching less than 4 million eggs annually each, to 16 facilities with annual capacities ranging from 5 to 30 million eggs each. During the same 30-year period the system went from having no rearing capacity to having sufficient capacity to rear at least half of the eggs that were hatched.[408] The hatcheries were supported by a direct tax on the commercial fishing industry.

Fishway technology had become sufficiently sophisticated in the 1920s to provide the means by which returning adult salmon could move upstream around dams as high as 200 feet. Keeping outmigrating salmon from diverting into irrigation canals and passing through power-generating facilities at the dams was still a problem, and there was increasing recognition that water pollution associated with activities of the pulp and paper industry was becoming a threat to fish. The screening of irrigation canal intakes would have gone far to solve the first problem, but the means to accomplish that were not available:[408]

The trouble no doubt lies in the fact that our statesmen, representatives in Congress and engineers in promoting and putting through these great developments never realized, appreciated, or knew they were destroying one of the state's greatest industries, the fishing industry. If they had ever realized the damage that was sure to follow, without a doubt funds would have been provided and the canals so designed with screens that the seaward migration would have been efficiently taken care of.

As an example of the severity of the problem, it was estimated that in excess of 90% of the water in the Yakima River, Washington, was diverted during certain seasons of the year into irrigation canals, with the result that millions of young salmon and trout were lost. The Western Food and Game Fish Protective Association was formed to seek legislative intervention and in the late 1920s was able to get support for the Bureau of Fisheries to begin screening irrigation canal intakes.[409]

Some dams were too high to allow the construction of fishways. The Grand Coulee Dam, completed in 1941, was one of those. In 1939, plans were implemented to relocate the salmon that would be obstructed by the Grand Coulee. Tens of thousands of fish were collected at the Rock Island Dam and trucked to the Wenatchee and Entiat Rivers, Nason Creek, and Lakes Wenatchee and Osoyoos, where they were released. The theory was that upon return from sea, the adult fish would seek out the release areas rather than attempting to move up the main-stem Columbia River to their spawning streams, which would be blocked by the new dam.[410] The effort to salvage fish was considered to be successful and was continued through 1941.[411] A similar technique was developed to deal with the lack of a fishway around the Shasta Dam in California.[412]

What of downstream migrating fish that went over the spillways of the dams? That issue was addressed in 1939 in an experiment that involved dropping individual coho salmon from the top of Ariel Dam to the stilling basin some 190 feet below, recapturing the fish, and observing them for several days. Salmon treated in that manner not only showed no mortality, but fed actively the day following the drop.[413] That experiment may have influenced the decision some years later to stock fish from airplanes in remote regions.

ACTIVITIES IN THE PRIVATE SECTOR

Private aquaculture, while not generally recognized by the public to any extent, had been a factor in the United States from the beginning. Recall that the early culturists employed by the U.S. Fish and Fisheries Commission came from the private sector and that that sector continued to provide fish, in particular trout, for recreational stocking. Demands for yearling fish for stocking by individuals and government agencies could be met with fish produced in private hatcheries, and there were appeals made for governmental encouragement of the continued expansion of private facilities.[414] Fish and game publications were advertising the availability of privately reared trout early in the twentieth century. Virtually all of the fish were being produced for stocking, and though the notion that fish could be produced for the food market had been put forward as early as 1913, there were no fish culturists targeting that market. Similarly, private fish culturists were in a position to provide fish for stocking farm ponds, which were being largely overlooked by both state* and federal stocking programs.[415] Radcliffe, writing in 1926, estimated that there were between 50 and 100 commercial fish culturists in the country who produced trout eggs valued at around $200,000 annually and also sold fish at what he termed "fancy prices." Other aquaculture activities were also being undertaken:[271]

> Our annual output of goldfish is probably not less than 20,000,000 fish valued at $350,000. Oyster farming† is now developing with about 150,000 acres of grounds under lease. Considering the suitability of our coastal waters to this crop, it is conceivable that we may double or treble our present harvest of 18,000,000 bushels with the further development and improvement of cultural practices.

Production of tropical fish for hobbyists was another aquaculture industry that was being developed in the 1920s.[416]

Conservation clubs in Indiana became organized in the 1930s. There were over 800 of them in existence by 1938, dedicated to increasing the numbers of fish in the waters of the state.[417] Hatcheries were constructed to augment governmental stocking programs, so another form of private aquaculture was born.

---

*Many states did adopt stocking programs for private waters, though that policy is coming under scrutiny currently and some states are moving away from providing free fish to farm pond owners, thereby returning that aspect of aquaculture to the private sector.
†Oysters were a notable exception to the basic fact that aquaculturists were not producing products that went directly into the marketplace.

CHAPTER 4

Aquaculture from World War II to 1970

THE HATCHERY SYSTEM GROWS

What seems to have been the last readily available survey of U.S. fish hatcheries was published in 1949.[418] That survey was written by Abram Vorhis Tunison and two co-authors. Before we look at the survey in some detail, a few words about Abe Tunison (1909–1971) are in order. A 1930 graduate of Cornell University with a major in agriculture, Tunison conducted trout feeding experiments with a commercial firm for a year, then returned to his alma mater for graduate work.[419] He took his M.S. degree in 1932 and then joined the research staff of the New York State Conservation Department. He was assigned to the Cortland, New York, U.S. Bureau of Fisheries hatchery, which had a cooperative relationship with the state. Tunison is credited with developing the Cortland facility into what became one of the premier aquaculture research laboratories in the nation.

In 1944, Tunison left Cortland and joined the U.S. Fish and Wildlife Service, where he spent eight years as assistant chief and three years as chief of the Division of Fish Hatcheries. In 1957, when the U.S. Fish and Wildlife Service was reorganized, Tunison became one of the administrators in the newly formed Bureau of Sport Fisheries and Wildlife. He ultimately became deputy director of the Fish and Wildlife Service in 1964.

Tunison is credited with having advanced the science of fish culture and with being a dedicated conservationist. He was presented with the Department of Interior's Distinguished Service Award in 1967. Tunison was heavily involved in the training of fish culturists and was a pioneer in the area of fish nutrition and in the development of feeding charts for fish.

The survey that Tunison and his colleagues conducted compared the number of state and federal fish hatcheries in 1937 with those in place in 1949.[418] A total of 383 state hatcheries existed in 46 states in 1937. The number had increased to 522 facilities in 1949. There were 87 federal hatcheries in 37 states during 1937. That had grown to 99 hatcheries in 43 states in 1949. The need to provide fish for stocking in farm ponds was part of the reason for the increased number of hatcheries. Expansion of programs geared toward the production of catchable-sized trout also increased the requirement for new hatcheries. While there had been a significant increase in hatchery numbers over the 12-year-period, 116 hatcheries that had been on the 1937 list were no longer in operation in 1949. Thus, the number of new hatcheries was actually considerably larger than the difference in total numbers between the two years. The move toward production of larger fish by the U.S. Fish and Wildlife Service is quite apparent from the data presented in the survey:[418]

Species	Fry Numbers	Fingerlings	Fish >6"
Largemouth bass	2,031,848	5,700,196	17,284
Smallmouth bass	851,000	185,653	10,656
Bluegill	—	4,514,092	22,492
Green sunfish	—	56,500	—
Redear sunfish	—	489,284	1,099
Striped bass	7,756,000	—	—
Buffalo	—	10,000	500
Channel catfish	130,750	136,852	1,041
Crappie	—	267,689	3,168
Flounder	1,136,737,865	—	—
Yellow perch	—	33,521	—
Walleye	13,250,597	58,014	10,666
Atlantic salmon	—	75,775	74,506
Chinook salmon	26,851,750	31,136,140	—
Coho salmon	—	1,775,865	—
Brown trout	75,860	1,623,871	220,593
Cutthroat trout	3,896,540	2,581,169	14,750
Eastern brook trout	903,987	3,764,622	459,444
Lake trout	108,100	508,296	857
Rainbow trout	207,650	5,224,621	794,394
Steelhead trout	—	263,291	—
Shad	14,311,000	1,500	—
Lobster	1,164,300	—	—
Terrapin	7,000	—	—

Note that the difficult-to-rear species such as striped bass, flounder, shad, and lobster were largely stocked as newly hatched individuals.

In 1949, the states produced 1,545,606,655 animals, while the Fish and Wildlife Service produced 1,312,860,117. Thus, state production had developed to the point where it exceeded that of the federal government. The nearly 3 billion fish that were being produced supported state payrolls of $3,650,869.08 and a federal payroll of $952,495.00. Hatchery operations (with the exception of payrolls) for the state and federal systems required nearly $10.5 million, and the cost of fish food was almost $2 million. A few additional states responded to the survey following initial publication of the data, and the new information appeared in a later paper.[420] The basic trends and values were not appreciably changed with the interjection of the new information.

Data from 1957 showed that the states produced 822 million fish and the federal government over 1 billion. In 1965, production in the states reached 1.5 billion while the federal government was responsible for over 1.7 billion salmon, trout, and pondfish.[5] Construction of reservoirs and farm ponds meant a continuous increase in the surface waters of the United States. It was not only the states that were involved with the stocking of farm ponds. In 1948, the U.S. Fish and Wildlife Service stocked 16,455 of those water bodies.[421] Even with the proliferation of hatcheries, the creation of new ponds and reservoirs, not to mention natural lakes and streams, was clearly beyond the capacity of both the state and federal government to provide the fish required for stocking. By 1965 there were 82 million acres of inland water, of which only half had been stocked. There was a deficit of 287 million hatchery salmon alone in 1965.[422]

Increased angling pressure meant that hatcheries had to continue to increase their production of catchable fish. While the notion of catching such fish as trout that had been released from a hatchery only minutes or hours before entering an angler's creel was found offensive in some circles,[423] there were also those we felt that trout fishing would vanish if it were not for maintenance of strong hatchery programs.[424] The quality of the fish released by hatcheries was considered to be inferior by some anglers, and in a comparative sensory evaluation of brook trout samples, wild fish were found more acceptable than hatchery fish in terms of such factors as aroma, flavor, texture, and appearance, though all samples were deemed acceptable.[425]

While trout fry stocking was being abandoned by federal hatcheries, not all the fish being stocked were immediately available to anglers. In many instances, fish below the legal catchable size, or during periods when the fishing season was closed, were stocked. Those fish were expected to survive, and in many cases grow, until they became available to anglers. Studies showed, however, that the catch of fish planted in advance of their availability to anglers was low, largely because survival was poor compared with that of wild fish of the same age. Some reasons postulated for poor survival of the hatchery fish were:[426]

- high levels of carbohydrates and fats in the diet
- overfeeding, which leads to detrimentally high growth rates
- a relative lack of exercise
- lack of foraging for natural food
- relative freedom from predators
- relative stability of water temperature
- domestication
- selection of broodstock for good performance in hatcheries, but not necessarily in the wild.
- relative absence of natural food leading to improper nutrition, since prepared feeds may not mirror the nutritional characteristics of natural foods
- methods associated with transporting and planting the fish

Schuck, writing in 1948, called for changes in hatchery practices to deal with the situation in trout hatcheries,[426] yet many of the same items continue to cause problems today. Calls for changes in conjunction with salmon hatcheries are still being made.[427]

Cooperative agreements between the individual states and the federal government helped meet the growing demand for sport fish without duplicative stocking efforts. In the Pacific Northwest, the states of Washington and Oregon, in association with the U.S. Fish and Wildlife Service and the U.S. Army Corps of Engineers, stocked hundreds of millions of fish to mitigate losses of salmon and steelhead trout in the Columbia River system caused by ongoing dam construction programs. The cooperative activity developed over a decade after passage of the Mitchell Act, and apparently, not a great deal of progress had been made by that time, as the problems discussed in Chapter 3 continued to exist.[428]

> Th[e] program included the clearance from . . . streams [below McNary Dam] of all obstructions to salmon passage, the laddering of impassable falls and small dams, the construction of hatcheries to develop more rapidly the lower river salmon populations, the screening of irrigation and power diversions to preserve the downstream migrating fingerling salmon, and a number of other associated activities. A crucial function of the States in the salmon maintenance program, and one in which they alone can function, is the judicious regulation of the commercial fishery.
>
> If the Congress provides the required funds in future years as programmed, it is anticipated that the greatest possible use of the lower river for salmon will be attained within about ten years after completion of all material facilities.

Much of the above program was implemented, and a large number of hatcheries were constructed. Yet, several decades later Columbia River salmon populations have reached historical lows. The Bonneville Power Authority has spent tens of millions of dollars attempting to return salmon runs to historical levels. A distinguished group of scientists developed a salmon recovery plan that was unveiled and approved in 1994. That plan is as controversial as any that has been proposed. Any plan is going to be controversial because of the various special-interest groups involved in activities in the Columbia River Basin. One of my former graduate students made a remark to me which may reflect one of the few truths surrounding activities on the Columbia river: "There is a direct inverse correlation between the number of biologists working there and the number of fish that return."

MISCELLANY—ITEMS FROM THE *PFC*

The *Progressive Fish-Culturist* *(PFC)* provided an interesting array of information on advancements in aquaculture during the period from World War II to 1970. Gadgets and new methodologies were among the subject matters covered. The *PFC* did take a break from publication during World War II to help the war effort by conserving on paper and zinc used in the plates from which the journal was printed.[429] Not too long after the war, a short paper was published in the *PFC* comparing the costs of some of the journals that were of use to fish culturists. Government publications, like the *PFC,* were available free of charge (a practice that ended not long after I got on the mailing list in the late 1960s). The *Transactions of the American Fisheries Society* was running $3.00 per year, *Copeia* (a publication of the American Society of Ichthyologists and Herpetologists) was also $3.00 per year, and the *Transactions of the North American Wildlife Conference* ranged from $1 to $2 per annual issue.[430]

Among the tips for fish culturists that appeared in the *PFC* were a recommendation to restore the soles of rubber boots that had become smooth from long use by cutting grooves into them. The technique could supposedly give an extra year of wear.[431] One little article described how to make a paint stirrer for use on an electric drill from a 1/4-inch-diameter metal rod.[432] Fish culturists are always painting something, so it seems to have been an appropriate subject for inclusion, and no more unusual than an article authored (not too surprisingly) by the Portland Cement Association on how to prepare and utilize cement in conjunction with fishery projects, including the construction of ponds and hatchery buildings.[433]

The stores of 1950 were not stocked with several hundred kinds of skin lotions and creams as they are today, though some products were available. That year the *PFC* published an item on the use of a small amount of lanolin to prevent rough, cracked hands:[434]

> Lanolin is packaged in tubes and is usually available at the corner drugstore under the name of lanolin hand cream or toilet lanolin, or some such title. Some preparations even have fragrant essences added. . . . perhaps the fish culturist of the future will reek of an essence of lilac instead of fish and meat scraps.

Today the question is, what corner drugstore?

Asbestos cement pipe was being employed in fish hatcheries after the war, though it did not become available in the eastern part of the country until 1950.[435] Now the mere mention of the word *asbestos* can invoke widespread terror, conjuring up the vision of men in moon suits, but it was all the rage for a number of years.

We haven't heard much lately about incidents in which it rained fish, but that seemed to be a phenomenon of the postwar years. For example, the following statement appeared in the *PFC* during 1950: "The rainfall of fish reported in October 1947 at Marksville, Louisiana, included largemouth black bass . . . , goggle-eye . . . , two species of sunfish . . . , several species of minnows, and hickory shad. . . . "[436] It is not known whether any of those were cultured fish.

Problems at fish culture facilities were not often of the type one might anticipate, as the following example demonstrates: "At the U.S. Fish-Cultural Station at Marion, Alabama, 150,000 fingerling and 312 adult bream were lost late in December when their 3-acre pond (which had been lowered for removal of the fingerlings) was invaded by a large flock of mallards."[437] The likely cause of mortality, while not explained (because this quote represents the entire story as printed in the *PFC*), was probably related to water-quality problems associated with duck droppings. Mallards are not highly piscivorous, though perhaps in this particular case they picked some up and dropped them over Louisiana.

A pet peeve of mine was mentioned in a single sentence in 1952: " 'Specie' is still heard occasionally for a single 'species' of fish."[438] The spelling error continues to be heard today, though everyone should know that *species* is both singular and plural when used in conjunction with living organisms. (The term *specie* is correct when it comes to monetary matters. The dollar is, for example, a specie of currency.)

A number of less mundane matters made up the bulk of what appeared in the postwar *PFC*. Here are a few other items of interest:

- The use of bentonite, a clay mineral used primarily by the petroleum explora- tion industry to lubricate drill bits, as a substance that can be used to seal leaky ponds was described in 1947.[439]
- The use of aluminum, rather than wood, for the construction of hatchery troughs was also described in the *PFC* during 1947.[440]
- Municipal water, which had been used by some fish hatcheries, became toxic to fish once chlorination systems were added to protect human health. The use of a sodium thiosulfate drip systems to neutralize ammonia in a fish hatchery was described in the pages of the *PFC* in 1951.[441]
- A shortage of glass hatching jars led to the development of plastic units in 1953,[442] a time when plastics were just coming into widespread general use.
- A vertical egg incubator, which was the precursor to the very popular Heath incubators used in trout and salmon hatcheries throughout the country, was described in 1955.[443]
- The first dissolved-oxygen meter was described in the pages of the *PFC* during 1955.[444]
- Cryopreservation of fish sperm was in its infancy. Early studies involving the use of very cold temperatures were reported in the *PFC* during 1956.[445]
- The use of pituitary hormones to spawn fish in Brazil was descried in the *PFC* in 1937.[446] That and other hormones found widespread use in the United States in the 1950s,[447] with the first use of human chorionic gonadotropin in catfish, goldfish, and crappie occurring in 1959.[448]

Kermit E. Sneed was one of the individuals involved with both sperm cryopreserva- tion and hormone-induced spawning. Sneed was also involved with the early work on grass carp, a species that was to become even more controversial than the common carp that had been introduced nearly 100 years earlier. Sneed is considered to be one of the pioneers of modern fish culture. His involvement with the introduction of new technology developed in other disciplines to aquaculture gives testimony to including him among the pioneers. Born in Pickshin, West Virginia, in May 1918, Sneed served in the U.S. Army Air Corps during World War II before obtaining his B.S. from the University of Oklahoma in 1948.[449] Following graduation, Kermit continued his association with Prof. Howard Clemens at the University of Oklahoma by assuming a position with the Oklahoma Fisheries Research Laboratory. Clemens and Sneed conducted pioneering work not only with cryopreservation and hormone spawning, but also in the areas of fish behavior, nutrition, toxicology, pathology, and intensive culture.*

In 1958, after having worked with private fish hatcheries as a management biologist and with private and state organizations as a consultant, Sneed joined the

---

*Additional information was also provided by Harry K. Dupree.

U.S. Fish and Wildlife Service. He was initially stationed at the Tishomingo, Oklahoma, station. When the Southeastern Fish Cultural Laboratory was created at the Marion, Alabama, hatchery in 1959, Sneed became its first director. Another research facility, the Fish Farming Experimental Station, was constructed at Stuttgart, Arkansas, in 1965, and Sneed was transferred to Stuttgart as director of that laboratory.*

Kermit was transferred to the Division of Fisheries Research in Washington, D.C. in 1973, from which he retired in 1976.[449] He worked with his sons, Roger and Buddy, who had gone into the commercial catfish farming business in Mississippi, and also served as a consultant in Kuwait and Bangladesh. He died in Belzoni, Mississippi, in May 1987.

Large numbers of pesticides and herbicides were developed beginning in the 1940s. Some of the herbicides were beneficial for controlling noxious water plants. One of them was 2,4-D, which was demonstrated as being effective in controlling water lilies in 1947.[450] It remains one of the few chemicals approved for use around fish ponds. Both pesticides and herbicides could kill aquatic animals if not used properly. One of the most commonly used pesticides was DDT, which has now been banned for over 20 years. In 1948 a writer to the *PFC* reported his experiences with fish mortality that were attributed to the aerial spraying of DDT.[451] He expressed the opinion that the problem would become serious as spraying activity increased and asked if others could provide details about their experiences.

Some of the early data on the toxicity of several pesticides on fish were published in the *PFC* in 1950.[452] Various other such studies followed. One of the problems with conducting toxicity studies in ponds was that the soils could be contaminated and, if not removed and replaced, could impact fish reared in those ponds for many years. In 1966, E. Wayne Shell at Auburn University compared fish production in plastic pools, concrete raceways, and earthen ponds.[453] He concluded that the first

---

*I got to know Kermit when I spent the summer of 1968 at Stuttgart. I had met Harold Webber, a consultant who was working with a couple of companies involved in shrimp culture in Latin America (and whose name is mentioned again in Chapter 5). As a part of his work, Webber had contracted with Florida State University (where I was in graduate school from 1968 to 1971) to undertake research on shrimp rearing. Webber also felt that catfish might be appropriate for culture in Latin America and asked if I would be interested in conducting my Ph.D. research on the potential for catfish farming in Honduras. Preliminary to the Honduras assignment I would be paid to gain some experience at Stuttgart. Since I was interested in catfish (primarily as a result of having written a term paper on the subject while at Missouri) and was looking for a summer salary, I jumped at the chance. Webber had a falling out with the companies who were underwriting his activities, so I never did get to Honduras. However, I did get to spend that summer at Stuttgart and to meet Kermit Sneed and several others who were at the cutting edge of research aimed at providing guidance for those involved in the newly developing channel catfish industry.

My wife, Carolan, and I arrived in Stuttgart with a total of $7.00. The money for my summer salary was, as they say, "in the mail." By the time we spent $5.00 for a hotel room and got our two young children breakfast the next morning, I was down to less than 50¢. The only credit cards we had were good for gasoline but little else.

After meeting Kermit and explaining my plight, he took me to his bank and arranged for enough money to keep us going until my summer salary money appeared (which happened about a week later). Not only did he provide financial assistance; he got us into an apartment right next to his in the nicest apartment building in Stuttgart. We got to know Kermit and his family quite well that summer, and their friendship was something we cherished.

two could be used for some types of studies, but cautioned that the results should not be extrapolated to earthen ponds without verification.

Wayne Shell is another individual who deserves additional attention. Born in June 1930 in Butler County, Alabama, he took his B.S. in Fishery Management at Auburn University in 1952,* and his M.S. from the same institution in 1954. After two years in the U.S. Army, Shell completed his graduate studies in 1959, when he received his Ph.D. from Cornell University. He returned to Auburn University as an Assistant professor and worked his way through the ranks to Professor. He was involved in a number of research areas, and was instrumental in initiating work in the United States with tilapia. Upon the death of H. S. Swingle in 1973, Shell assumed the position of department head in the Department of Fisheries and Allied Aquacultures. He retired in 1994 and was succeeded by Wilmer (Bill) Rogers, who became only the third department head of the organization since its inception. Wayne Shell was actively involved in the international fisheries activities of Auburn University. He participated in programs conducted in Brazil, Guyana, Honduras, Indonesia, Italy, Panama, Peru, the Philippines, Rwanda, Sudan, and Thailand. He undertook a total of 23 short-term international assignments in those countries.

FISH TRANSPORTATION

As automobile and, later, air travel became the dominant means of moving people long distances, the method by which fish were transported also changed. The fish cars that had operated effectively for decades were gradually replaced by trucks, including as early as the 1920s, those designed specifically for hauling fish.[454] Aeration technology advanced as well. The feasibility of using bottled oxygen to provide aeration was demonstrated by 1940.[455] By the late 1940s, virtually all of the techniques currently in use to transport fish had been developed: "Live fish have been moved by pail and can, by tank (plain and complex), by specially designed truck and railroad car, and by airplane."[456] The technique for transporting fish in polyethylene bags charged with oxygen, which is widely used today for moving fish, particularly in the ornamental fish trade, was developed in the late 1950s,[457] and adapted to the shipment of large numbers of channel catfish fry in the 1960s.[458]

Aeration in the 1940s could be provided in at least four ways:[456]

1. Aeration by circulation of water through a Venturi valve.
2. Aeration by spraying.
3. Aeration by forced air, i.e., by use of air compressors.
4. Aeration by introduction of "canned" oxygen.

About the only major additions to the list since the 1940s would be electric agitators that run off batteries and the use of liquid oxygen.

A glass fish tank mounted on the bed of a truck was developed by the Colorado Fish and Game Department. The truck was used primarily as an exhibit to demonstrate to the public the problems associated with distributing fish. State and county fairs, sports shows, and other venues were visited by the truck carrying the glass fish tank.[459]

---

*Information from Wilmer Rogers, Auburn University.

EXOTIC INTRODUCTIONS

Agriculture in the United States is heavily dependent on the use of exotic species of both plants and animals. Important native species include turkeys, tobacco, and corn (maize) but little else. Horses evolved in North America but became extinct on this continent thousands of years ago, only to be reintroduced by the Spanish.

The introduction of exotic species was not undertaken in the absence of expressions of concern.* We are all familiar with appeals to determine potential environmental impacts prior to introducing new species today, whether those organisms are going to be moved from one part of the country, where they are currently present, to another region that lacks the species, or in instances where a species that does not exist in the United States is being considered for introduction. Similar appeals were heard in the past. Speaking of fish introductions to Oregon, Francis P. Griffiths wrote the following in 1939:[460]

> Many introductions of exotic species have been made without considering the possible effects on native species. Of eighty fresh-water and anadromous fishes of Oregon, nineteen have been established by introductions.

By 1945, at least 39 exotic fishes had been introduced into the state of Nevada beginning in 1873.[461] Various other states could undoubtedly make similar claims.

The federal government, which had been very free in its use of exotics during the early years, changed its position by 1940, a year in which two chinook salmon stocked in 1934, and weighing over 30 pounds, were caught in Maine:[462]

> As a result of the interest aroused by these salmon, it is likely that requests will be made for the [U.S. Fish and Wildlife] Service to ship additional quantities of eggs from the West coast for further plantings. Because of the rather general unsatisfactory experience with transplanting exotic species, Service officials announce that no eggs will be transported for planting in East coast streams until the necessary biological studies have been made.

Whether such studies were ever conducted is not clear, but the federal government does seem to have retained the policy of not shipping Pacific salmon to the East Coast.

An exotic species considered highly desirable in one region may be totally or largely unacceptable in another part of the country. One example is the brown trout:[463]

> . . . in the West a general feeling of dislike is expressed, whereas in the East there are some who seem to prefer brown trout above all other species. Some of the points at issue in this controversy are not without foundation in fact.

Other examples are grass carp and tilapia.

Grass carp, native to Asia, were recommended in 1957 for introduction into the United States by H. S. Swingle.[464] The fish were supposed to be strict herbivores and were seen as a means of controlling aquatic weeds without resorting to herbicides. In

---

*This word *concern* is one of the most overused words in the English language and one that I make an effort to avoid. The term today has little meaning, other than to label the utterer as a person with an agenda. I decided to use it here, however, because it is appropriate in the context of the historical period being discussed.

1963, a shipment of grass carp was received by Auburn University. Those fish were subsequently spawned in 1966.[465] The Fish Farming Experimental Station at Stuttgart, Arkansas, also produced a small number of grass carp fry in 1965. Subsequent distribution by state and federal agencies led to the spread of grass carp to 35 states by 1978.[466]

The herbivorous nature of grass carp was of interest to fish farmers, and studies were initiated in the 1960s to examine which plants grass carp would eat.[467,468] Not everyone was convinced that grass carp were strict herbivores,* and there were calls for caution in stocking them until additional studies could be conducted.[469] Opponents of grass carp expressed views that their voracious feeding habits would mean the destruction of valuable wetlands should the fish be allowed to escape into the river systems of the nation. Such fears led to the outlawing of grass carp in over 30 states.

Tilapia, which look a good deal like sunfish, are native to Africa and the Middle East. They are tropical fishes that cannot tolerate water temperatures approaching freezing, so they cannot survive outdoors in temperate climates under normal circumstances. They were widely introduced throughout the tropical world as culture species, beginning in the 1930s. They have firm, white flesh, are extremely hardy when not exposed to cold water, and are generally accepted as an excellent foodfish by most human cultures.[470] Successes with tilapia culture in Asia created interest in the fish by U.S. fish culturists during the 1950s. The U.S. Fish and Wildlife Service received a number of inquires with respect to the possibility of tilapia culture in the United States, but took the position that they could not advocate use of the fish as food or sport fishes until appropriate research had been conducted.[471] The service indicated that tilapia were often available from aquarium fish dealers, but that those interested in rearing tilapia should obtain appropriate permits from their state fish and game departments.

Early studies on the suitability of tilapia for ponds in the southeastern United States demonstrated the inability of the fish to survive during winter.[472] Swingle felt that if broodstock or young-of-the-year fingerlings were overwintered and stocked during spring, they could provide sportfishing opportunities by fall.[473] Much of the early work was conducted with Java tilapia,† though other species, such as the Nile tilapia, were also under scrutiny.[474] There was little discussion of tilapia as a foodfish during the 1950s and 1960s, but there was interest in using some tilapia species for weed control.[475]

MORE ON NUTRITION, FEEDS, AND DISEASE

"Real fish don't eat pellets," was seen on bumper stickers a couple of years ago. The message represented one in a long series of criticisms concerning the stocking of hatchery-reared fish (in this case salmon) in the natural environment. The criticism in the 1990s is in some cases similar to that of the 1890s, to the extent that the use of feed other than that which the fish would obtain from nature was felt to be less than

---

*Grass carp fry do consume zooplankton, but the preponderance of available information indicates that they are strict herbivores throughout most of their lives.
†This species was introduced to Asia during the 1930s but is native to Africa.

desirable. In the beginning the fear was that fish fed prepared feeds would not perform properly when released.* Now the focus is on misconceptions that cultured fish are constantly fed antibiotics, hormones, and other food additives that should not be present in the fish, the environment, or the people who might eat those fish. In reality, antibiotics are used only as necessary, and the withdrawal period (with subsequent clearing of the chemicals from the bodies of the fish) is sufficiently long before the fish are marketed that no resides are present. Hormones are not routinely fed, though they have been used to induce spawning and to sex-reverse certain fishes. Background levels of hormones, which are chemically indistinct from the naturally occurring ones, are well below what the fish are producing at the time of marketing. Additional food additives, other than vitamins and minerals, are not used by U.S. fish culturists.

The development of dry fish feeds was only possible following World War II when pellet mill technology for their production was developed in conjunction with the terrestrial livestock industry.[476] In the case of extruded products, the human food industry was the primary target for products such as extruded breakfast cereals. (I vividly recall listening to Sky King, the Lone Ranger, and other programs on the radio around 1950 and hearing how certain breakfast cereals were "shot from guns." It was perhaps 25 years later that I learned that those products were produced in extruders.)

Donald Brockway discussed the development of an early pelleted fish-feed formulation in 1953.[477] During the same year, Harvey Willoughby described his research on pelleted feed used in conjunction with trout production. Willoughby's formulation was as follows.[478]

Dried skim milk	11%
Whitefish meal, 60% protein	16
Wheat flour middlings	24
Cottonseed meal	24
Distiller's dried solubles	11
Dried brewer's yeast	10
Salt (sodium chloride)	4

The use of pelleted feeds became quickly accepted in trout hatcheries, but the typical protocol involved feeding meat to fry and eventually training them to accept pelleted feeds as they aged. There was some feeling among aquaculturists that trout could be converted to prepared feeds earlier in their life cycle if the feed was properly presented. Early studies on how color affected the acceptance of pellets by young trout led to the conclusion that red pellets were the most attractive.[479] Size was probably more important. Today all the starter feeds are dark brown to almost black, with the color attributed to the ingredients, not additives.

Between 1945 and 1955, the amount of meat products purchased for use in federal fish hatcheries had grown from 995,323 to 2,695,633 pounds. Dry-ingredient usage had increased from 401,588 to 2,128,605 pounds over the same period.[480] The trend was toward increased use of dry materials. Nearly 500,000 pounds of pellets were used in the federal hatcheries during 1955 and 1.6 million

---

*Up until the 1950s such feeds were largely in the form of fresh organ meats.

pounds in 1957,[481] so dry pelleted feeds were clearly becoming accepted, though with some reservations:[480]

Pellets appear to be well on the way to becoming a poplar food, requiring a minimum of labor and of capital investment. Pellets are not the whole answer, but they can be expected to improve and to provide the basic nutritive requirements in the next few years. Perhaps the use of meats, as such, will not longer be required.

The move toward pelleted feeds was predicted to accelerate with the construction of additional trout and salmon hatcheries. (Such hatcheries were the largest users of prepared feeds.)[480]

Formulations were crude in the 1950s, and fish such as trout, which are currently fed from swim-up to stocking or market size on dry pelleted rations, often took several days or even a few weeks to convert from wet ingredients to pellets. In Michigan, dry pelleted rations were adopted by state fish hatcheries by 1957, though the recommendation was to feed raw beef liver at least once every three weeks.[482] Since the nutritional requirements of trout were just beginning to be determined, the beef liver provided additional protein, and probably more importantly, an array of vitamins that were generally deficient in the early fish feeds.

The idea of feeding medicated feeds was also developed in the 1950s. The adoption of dry pellets allowed for even distribution of antibiotics within the ration and allowed long-term storage of medicated feeds. One suggestion was that medicated feeds might be feed continuously to control certain types of bacterial diseases.[483] The possibility of developing antibiotic-resistant strains of bacteria had yet to be recognized, at least by fish culturists. Today it is recommended that antibiotics be fed only when absolutely necessary and then only for a period of 10 days at recommended levels. Because of additional costs for medicated feeds, aquaculturists typically follow the recommendations religiously.

In the 1960s, commercial feed companies began to produce feeds specifically formulated for the fish farming industry. Designed for trout and salmon, the feeds featured high levels of protein (approaching or even exceeding 40%). Fish culturists soon began testing supplemental feeds of their own formulations as well as commercial feeds on other species of sport fishes. The state of Kansas first reared channel catfish from fry to stocking size on dry granules and pellets in 1960.[484] Attempts were made to feed a commercial trout ration to sunfish in 1967[485] and to northern pike and muskellunge by 1970.[486]

Diet development research moved rapidly in 1960s. Supplemental rations, which met many of the nutritional requirements of fish but not all of them, began to be replaced by complete formulations as the requirements of fish were defined and the ingredients that would meet those requirements were identified. The first successful complete rations were developed for trout.[487488]

By the early 1970s, a standard dry-pellet formulation had been developed for salmon. That ration, called the Abernathy Diet, was used as a control diet for studies aimed at further delineating the nutritional requirements of the fish.[489] The diet, which underwent some modification in later years, initially contained fish-carcass meal, dried whey, wheat germ meal, cottonseed meal, soybean oil, and a vitamin supplement.

Earlier, a new approach was developed by aquaculturists in Oregon that incorporated fresh ingredients together with dry products to form what came to be known as the Oregon Moist Pellet (OMP). First used in 1959, OMP was formulated as a salmon and steelhead trout feed. Its formula was modified from time to time over the years. In 1963 it was as follows:[490]

Dry Mix	
Cottonseed meal	23.00%
Herring meal	21.00
Crab or shrimp solubles	6.00
Wheat germ meal	3.60
Distiller's solubles	2.40
Vitamin premix	1.50
Wet Mix	
Albacore tuna viscera	20.00
One or a mixture of:	20.00
Turbot, Pasteurized salmon viscera, dogfish	
Corn oil	1.80
Choline chloride liquid	0.65
Antioxidant	0.05

OMP contained nearly 35% protein and was being manufactured by three commercial firms by 1963.[490] Because of the high moisture content (over 34%, as compared to about 9% for dry pellets), OMP had to be used immediately after manufacture or stored frozen. In either case, it was necessary to use the feed in the immediate vicinity of where it was produced,* which continues to be the case today. OMP is still available in the Pacific Northwest and is used almost exclusively for feeding salmon, though it was used experimentally for such species as largemouth bass as early as 1968.[491]

The nutritional requirements of channel catfish were being investigated by the late 1950s. In 1963, the ingredients for a series of experimental dry-pellet formulations were adopted by the Fish Farming Experimental Station in Stuttgart, Arkansas. The results of studies at Stuttgart, and elsewhere, provided commercial feed companies with the information they needed to develop some of the early commercial catfish feeds. The Stuttgart formulas of 1963 included mixtures selected from the following ingredients:[492]

Fish meal	Rice hulls
Soybean meal	Rice by-product
Feather meal	Sorghum grain
Blood meal	Alfalfa meal
Distiller's solubles	Soybean oil
Rice bran	Fish solubles

A vitamin premix was added to each experimental diet, and purified amino acids were added in a few cases. Several of the ingredients listed above see little or no use in modern catfish rations. Those not used much in present feeds include feather meal, blood meal, various rice products, sorghum, and alfalfa meal.

---

*High shipping costs precluded sending frozen OMP very far from the point of manufacture.

I was asked to assist with the reformulation of a commercial catfish ration in the late 1970s and recommended the elimination of alfalfa meal from the existing formulation, where it was included at 50 pounds per ton. Several months later I was told that there had been a complaint from a customer who was upset because the feed was no longer green. I recommended that the customer be told that the change in color was going to result in better performance of the fish on a new and improved feed.

Salmon, trout, and catfish were certainly not the only species being offered prepared feeds by 1970. Shrimp culture was taking off, at least in the minds of those who thought it had potential, and a number of fishes other than the ones mentioned were being offered pelleted feeds. Floating pellets, produced with extruders, were beginning to appear on the market by 1970, though they were not widely adopted until a bit later, partly because they were more expensive than pressure pellets. A few additional comments on feeds and feeding of various species appear in the sections pertaining to the development of commercial aquaculture that follow.

Now let's turn our attention, briefly, to aquatic animal health. One of the widely accepted tenets of aquaculture is that the progress made toward increased production is accompanied by the proliferation of diseases. Aquaculturists have never had a large arsenal of treatment drugs, and in recent years that arsenal has been severely restricted due to government regulation. For example, while dozens of antibiotics are available for use in mammals (including humans), only two are currently approved for use in foodfish.

Fish diseases have been widely studied since World War II, and a number of individuals have become widely recognized for their contributions to the field. H. S. Davis was one such person. He was actively involved in describing the diseases of various fishes during the postwar period. Others who have contributed significantly to our understanding of fish diseases are Stanislas F. Snieszko, Glenn L. Hoffman, Ken Wolf, and Roger Burrows.

Snieszko was born in Poland in 1902.[493] He was trained as a bacteriologist at Jagellonian University in Cracow, where he received his Ph.D. in 1926. He took a position in bacteriology at Jagellonian University after graduation and was provided a fellowship by the Polish government at the University of Wisconsin in 1929. He returned to the United States again, this time on a Rockefeller Fellowship, in 1939. That time he remained in the States since his homeland was overrun. He received the rank of captain during World War II and worked in the bacteriological warfare unit at Fort Detrick, Maryland, during that conflict. In 1946 he joined the U.S. Bureau of Sports Fisheries and Wildlife and was stationed at Leetown, West Virginia, where he was responsible for diagnostic services for the eastern United States through 1957. In 1958 Snieszko developed a training program in fish disease diagnosis for federal hatchery biologists, which he conducted until his retirement in 1972. Snieszko published over 200 professional papers and was one of the most highly regarded fish pathologists of his time. He died on January 12, 1984.

Roger Burrows, a U.S. Fish and Wildlife Service employee, developed a prophylactic treatment for fungus infections of fish eggs using malachite green and formalin.[494] While others had used the same chemicals in earlier studies, Burrows was the first to develop, in 1949, an effective control protocol for salmon eggs. The malachite green protocol was extended to northern pike in 1952,[495] and was subsequently used on a variety of fishes until the chemical was banned in the 1980s because of carcinogenicity. By 1954 formalin was the recommended treatment for many external parasites.[496] Formalin was not being recommended as a fungicide in 1954, nor is it today. A few other chemicals were being tested for their effectiveness against parasites in 1954, and the notion of using selective breeding to produce disease-resistant fish was introduced.[497] That approach continues to be used today, though the results have not been particularly encouraging.

A review of advances in the study of fish diseases between 1954 and 1964 showed that the development of resistant strains of pathogens had made some formerly efficacious drugs ineffective.[498] By 1964 some of the drugs that had been added to the feed were no longer approved for that purpose and could not legally be offered orally. Mercurial compounds were in use, though that practice was discontinued a few years later when national awareness of the environmental impacts of mercury developed. Formalin and malachite green were widely used, and a host of chemicals that are no longer approved played roles in the treatment of parasitic diseases. Examples are Lignasan, Gammexane, Lindane, and Enheptin, none of which are available today. Copper sulfate and ferrous sulfate were used to treat both external bacterial and parasitic problems. A metal-based product, di-n-butyl tin oxide, was used to control fish tapeworms.

A number of antibiotics had been screened for efficacy against various bacterial infections by 1964. As early as 1959, Snieszko had initiated work with Chlormycetin and Terramycin.[499] Terramycin was approved for use in conjunction with foodfish in the early 1970s.

Progress in disease identification and treatment was clearly occurring in the 1960s, with researchers concentrating on a few species of fish and a relatively small number of known diseases. When aquaculture began to expand in the private sector at an exponential rate in the 1970s, the number of species being reared increased, as did the numbers of new diseases and of old diseases that had to be treated differently when they impacted new hosts. Fish health specialists tended to be located within the federal government system. They became swamped with requests for diagnoses and treatment recommendations. A few universities established diagnostic facilities for fish diseases. In a few cases research programs that dealt with aquatic animal diseases were housed in schools of veterinary medicine; the University of Georgia, Texas A & M University, and Mississippi State University are examples. In most veterinary schools, courses in aquatic animal health are either not available or are offered as electives. Fish and invertebrate disease diagnosis and treatment continues to be centered outside of the veterinary community in the United States. That in no way detracts from the excellent work that has been accomplished. Given the handful of available treatments* and limited research budgets, great strides have been made. In recent years vaccines have been developed for some fish diseases. Alterations in management practices, including stress reduction and proper sanitation, have gone a

---

*Many drugs of demonstrated efficacy have not been cleared for use with aquatic animals by the Food and Drug Administration. Clearance can require years of effort and the expenditure of millions of dollars. Drug companies are reluctant to expend the time and effort because the size of the U.S. aquaculture industry is too small to provide them with a reasonable return on investment.

long way in preventing the onset and spread of aquatic animal diseases. Some diseases, however, including some of those first described in the early days of U.S. aquaculture, continue to cause major problems.

PRIVATE AQUACULTURE BEGINS TO EXPAND

It is unlikely that most U.S. citizens in the period from World War II to 1970 were aware that private-sector aquaculture was not only a reality, but was quite widely diversified. Prior to 1970 many of the species or species groups currently being produced commercially were either actually available from private farms or were being actively researched to determine their potential and develop appropriate technologies that could help ensure success. The small aquaculture industry, typically a family operation, was poised for rapid development, spurred to a large extent by interest within the research community.

Agricultural research, which involves billions of dollars a year in federal expenditures, was developed as a result of expressions of need for information from the farmers. The land-grant college system was founded in response to the need for regional research on real problems being faced by terrestrial farmers and livestock producers. I would submit that aquaculture research developed in a rather different manner.

Researchers had long been paying attention to problems associated with the production of sport and capture fisheries but had done relatively little to assess the needs of the small private-aquaculture community. When private aquaculture began to expand, the government responded by dedicating facilities like those at Stuttgart (Arkansas) and Marion (Alabama) to research aimed at supporting the fledgling industry. University programs were developed, largely at land-grant universities, but the researchers in those institutions often fought uphill battles for funding with traditionalists who did not feel that aquaculture was a legitimate part of agriculture. Many, such as me, who became interested in aquaculture research beginning in the 1960s, seemed to be fascinated by the activity itself and may not have recognized that they were laying the groundwork for the rapid expansion that was soon to come. Most of the people who became agricultural researchers came from farm backgrounds. A much lower percentage of aquaculture researchers had that heritage, and sometimes the fish farmers were skeptical of the so-called experts in the research community who had never driven a tractor or actually grown a fish to market size. In any event, it seems safe to say that the ratio of researchers to farmers has been, and continues to be, considerably higher in aquaculture than in terrestrial agriculture.

Inland aquaculture in the United States is dispersed throughout the nation, but there are some areas of concentration that, because of suitable environmental conditions (including plentiful supplies of good-quality water), make them far and away the largest production regions in the nation. With respect to trout, such a location is the Thousand Springs area in the Hagerman Valley of Idaho, not far from Twin Falls. The underground rivers that emerge as the Thousand Springs and flow into the Snake River in the Hagerman Valley provide an enormous supply of water that is nearly perfect for the culture of trout. Large-scale commercial fish farming began in the area immediately after World War II,[399] and the industry grew to the extent that at one point in the 1980s, something like 95% of the commercially raised trout in the nation were produced within 20 miles or so of Twin Falls.*

In terms of foodfish production volume, trout was virtually unchallenged prior to the 1950s when buffalo (the fish, not the big land mammal) and catfish farming were being developed but were not being marketed in significant quantities.[500,501] The early work in the Southeast involved rice-fish farming, whereby diked fields were reflooded after the rice had been harvested. The fields were then stocked with fish such as buffalo, carp, or channel catfish,[501, 502] which were allowed to grow to market size (about two years). Through the middle 1950s buffalo and channel catfish were considered difficult to propagate,[501] though that was about to change. Some success with buffalo spawning and larval rearing had been achieved early in the 1950s,[503] and refinements improved the chances for commercial culture success throughout that decade.[504–506] Commercial culture feasibility was demonstrated in 1957.[500]

Aquaculture in Arkansas prior to 1970 was not restricted to foodfish production. Following World War II, the industry began with commercial rearing of fathead minnows, golden shiners, and goldfish.[502] Interest in the production of game and foodfish followed, with the farmers learning through experience since little research information was yet being distributed to them. By 1959 buffalo was the predominant foodfish species under culture. It was usually produced in combination with sport fishes such as bass, crappie, and catfish.[507] Buffalo prices were so low by 1960 because of the availability of wild fish that the farmers could not make a profit. Also, the presence of wild fish in ponds stocked with buffalo, channel catfish, or large-mouth bass created competition that reduced production of the desired species while the so-called trash species such as gizzard shad, carp, green sunfish, and bullheads were unmarketable.

By 1966 fish farming was the fastest growing segment of agriculture in the state of Arkansas.[508] Channel catfish had been moved from the sport-fish to the foodfish category and had replaced buffalo as the leading foodfish being cultured. Monoculture of catfish had largely replaced the polyculture of earlier years. Sport-fish production was centered on bass, crappie, and sunfishes. The largest contributor to income from fish farming in Arkansas was still associated with baitfish facilities. A total of 3,800,000 pounds of minnows valued at $5,325,000 were produced in 1966. Total income from all species that year was placed at $9,165,000. A new form of fishing enterprise was also becoming popular:[508]

> A need for recreational fishing by Arkansans is reflected in the success of numerous fee fishing operations across the state. Anglers are charged a base fee for the privilege of fishing and, at intensively managed installations, are charged for the fish taken.

To supplement income at fee-fishing locations, operators typically sold bait, cold drinks, snack food, and in some cases rented fishing equipment.

---

*That figure, cited by various trout producers, seems to have decreased to some extent in recent years as trout culture has expanded in various other states.

By 1970, Mississippi had jumped into the catfish production leadership role, followed in order by Arkansas, Louisiana, Alabama, and Texas.[509] While most farms were family enterprises, there was some movement by big business to explore catfish farming. For example, The Pennzoil United Company was operating a farm in Louisiana before 1970.[510]

Trout were also being widely cultured around the nation by 1970, though the leadership role that had established Idaho was becoming apparent. There was some research activity aimed at rearing salmon to market size in captivity, and shrimp culture research was receiving a great deal of attention. Oysters were still being cultured along the coasts using technology that had changed very little in decades. Interest in culturing crawfish* had developed in Louisiana, where it was thought that aquaculture could be particularly important in filling demand for that crustacean during years when the capture fishery was weak. The estimated production levels and retail values of several aquaculture species in 1969 were as follows:[511]

Organism	Production (lbs)	Retail Value ($)
Trout	13 million	18–20 million
Catfish	25 million	29–31 million
Bait minnows	40–50 million	125–130 million
Crawfish	4 million	1.4–1.6 million
Oysters	31 million (meats)	45–51 million

We will look at each of these species and at some others in the sections that follow.

Aquaculture was often lauded as a means by which the increasing demands of people for animal protein could be met. On the other hand, potential problems associated with commercial aquaculture were being heard. Among the issues being raised were the use by aquaculturists of chemicals that might impact nontarget species, and the use of exotic species. The question was also raised as to whether using aquaculture as a means of feeding a growing human population was actually wise:[512]

> Regardless of its benefits, aquaculture should not be used for permitting a continuation of the increase of human population densities. Making more food available to permit survival of more people will eventually (or should we say shortly?) force an increase in mortality. When that occurs, the larger the population there is, the more people die.

The fact is that, with few exceptions, commercial aquaculturists are in the business of making a livelihood, and are not dedicated to feeding the starving masses of the world. Aquacultured fishes of the types produced in the United States and most other nations are too expensive for the pocketbook of the world's poor.

---

*Depending on where one lives in the United States, these crustaceans are called either *crayfish* or *crawfish*. The American Fisheries Society now recognizes the *crawfish* spelling in conjunction with the commercially important species, so I've used that spelling.

Development of commercial fish farming was not accomplished in a vacuum. While, as previously mentioned, a good information transfer network between the research community and private culturists was only beginning to be developed by 1970, progressive fish farmers sometimes had university training[399] and were able to obtain information from nearby state or federal hatcheries and a few federal research laboratories that were actively involved in aquaculture research. There were also a few magazines and a number of scientific publications that concentrated on, or at least contained some information on, aquaculture.

The World Aquaculture Society, which was dedicated to information transfer, was formed in 1969. Additional universities were becoming involved in aquaculture during the postwar period as well. We'll look at some of these topics in more detail after taking a closer look at some of the species that were being farmed by 1970.

The information presented on the various species comes from literature that seems particularly important and pertinent. Not all of the technological advances are chronicled by any means, nor are the names of all the important players included. That becomes increasingly true in the chapters that follow, since information was being developed so rapidly that summarizing all of it within a reasonable number of pages would be impossible.*

A final word is needed before we begin looking at individual species that were being considered for commercial culture before 1970. Toward the end of this section I've included a few that received consideration for reasons other than commercial culture but which are of some interest. In some cases (for example, striped bass), the way for commercial culture was paved by work that was not aiming at that goal at the time the research was conducted. Other species are discussed as oddities. One example is sponges, which don't seem to be of interest today but once received a good deal of attention. Also, in some cases I've gone back in time to the years prior to World War II to obtain background information. A good example is, once again, sponges. (That information could have been incorporated into Chapter 3, but since this chapter takes a closer look at a diversity of species, I elected to keep the material back until now.)

Catfish

Studies on catfish ultimately became focused on the channel catfish, but there were also investigations on blue catfish, white catfish, and ultimately, hybrids among them. John Giudice had just completed a good deal of work in which he examined the growth of catfish hybrids at the time I got to Stuttgart, Arkansas, in the summer of 1969. He had crossed all three in various combinations, including backcrosses. A comparison between albino and normal channel catfish was made in 1961.[513] Flat-head catfish were also a topic of research in the 1950s and 1960s,[514–516] but the

---

*To illustrate that point, the following little story is appropriate. I recently attempted to review the literature published between about 1982 and 1993 when I revised and expanded my book, *Principles of Warmwater Aquaculture* (which was renamed *Principles of Aquaculture* to more accurately depict content). The CD-ROM literature search software that was available to me at the University of Washington library contained over 40,000 references for that period, and that certainly was not a complete listing.

cannibalism exhibited by that species ultimately led culturists in other directions. Bullhead culture was also evaluated in the 1950s,[517] and while the culture of those fish has been advocated periodically over the years, no commercial culture ever developed.

A number of review papers on the culture of catfish became available in the years immediately following World War II as government fish culturists reported to their colleagues on their experiences. Gehard Lenz, Superintendent of the Gretna state hatchery, Gretna, Nebraska, wrote such a summary in 1947.*[518]

Both channel and blue catfish were being reared at Gretna in the 1940s. Lenz described pond preparation, nest placement, and spawning activity. He went on to discuss egg collection and incubation in the hatchery, fry feeding, early rearing, and disease problems and their control. The paper was published in *Outdoor Nebraska* magazine and reprinted by the *Progressive Fish-Culturist*. Lenz fed a mixture of beef livers and melts (spleen), dried skim milk, white fish meal, cottonseed meal, and wheat middlings.[518] During the same year, the U.S. Fish-Cultural Station at Neosho, Missouri, was feeding a mixture of $1/2$ crayfish abdomens, $1/6$ fish meal, $1/6$ cottonseed meal, and $1/6$ dried skim milk.[519] Beef livers and melts were sometimes substituted for the crayfish, so the ingredients were quite similar to those used at Gretna.

A second review on catfish propagation, this time restricted to channel catfish, was also published in 1947 by H. L. Canfield, an employee of the U.S. Fish and Wildlife Service at LaCrosse, Wisconsin. Canfield also described pond preparation, placement of nests, and spawning behavior.[520] He noted that nail kegs made good nests and that they could often be obtained from hardware stores for 10¢ apiece. Canfield suggested feeding fry such things as beef liver, beef heart, and sheep liver and indicated that cereal grains were not satisfactory feeds (which is true if they are fed in the absence of certain other feedstuffs).

Yet another review of propagation techniques appeared in 1951.[521] That one, by Marion Toole with the Texas Game, Fish and Oyster Commission (now Texas Department of Parks and Wildlife), went through much of the same information that had been presented in earlier reviews. Of some interest was the comment that flathead catfish culture had been attempted but had generally been unsuccessful. Also of note is that the Texas procedure involved the use of spawning pens (which were also being used in Arkansas by the late 1950s[522,523]). Spawning pens allowed the culturist to pair up fish rather than allowing random mate selection. The use of milk cans in Texas was said to have resulted in the production of millions of eggs, but crockery jars had been substituted for the milk cans by 1951 because the latter had become badly rusted.[521] There was no explanation as to why the culturists didn't just go out and find some new milk cans since they were certainly available at that time (though it's difficult to find one in good condition now).

The methodology for successful catfish spawning and rearing was in place by the

---

*I know virtually nothing about Mr. Lenz, but there is a soft spot in my heart for his hatchery, because it was the first one I ever visited. In about 1966, students in the ichthyology class that I was taking had the good fortune of taking a field trip to the Gretna hatchery. The facility appeared to have been constructed in the 1920s or 1930s and did not appear to have undergone a great deal of renovation in the interim. It was a classic facility that was involved in the culture of trout and warmwater fishes, including catfish.

1950s, but commercial ventures were almost nonexistent. H. S. Swingle added the final piece needed to confirm the potential for commercial production when he presented a paper in 1957 to the Southeastern Association of Game and Fish Commissioners on the economics of growing channel catfish from fingerling to market size.[500] He concluded that with the production levels achieved during research at Auburn University (1,242 pounds per acre) and a market price of $0.50 per pound, the farmer could realize a modest profit. Swingle included recommendations on pond fertilization and included his formula for the Auburn No. 1 prepared feed. That ration contained 35% soybean meal, 35% peanut meal, 15% fish meal, and 15% distillers solubles. The paper is a classic because it demonstrated for the first time that commercial catfish culture was a possibility. Additional research on catfish production experiments was reported on at the same meeting in 1959 by Swingle.[524] By that time it had been discovered that the peanut meal used in the Auburn No. 1 formulation contained shells that were indigestible. As a result, the Auburn No. 2 formulation was developed. It was identical to Auburn No. 1 with the exception that peanut cake (which contained no shells) was used in place of peanut meal. Improved performance was obtained when the feed was pelleted.

Swingle concluded that if fish were stocked in the spring at the proper density and were marketed at $0.55 per pound, a profit of $266.40 per acre could be realized. Production levels varied as a function of stocking density, with an initial density of 2,000 fingerlings per acre yielding fish of nearly one pound at the end of the growing season.

Catfish culture in Alabama has been largely confined to so-called runoff ponds; that is, ponds that are constructed in areas where they can be filled with rainwater runoff. That is a requirement in much of the state because of a scarcity of groundwater. The industry in Arkansas was established in the central part of the state, where rice was being raised and groundwater was plentiful. As the Arkansas industry developed during the 1960s, the water table began to fall and farmers looked for alternative production sites. They found exactly what they needed in the flat Mississippi Delta region, where groundwater was seemingly inexhaustible. Mississippi quickly dominated the industry and continues to do so today.

Basic pond design and construction techniques were modified as the industry developed. To simplify management and harvest, the industry went to ponds that were rectangular or square rather than those that followed natural contours and tended to be irregular in shape. Uniform levee side slopes, bottoms that gradually sloped toward a drain, and uniform maximum depths of 5–6 feet became standards within the industry. By using blowers to distribute feed, employing trucks and tractors to pull large seines, and adopting live-cars to help hold the catch from seine hauls until the fish could be loaded on trucks, it was possible to manage increasingly larger ponds. For a while the industry moved toward ponds of 20 and even 40 acres. Managing water bodies of those sizes was difficult, and when a disease or water-quality problem occurred, the potential loss in revenue was unacceptably high. Thus, most of those very large ponds were eventually cut in half or in quarters, and new ponds were typically not over about 10 acres in size.

Assistance in selecting suitable pond sites was often available from the Soil Conservation Service (SCS) of the U.S. Department of Agriculture. SCS personnel, who had long been involved in advising farmers on farm pond construction, developed expertise in fish farm requirements as well. In 1968, Roy Grizzell, a SCS biologist in Arkansas, summarized information on site selection, pond construction, and other aspects of catfish farming.[525] He included cost analyses for fingerling production ponds and growout ponds and provided balance sheets based on various pond sizes and production strategies (for example, fingerling production with or without use of a hatchery, growout of fingerlings, and fee-fishing operations). Among the assumptions he made were that earth moving would cost $0.15 per cubic yard for small ponds and $0.20 per cubic yard for larger ones, that money could be borrowed at 6% interest, and that annual land taxes would be $2.00 per acre. His various scenarios used food conversion ratios ranging from 1.65 to 5.7.* Another paper on the economics of aquaculture appeared in 1970.[526] That paper used a cost of $157 per acre for pond construction based on a survey of fish farmers.

While small raceways and aquaria were typically found in the laboratories of aquaculture researchers, those units were usually used for holding fish and conducting bioassays, physiology experiments, and other studies that did not directly deal with production. Performance trials, including the testing of new diet formulations, were almost always conducted in ponds. Pond trials were expensive because of the number of fish and the amount of feed involved, and they required a large number of experimental units if several variables were to be simultaneously evaluated in replicated experiments.

Catfish researchers in government laboratories and universities could be generally characterized as having been trained as fisheries biologists who moved into aquaculture and as scientists who conducted their experiments in ponds. There were obviously exceptions to that characterization. I, for example, was trained in limnology and oceanography at the graduate level, never had a fish culture course, and did most of my early work in tanks (except for the summer at Stuttgart, where I spent a lot of time pulling a seine in ponds with Bill Simco). Another exception was a nutritionist named James W. Andrews.

Beginning in the spring of 1969, Andrews began conducting studies on the nutritional and environmental requirements of channel catfish at the Skidaway Institute of Oceanography in Savannah, Georgia. Whereas many who had been involved in fish nutrition research before him had been trained primarily as fisheries biologists, Jim knew little about creatures with gills, but he certainly knew a lot about basic nutrition. He held a B.S. degree in chemistry, an M.S. in poultry nutrition, and a Ph.D. in nutrition from the University of Georgia. Before setting up shop at Skidaway he had obtained experience with both poultry and human nutrition.[527]

Jim had no prejudices about conducting experiments in ponds, which was good since he had no ponds at his disposal and there was little likelihood that ponds would be constructed at the institute. Jim conducted his work at Skidaway in a

---

*The food conversion ratio is calculated as the dry weight of feed required to produce a pound of fish. Currently, catfish farmers typically grow fish to market size with overall food conversion ratios of less than 1.5. Better feeds and feeding strategies have led to more efficient use of feed and have helped keep production costs down.

circular cattle-show barn that had been donated to the state along with about 1,000 acres of land and a few buildings by the landowner, Robert Roebling, whose father had constructed the Brooklyn Bridge. The Skidaway Institute of Oceanography was established as an independent research unit within the state of Georgia's university system.

The cattle barn contained about 20 bull stalls, each about 10 feet wide. Jim purchased 20 circular fiberglass tanks 8 feet in diameter and had them set up in the stalls. He also set up an aquarium room that contained batteries of 30-gallon tanks in a portion of the barn that has once been used for storage. A number of 4-foot-, and later 3-foot-diameter fiberglass tanks were also installed in the building's halls, hayloft, and viewing gallery. Water was supplied from a well that delivered 1,000 gallons per minute, but required heating to bring its temperature to that considered optimum for catfish growth (about 80°F). A gas-fired heater did that job quite nicely, though not cheaply.

I met Jim a few months after returning to Tallahassee, Florida, following my summer at Stuttgart. He had just gotten his catfish work started and was interested in initiating some research with marine shrimp. To learn more about shrimp, he drove down to Florida State University to speak with a few of the researchers there. During those conversations he mentioned his work with channel catfish and was informed that one of the graduate students (me) had some experience. We met, and Jim asked me what my research plans were in conjunction with my Ph.D. At that time, funding for my assistantship was running out—there was a two-year limit on state support, and my major professor had been unable to find additional funding for me—and my G.I. Bill benefits had almost been used up. It was a desperate time for me because there didn't seem to be any way I could complete my degree, with a wife, two small children, and no financial support.*

Jim asked if I'd be interested in conducting my Ph.D. research at Skidaway. He would pay me as a research technician and would provide housing for me at the institute. Following a quick trip up to Savannah to survey the situation, and having obtained approval from my advisory committee, we moved to Skidaway.

The next year, 1970, was one of the most exciting in my career. Jim set me up with 54 fiberglass tanks 3 feet in diameter in which to conduct my studies on the lipid requirements of channel catfish. At the same time, he designed a number of other studies that I conducted and we jointly published. Included were studies on the effects of environmental temperature, photoperiod, and feeding rate on growth. Jim had been looking at loading rates in flow-through systems and had also done some pioneering work on recirculating systems. One of his closed systems sustained catfish at 10 pounds per cubic foot of water for several months.

While I was working with catfish, Jim initiated his work on marine shrimp. Once I had graduated, I stayed at Skidaway and became involved with flounder culture and in field studies in conjunction with various scientists at the institute. Jim turned

---

*Carolan had worked throughout our marriage, but we agreed she would stay home with the children when they were very young. She was babysitting, but we were obviously not going to be able to support our family without making a significant and rapid change in our financial status.

the catfish work over to Yoshiro Matsuda and Takeshi Murai, two Japanese scientists who received advanced degrees at the University of Georgia during their tenure in the United States. Taki worked for the U.S. Fish and Wildlife Service for a few years and is now back in Japan, having become one of that nation's leading fish nutritionists. I lost track of Yoshi, but I believe he is also back in Japan.

When the director of the institute asked Jim to move his operation to Tifton, Georgia (where the Georgia Experiment Station research center is located), Jim appealed to be allowed to remain in Savannah. The director, who opposed the catfish work because it did not involve a marine species, was adamant, so Jim resigned and went on to establish a number of water-quality analytical laboratories. His legacy in aquaculture lives on, however. It is virtually impossible to find an aquaculture research facility that doesn't have a number of circular tanks with center drains similar in all respects to those used by Jim Andrews in the Skidaway catfish barn (also known as the "world's largest cathouse" in some circles). Today tank experiments are the standard among aquaculturists and ponds are used less and less for replicated studies.

Another departure from pond culture was first described in the late 1960s by Richard Collins, a college faculty member in Arkansas. Collins appears to have been the first person in the United States to rear catfish in cages.[528] The idea was not a new one, having been employed by aquaculturists in Asia, but it was a new application. Collins recovered quite a lot of notoriety in the aquaculture community, and the idea was soon picked up by others.[528,529] There was even an article that related cage culture to a vision expressed President Franklin D. Roosevelt: ". . . who visualized in the 1930s a fish pond on every farm large enough to grow a quantity of fish sufficient to feed a family."[530]

I met Richard Collins a few years after the furor died down and learned that he had used cages initially as a means of holding fish, not for growout. He was not really interested in aquaculture when he began utilizing cages, but he is one of the people who will be connected with the field because of the technology transfer he undertook.

Cage culture is undertaken in relatively small containers (usually no more than a cubic yard or two) with sturdy frames over which nylon netting, or more commonly, wire or plastic-coated wire mesh is stretched so the fish will be confined. Floats are used to keep the cage at the water surface. Cage culture is rarely practiced in the United States, but a similar technique that involves much larger units (often 40 × 40 feet and up to 60 feet deep) has been employed quite successfully in the marine environment. These marine units, called net-pens, were developed in 1969[531] and were used by researchers to demonstrate the feasibility of commercial salmon farming.[532]

Commercial channel catfish culture was a reality and growing rapidly by 1970, but there were significant challenges facing the industry. One of them was consumer acceptance outside the South. Another was product quality, since catfish were notorious for having off-flavors on occasion. Finally, there was the problem of profitability in the face of increasing costs of production during periods of stable or slowly rising prices paid by processors. The solutions to those problems occurred during the 1970s and 1980s and are discussed in the next chapter.

Trout

A survey of the commercial trout industry in the West was conducted in 1952 and 1953. The states surveyed were California, Colorado, Idaho, Montana, Nevada, Oregon, Utah, Washington, and Wyoming.[533] The most common facility was of the fee-fishing type. More than a third of the respondents indicated that their facilities included a hatchery, rearing ponds, and a public fishing pond. Many operators obtained their eggs from large hatcheries and did not produce them on site.

Most facilities were owned by individuals, though the larger operations were set up as partnerships or corporations. Several state and regional fish culture organizations had been formed by the early 1950s, and a national organization was under consideration. (The U.S. Trout Farmers Association was, in fact, formed prior to 1960.)

States varied as to when the bulk of their active farms had been established. For example, 80% of the California trout farms were less than five years old at the time of the survey, whereas the figure was 27.8% in Colorado. Nearly 60% of the Colorado enterprises were at least 10 years old.

The number of reported fish breeders in the various states (which included all types of fish, not just trout) in 1952 ranged from highs of 325 and 150 in Colorado and California to 3 in Nevada.

Reported investments in the various enterprises ranged from $500 to $1,620,000 (only four farms reported having invested over $200,000). Incomes reportedly ranged from $50 to $300,000 annually. Idaho, which reported only 14 fish breeders in 1952, was demonstrating its rise to the top in terms of production. Gross income in Idaho was higher than in any other state ($678,200 a year based on a reported investment of $766,500). California was a close second ($612,000 a year with an investment of $855,500). Colorado, with reported investments of over $3.2 million, had a gross annual income of only $379,500.

In addition to fish stocked directly as catchables or sold to people who operated fee-fishing lakes, fingerlings were sold for stocking lakes and 8–12-inch fish were marketed to restaurants, resorts, and grocery stores. The cost of producing catchable trout was estimated at $0.575 per pound.[533] By 1958, commercially raised trout were bringing from $0.85 to $1.25 per pound,[534] which is similar to the current range.

Water shortages were reported by many trout culturists in the early 1950s. Water sources included springs, streams, wells, and storage reservoirs. Competition from imported fish was seen as a problem throughout the 1950s,[533,534] and dumping of trout by marginal producers was helping keep the price down.[534] Difficulties associated with obtaining financing, a shortage of trained personnel, and the need for additional research were also seen as pressing needs.[534] Researchable problems included those associated with disease, nutrition, and environmental requirements.[535]

The need for regulating commercial trout farms was seen as early as 1958. Licensing of trout farms and control of the distribution of fish from such farms was seen as[536]

. . . necessary to avoid indiscriminate and unplanned stocking of public waters and introduction, accidental or otherwise, of undesirable species of fish. Such control also aids in curtailing the spread of fish diseases.

One assumption about the perceived future of trout farming that was made in the early 1950s is particularly interesting and prophetic since it was made early in the era of commercial aquaculture development:[533]

> It is not expected that the development of private trout culture will ever approach the magnitude of the development of the warm-water fishes, as the environmental requirements of trout are much more exacting. Suitable water supplies are difficult to locate, and the investment that is required to establish such a business is prohibitive for many individuals who might otherwise enter into such an activity.

In 1960, the U.S. Trout Farmers Association conducted a survey in all the states and the provinces of Canada.[537] Responses were obtained from all but North Carolina, Georgia, Kansas, New Jersey, and Vermont. The survey revealed that no trout were being stocked in Alabama, Florida, Louisiana, Mississippi, Oklahoma, or Texas. Commercial trout farmers sold 146,381,000 eggs in 25 states and provinces. There were also sales to England, Scotland, and Denmark.

Much of the activity by commercial trout farmers continued to be associated with fee fishing in 1960. Thirteen states required a valid state fishing license for people who frequented fee-fishing establishments, and the farmers felt that was an unfair requirement since it applied not only to anglers but also to tourists who passed by to purchase fish.

All of the states surveyed allowed the sale of dressed trout except Delaware. Only frozen dressed trout could be marketed in Vermont. Virginia allowed the sale of only fish originating from outside the state.

Relationships between the state agencies and commercial trout farmers were reported to be exceptionally good in 33 states, and there were only minor conflicts in 5 additional states. Reported sources of friction were ". . . stocking of diseased fish in waters where the disease was previously unknown, planting fish without a permit, restricting fishing in private waters immediately below a trout farm, and the reduction of trout waters by the commercial use of big springs."[537] Regulations controlling the stocking of fish in public waters had been developed in 19 states.

The rearing of trout by governmental and private producers was largely conducted in ponds during the 1950s. Earthen raceways were also in use,[538] but the era of the high-volume concrete raceway was yet to come. And come it did. The Snake River Trout Company in Idaho, purchased in 1952 by Robert Erkins, became the world's largest trout farm by 1969, though it sat on only 10 acres of land.[539] Unconvinced that trout needed large areas in which to swim about, Erkins constructed concrete raceways on much of the facility, stocked them heavily, and employed large flow rates to maintain water quality. By stocking 100,000 pounds per acre of fingerlings, he was able to produce 400,000 pounds per acre of marketable fish per year.

Keen Buss, an innovator in the design of trout culture facilities, began tinkering with raceway design in 1953 and felt that long raceways provided better conditions than short raceways.[538] Buss was also trying to obtain better utilization of the vertical space in hatcheries by rearing trout in deeper water. As a start in that direction he experimented with fry rearing in 55-gallon drums. The next step involved expanding the size of the vertical units. Buss and his colleagues with the Pennsylvania Fish

Commission developed a silo that was 17 feet high and 7.5 feet in diameter and could be used to rear 20,000 rainbow trout.[540,541]

Silo culture never did catch on among trout farmers, but the idea reappears occasionally. During the 1970s, for example, many dairy farmers in the upper midwestern United States were experiencing economic problems and were looking for ways to move into new types of farming enterprises. The presence of large grain silos on defunct or failing dairy farms appeared to be ready-made fish culture units. Consultants, sometimes with questionable credentials, hit the road with sure-fire proposals to turn some of the dairy farms into yellow perch culture facilities in which recirculating water systems would be employed with the silos as rearing units. A number of farmers staked what few resources they had left to embark on the new enterprise, but insofar as I know, none of them ever turned a profit.

Carp

Interest in common carp culture and stocking had evaporated by the 1940s. Commercial fisheries adequately satisfied the ethnic market demand, and at least one state was incorporating carp into its trout feeds by 1950.[542] In fact, carp made up 80–90% of the fish used in feed formulations at most agency hatcheries. Cooked carp was mixed with cereal grains and a small amount of salt. Even that use of carp largely disappeared when pelleted trout feeds became popular.

Most of the aquaculture action with respect to carp was confined to research on grass carp as previously discussed. During the 1960s research with grass carp was conducted by the U.S. Department of Agriculture in Fort Lauderdale, Florida; the Fish Farming Experimental Station in Stuttgart, Arkansas; and at Auburn University.[543] While much of the research was favorable in terms of the ability of grass carp to control aquatic vegetation, thereby reducing the need for chemical herbicides, opposition soon developed. Among the concerns were a lack of information on how grass carp would affect native fish populations if the carp were to escape and a belief that grass carp would consume beneficial aquatic vegetation, thereby degrading waterfowl habitat.

Arkansas began stocking grass carp, and their appearance in the Mississippi River system meant that they could potentially spread through a considerable portion of the nation. Many states outlawed the stocking of grass carp. One Texas game warden once told me that the Red River on the border between that state and Arkansas was a natural barrier to grass carp so the fish couldn't get from Arkansas, where they were legal, to Texas, where they were banned. (I'm not sure why the fish would fail to negotiate a slowly flowing river, but everyone is welcome to their own opinion.) There were also reports that every truckload of bait minnows that left Arkansas contained a few small grass carp. That rumor is one that I can easily believe.

We won't spend any additional time on grass carp in later chapters. To sum things up, successful captive spawning was reported in 1971 by investigators with the Arkansas Game and Fish Commission.[544] Commercial fish farmers were soon producing the fish. Jim Malone in Arkansas was one of the leaders in the production of juvenile grass carp, beginning his activities in 1972.[545] As criticism of the potentially negative environmental impacts of grass carp grew, an increasing number of states banned the fish. In response, Malone learned how to produce hybrid grass carp by crossing them with bighead carp. The hybrids consumed vegetation but were sterile. With the development of the hybrid, states began to reconsider their bans. Many states ultimately changed their regulations to allow the planting of sterile hybrids. Grass carp continue to have a place in commercial fish ponds, farm ponds, and even some large reservoirs, and there is a foodfish market for them as well. (From personal experience I can attest to the delicious nature of a smoked grass carp of sufficient size that the bones can be removed easily; fish of 20 pounds or so easily meet that requirement and can be grown within two years in the South).

Striped Bass

In the early part of the twentieth century the U.S. Fish and Fisheries Commission was beginning to perfect the techniques of spawning striped bass, hatching their eggs, and releasing fry in what were probably ill-fated attempts to enhance the commercial fishery.[546] By 1910 the basic technology was in place.[547] Plans to augment marine stocks of striped bass were ultimately abandoned, but renewed interest developed in striped bass culture during the 1950s after a resident striped bass population was established in the Santee-Cooper Reservoir in South Carolina.[548] A primary purpose for striped bass stocking was to establish populations that would control gizzard shad in southeastern U.S. reservoirs. Striped bass would also provide recreational fishing. Since the fish typically would not spawn in reservoirs, it was necessary to establish hatcheries. One hatchery at Weldon, North Carolina, had been established during the earliest years of striped bass culture. A new hatchery was constructed much later at Moncks Corner, South Carolina, and attempts to induce striped bass spawning with hormones began in 1962.[549,550] Within two years the basic procedures had been worked out, and tens to hundreds of millions of striped bass fry were being produced at both hatcheries.[551]

Interest in striped bass spread beyond the southeastern Atlantic coastal area as other states began, once again, to look into stocking reservoirs to provide an additional sport fish. Adults had been stocked in the inland waters of several states at one time or another, but they did not reproduce.[552] Fry stocking was possible but might result in poor survival because native fish populations could forage on the small striped bass. In 1965, biologists in Oklahoma obtained striped bass fry from North Carolina and reared them to fingerling size in ponds, with survival rates after two months of 15–60%. The Bureau of Sport Fisheries and Wildlife facility at Edenton, North Carolina, became involved in the development of techniques to produce striped bass fingerlings in the 1960s.[548] The Edenton National Fish Hatchery obtained their eggs from the Weldon, North Carolina, hatchery.

Perhaps the first attempt to rear fingerling striped bass to subadult size in ponds and hatchery troughs was reported in 1967.[553] The study, conducted at Auburn University, revealed that striped bass fingerlings could be reared on prepared feed and helped set the stage for the commercial culture that was to develop several years later.

Fish culturists always seem to be looking for some new species or way in which to manipulate the species with which they have been working. In 1965, before much of the technology associated with producing striped bass fry and fingerlings had been standardized, investigators at the Moncks Corner hatchery were already producing hybrids between striped bass and white bass. The first attempt to produce finger-lings failed, but in 1966 some success was achieved.[554] Preliminary indications were that the hybrids grew more rapidly than either parental species. The most successful hybrid was produced with striped bass eggs and white bass sperm. Growth was considered to be excellent with some individuals weighing as much as 1.7 pounds at the end of their first year and as much as 4.8 pounds after two years. The hybrids were found to be fertile and were first backcrossed with striped bass in 1968.[555] Much of the attention on production of fish for both stocking and later for commer-cial culture has focused on hybrids.

Minnows

The use of minnows as bait undoubtedly goes back to the time during which recreational angling was initiated in the United States. Minnows were either col-lected by the fishermen or purchased from dealers who had seined them from the wild. By the early 1930s it had become apparent that minnow populations were being severely impacted by the heavy seining.[556] Not only was the forage base for predatory fish being impacted, but those involved in seining minnows often collected fingerlings of sport fishes and sold or used them directly as bait. Appeals from officials with the U.S. Bureau of Fisheries for the development of minnow culture began to appear in the 1930s.[556,557] Species recommended for consideration included golden shiners, bluntnose minnows, blackhead minnows, horned dace, stonerollers, topmin-nows, and mudminnows.* Suggestions for how propagation might be achieved, in a general sense, were provided by Lewis Radcliffe in 1931.[557] Experiences associated with the culture of golden shiners and bluntnose minnows were presented to the American Fisheries Society (AFS) in 1932 by the noted ichthyologist Carl L. Hubbs.[558] He indicated that the golden shiner was easy to breed and produced large numbers of offspring, but that when allowed to breed in bass brood ponds the minnows interfered with the production of bass offspring. That did not seem to be a problem when bluntnose minnows were used. Eugene W. Surber, another noted fishery scientist, and his colleague George Klak reported in 1939 that the addition of minnows, including golden shiners, to bass brood ponds increased the production of young-of-the-year fish.[559] There the debate seemed to end.

Circulars on bait minnow production were published in 1939 and 1940.[560,561] Following World War II, another circular was published by the U.S. Fish and Wildlife Service on minnows and other baitfishes.[562] That circular was revised and updated in 1956.[563] Research with golden shiners and fathead minnows was being conducted at Auburn University beginning in the 1930s, and continued into the 1950s.[564–566]

In his recent review of baitfish culture in the United States, James T. Davis indicated that there are currently three species of general commercial importance:

*You who are up on your taxonomy will note that some of those fishes are not members of the minnow family but were considered appropriate bait species in any case.

golden shiners, fathead minnows, and goldfish.[566] Those were the same species reported to be most important in 1967.[502] Of local importance are such species as killifish, chub suckers, stonerollers, tilapia, topminnows, and shiners.[566]

By 1970, minnows were being produced in large quantity. The largest producer, I. F. (Fay) Anderson, who farmed outside of Lonoke, Arkansas, had 6,000 acres of ponds in production by that time, most of which were dedicated to minnows.[567] The principle species under culture was the golden shiner, but Anderson also raised fathead minnows and goldfish. Anderson's annual production was about 660 million fish.

Anderson had opened a sporting goods store in Yazoo City, Mississippi, after being discharged from the military following World War II. Since demand for minnows was high and the supply undependable, he and his brothers began growing a local minnow species. Problems with the species they had selected, the redfin, led Anderson to eventually purchase a farm outside of Lonoke, Arkansas, where he felt conditions for growing minnows were more ideal. Golden shiners were selected as the initial species for rearing since they were in demand by anglers. He began working with that species in 1950, a few years before university research was initiated. His success undoubtedly enticed others to enter the business, and Arkansas became the leading state in minnow production.

Miscellaneous Fishes

Tilapia. Tilapia were introduced into the United States during the 1960s, when the interest in finding a biological control for aquatic vegetation was initiated.[543] At least some species of tilapia, like grass carp, were reputed to be herbivores, though most tilapia species do not consume macrophytes to any degree. Studies on tilapia were conducted during the 1960s,[568] including some work with the growth of hybrids, but interest in the commercial culture of tilapia did not develop until the 1970s and is discussed in more detail in the next chapter.

Paddlefish. The paddlefish is one of the more unusual freshwater fishes of North America. The fish is native to the Missouri River Basin and is prized both for its flesh and its roe.[569] Interest in culturing the fish began in the early 1960s.[570] Charles Purkett with the Missouri Department of Conservation was the first to describe natural spawning in paddlefish,[571] and was the first to spawn them in captivity during 1961.[570] Hormone injections were successfully used in 1962 by researchers at the Fish Farming Experimental Station in Stuttgart, Arkansas.[569] Biologists with the Missouri Conservation Commission continued conducting research on paddlefish thereafter, but little in the way of captive culture was undertaken. Interest in paddlefish aquaculture was revived in recent years because of dwindling populations to the point that the paddlefish may be listed under the Endangered Species Act. Aquaculture could provide part of the solution for stock recovery.

Pompano. One of the few marine fish species being evaluated for culture prior to 1970 was the Florida pompano. Research on that fish was initiated in 1967 by John H. Finucane, a biologist with the Bureau of Commercial Fisheries laboratory at St. Petersburg, Florida.[572] The perceived good market price ($0.60–1.10 per pound wholesale) and high demand for pompano led Finucane to believe that it would be a good aquaculture candidate. Advantages of pompano included the fact that it were not cannibalistic, adapted well to captivity, and grew well on commercially available feeds. Juveniles for stocking were not produced in hatcheries prior to 1970 but had to be collected from nature. Finucane obtained his fingerlings by beach seining but felt that controlled spawning was possible. Problems with the fish included its inability to tolerate cold. Pompano experienced mass mortality when the water temperature fell to about 50°F, and their growth was reduced at temperatures at or below 63°F. Feeds specifically designed to meet the nutritional requirements of pompano were not available (because the requirements had not been determined), but would be required before commercial culture became a reality. Finucane's work represents the bulk of what was accomplished with pompano, and while the species receives some mention from time to time, private aquaculture has never developed.

Sponges

The idea of culturing sponges developed during the latter half of the 1800s. Early studies were conducted beginning as early as the 1870s, but few data were collected.[573] By 1900, demand for sponges had grown to the point that supplies were diminishing because of overharvesting. In response to the problem, the Fish and Fisheries Commission launched studies in 1901 aimed at culturing sponges by planting cuttings of about a cubic inch taken from larger specimens. At three years of age the "cultured" sponges were worth $3.50 per pound. At four years their value increased by $1.00–1.50 per pound (about a dozen sponges to the pound).[574] Additional information on sponge rearing was published in 1910,[575,576] but there doesn't appear to have been any further activity on the subject until the 1930s.

Some private sponge culture was reportedly occurring in the Bahamas during the 1930s, but nothing like that was occurring in the United States, even though supplies were becoming scarce and the price was rising. A call for aquaculture was made in 1938: "As the world's largest consumer of sponges our hope of permanence and stability appears to depend upon the development of sponge cultivation."[577] Sponge culture could perhaps have become big business following World War II, but technology solved the problem of sponge scarcity by developing washcloths for bathing and synthetic sponges for car and dish washing. A potentially thriving aquaculture venture was nipped in the bud. Now it is rare to see a real sponge in the marketplace.

Oysters and Clams

Oyster production during the postwar period did not change significantly in most respects. It remained largely a hunting-and-gathering activity in which management was associated with the replacement of dead shells to provide cultch and the leasing of private oyster beds in certain states. There were at least some attempts made to modernize oyster culture, with the most well known one being an operation in Long Island Sound off Connecticut.[578] Long Island Sound had produced 3–4 million bushels of oysters a year early in the century, but production had fallen off to 50,000 bushels or less during the 1950s and 1960s. Pollution, storms, poor reproductive success, and predation by starfish were all involved with the decline. In 1966, the approach to oyster culture in Long Island Sound was altered. Commercialization involved the production of spat in hatcheries, control of starfish and oyster drills, modernization of equipment, and the use of Scuba divers to monitor the oyster beds. Historical production equated to 1 bushel of market oysters (200 oysters) to 1 bushel of seed oysters (5,000 oysters). With the use of the new technologies, 10 bushels of market oysters were being produced from a bushel of seed oysters by 1970.

While oyster culture had a long history, the ability of aquaculturists to cultivate clams was elusive. Victor Loosanoff, writing in 1959, had the following to say about the subject:[579]

> The Greeks couldn't do it, the Romans couldn't do it, and the Chinese couldn't do it. Even the predominating majority of present-day biologists were skeptical of the idea that such mollusks as the hard-shell clam . . . could be artificially cultivated, in large numbers and as a matter of routine, from the egg to a size large enough to be planted as seed. However, this has been accomplished . . . by the U.S. Fish and Wildlife Service's biologists. . . .
>
> During the past winter, with the small personnel and limited laboratory space available, we managed to grow millions of seed clams.

Loosanoff indicated that approximately 1 million seed clams were delivered to biologists and people in the private sector for experimental planting. He also reported that the first commercial clam hatchery in the world, which was planning to produce 300 million young clams annually, was under construction. Regardless, there was not any major movement toward clam culture prior to the 1970s.

Shrimp

Placing a specific date on when aquaculturists became interested in producing marine shrimp is difficult, but that interest seems to have first developed in the 1960s. Shrimp were well known and popular along the Gulf of Mexico and southeastern Atlantic coasts of the United States long before World War II, but it was only after the war, when the ability to ice and freeze seafoods had become common practice, that shrimp could be shipped inland.*

With the expanded availability of shrimp, demand certainly increased, though it seemed to be met relatively easily by the commercial fishery. Perhaps it was visionary that interest in shrimp aquaculture initially developed (though I think there was also a good deal of greed involved). Visionaries might be credited with seeing that demand would eventually outstrip supply and could argue that aquaculture provided a reasonable solution. Greedy individuals, on the other hand, can be blamed for promoting shrimp culture in the absence of the technology required to achieve

---

*My father, who has never been known to consume anything that lives in water, was out of town three to four days a week in conjunction with his job, and my mother would sometimes purchase fish sticks or shrimp in his absence. Both types of seafood came breaded and frozen, and since we lived in Nebraska, and later Kansas, perhaps I can be excused for thinking that one picked up a shrimp by its foot.

success. As I have previously mentioned, Florida State University scientists were conducting research on the rearing of shrimp when I arrived on campus in 1968. Their research was on species indigenous to the Gulf of Mexico: white, pink, and brown shrimp. The Bureau of Commercial Fisheries laboratory in Galveston was also conducting research on shrimp and had managed to spawn brown shrimp and produce postlarvae by 1966.[580]

Pond culture of shrimp in the United States was first conducted by Robert Lunz at the Bear's Bluff Laboratory in South Carolina in the 1950s.[581] Young white shrimp were reared in ponds by the Bureau of Commercial fisheries in 1966, and similar studies were undertaken with brown shrimp in 1967.[582] Clearly, shrimp culture was in the very early phases by 1968, yet there was already a lot of activity in the private sector.

I was told by colleagues at Florida State University that in 1967 there had been 50 commercial shrimp farms in Florida, and that there were also 50 such farms in Florida in 1968. Only one of the names on the list was the same during both years. The single farm that remained in operation (or ever went into production, for that matter) was Marifarms, Inc., located on 3,000 acres near Panama City, Florida.[583] Their first crop was harvested in 1970, and they remained in business for something like 15 years before ultimately giving up.

What seemed to be happening was that individuals or groups of investors would either be talked into initiating a shrimp culture venture or would decide to put their money into such a venture without external influence. They would, in many cases, already have the land, so what they needed was someone to actually grow the shrimp. Surely, any good biologist could accomplish that task. Florida State, and I'm sure other universities, received frequent calls from prospective shrimp farmers looking for biologists. Offers of up to $20,000 a year (an exorbitant amount of money in the late 1960s) were made, and sometimes accepted by students, none of whom had ever raised a shrimp, and most of whom did not alter their track record after accepting jobs as shrimp culturists.

The relatively small time investors that dominated the scene in Florida were not the only ones interested in shrimp culture. Large corporations such as United Fruit, Dow Chemical, Ralston-Purina, Monsanto, and Texaco also invested in shrimp culture.[581]

Shrimp were being raised successfully in Japan, and that success undoubtedly had a lot to do with the interest that developed in the United States. Also, the commercial fishery seemed to be experiencing difficulties by the 1960s:[581]

> Of all U.S. fishing industries . . . the shrimp industry [is rated] as the most valuable fishery. Shrimping in 1969 represented 24% of the total U.S. value for all species. For the past 2 years the consumption of shrimp in the United States has averaged almost 1 million pounds a day (heads-off weight).
>
> Despite high demand and record high prices for shrimp, many offshore trawlers have had difficulty operating the year-round at a profit because of rising labor and maintenance costs. Due to the high cost of outfitting and operating offshore vessels, there was a decrease in the number of shrimp trawlers added to the Gulf of Mexico fleet in 1969.

So, the prospective shrimp culturists had reason for optimism. The basic pieces of the puzzle were in place by 1970 due to research at various governmental laboratories and that being conducted at such institutions as Texas A & M University, the University of Florida, Nicholls State College, and Louisiana State University.[581]

As we shall see, research on shrimp grew exponentially after 1970. A major problem that faced U.S. scientists was the difficulties involved with rearing native species. Ultimately, the industry that developed employed species that are exotic to the United States. It was exotics that were ultimately selected for further study by U.S. scientists and that ultimately led to development of a successful industry, though not one that has achieved any visibility within the boundaries of this nation.

Lobsters

As we have seen in previous chapters, American lobsters were a subject of interest from the earliest years of U.S. aquaculture. You will recall how Livingston Stone inadvertently stocked some of them in the waters of Nebraska during his first ill-fated expedition to the West by train in 1873. Serious attempts to spawn and release lobsters began in 1888 in response to the dwindling catches.[584] Overfishing and disregard for laws that were designed to protect the lobster fisheries were blamed on the decline. Releases of larvae from 1888 through 1903 by the U.S. Fish and Fisheries Commission in its attempts to augment the stocks totaled 879,511,565.

The realization that larval lobsters were not well equipped to survive after release was recognized early on as described in 1901 by A. D. Mead:[585]

> There is, in the life of the lobsters, a definite, well-marked period beginning when the eggs are hatched and ending when the young have shed their shells three times and have reached the fourth stage of development. During this period the young are very poorly equipped, wither in structure or habits, for protecting themselves against their enemies or from escaping from them. They swim about slowly and aimlessly in the water, an easy prey to shrimps, fishes, and other animals; they lack the hard shell, the protective coloration, and the swift movements common to most small crustacea; indeed, they do not have even the sense of fear which might lead them to avoid danger. During this period of life there is, as might readily be inferred, a very great mortality.

Experiments by H. C. Bumpus at Woods Hole resulted in limited numbers of larvae being reared to the juvenile stage in 1898.[584] In 1901, Bumpus developed a culture system in which large numbers of postlarvae could be produced. The culture system kept the water in motion by means of mechanically operated paddles. This was felt to be important for keeping the larval lobsters suspended in the water column.[585] While mortality was high in the laboratory, it was the contention of A. D. Mead that releasing 10 juvenile lobsters would result in better overall survival than the release of 100 larvae.

The studies were significant in that they provided the means by which to rear and feed lobster larvae. A variety of foods were tried, and it was found that lobster liver* was one of the best, but it was too expensive for routine use. Shredded fish flesh was

---

*It would be technically more correct to say hepatopancreas rather than liver.

acceptable, but clam meat, chopped into fine pieces, was the best choice overall. A little problem, and one that has plagued lobster culturists ever since, was uncovered during the experimentation:[585]

> There is one habit of the fry which makes the question of ample food supply especially important, their atrocious cannibalism. From the moment they are hatched, throughout the early stages of life their affection for one another take this disgusting form. The only way to prevent them from destroying one another is to give them an abundance of food, and in such a manner that they will take it in preference to other lobster fry.

As the researchers gained experience, the numbers of juvenile lobsters produced generally increased: 8,974 in 1901, 27,300 in 1902, 13,500 in 1903 (a reduction in that year), 50,597 in 1904, and 102,000 in 1905.[586] Of newly hatched larvae that were placed in the rearing apparatus, survival even during the early years of the work ranged from 16% to over 50%.[587,588] The young lobsters were being released into the waters of Rhode Island by that state's fish commission, and while Mead acknowledged that it would take some years to ascertain if the plantings were effective at increasing the lobster stock, there was early indication of success:[584]

> . . . in the vicinities where these lobsters were liberated, the lobster fishermen report that, for the last two winters young lobsters of about eight inches in length were abundant along the shore many of them being dug up by the clam diggers and by ourselves, and it is said that small lobsters have not been seen in abundance in these localities for twenty years. The lobster fry planted by this Commission should be about as large as these young lobsters by this time, and it seems probable that this large supply of young lobsters is the result of the efforts of the Commission.

Little more was heard of lobster culture and release for several decades, though there were apparently some 20 or more hatcheries operating in North America at one time or another until about 1950.[589] In 1951, Massachusetts established a state lobster hatchery at Martha's Vineyard and got involved in the rearing and release of juveniles. The hatchery released over 7 million juvenile American lobsters by 1970. The federal government became involved in lobster research at the Bureau of Commercial Fisheries laboratory at West Boothbay Harbor, Maine, in 1969. In California, research aimed at employing the heated effluent from coastal power plants to accelerate production rates of lobsters was initiated at the San Diego Gas & Electric Company power plant during the late 1960s. Oregon State University initiated studies during the same period with the goal of introducing lobsters to the waters of that state.

In 1970, the lobster fishery was number 6 in terms of value among the U.S. fisheries, and lobster meat was wholesaling at $8 per pound.[589] The high price and high demand for lobsters made it appear as though commercial aquaculture deserved renewed attention.[589] The technology for spawning and larval rearing was well established, and some success in rearing lobsters through their entire life cycle in captivity had been achieved. Surely, it seemed, lobster culture could develop into a major industry.

A considerable amount of research was conducted on American lobsters after 1970, both in the United States and Canada. There was even some pseudoculture activity in the private sector involving the holding and feeding of lobsters captured from the wild in cages. True commercial aquaculture was difficult to establish, even though research indicated that estimated costs of production in 1973 were on the order of $2.00–2.50 per pound.[590] By increasing growth through environmental and other types of manipulations it was thought that costs could be further reduced.[591] Projections from small-scale research in the laboratory and computer simulations never seem to have been put to the test on a commercial scale, however.

Cannibalism continued to be a major problem. The solution was to raise lobsters in individual chambers—lobster condominiums—which greatly affected production costs. Not only were facilities costs increased dramatically, but the time and labor involved in feeding and cleaning up after lobsters reared in individuals units were unrealistic. At the moment, commercial culture of the American lobster appears to be moribund, if not dead in the United States.

There was some talk about dosing American lobsters with tranquilizers during the growout process, but that doesn't seem to have been pursued. Had it been, there would undoubtedly have had some interesting ramifications relative to the U.S. Food and Drug Administration.

The aggressive nature of the American lobster and its penchant for cannibalism was a problem that some people felt could be avoided by raising lobsters that lacked large claws. The Florida lobster was just such an animal. That lobster also seems to be quite gregarious. Anecdotal information during the 1970s indicated that some success had been achieved by prospective commercial culturists who collected juveniles from nature and reared them in captivity. For commercial aquaculture to become a reality, culturists would have to close the life cycle; that is, they would have to achieve captive rearing from egg to market and would also need to establish and maintain captive broodstock. Obtaining juveniles from heavily and often overfished natural populations could not be relied upon.

The spawning of Florida lobsters had been achieved as early as 1920. That fact was documented in a paper that also had a description of the early larval stages.[592] The larvae, called phyllosomes, do not look anything like the adults. Phyllosomes have long appendages and have been described as being leaflike. As they go through the various developmental stages, which takes several months, they reach sizes measuring a few inches across. The problem associated with rearing the larvae begins even during the earliest stages in the process.[592]

It was observed that these long legs became entangled when the larvae were crowded and that it was impossible to separate the larvae. Consequently, they sank to the bottom in tangled mats and soon died.

The problem of larval survival is compounded as they become larger and swim more strongly, since the chances of encounters with one another increase even at low densities. No way to overcome the problem has been developed that doesn't involve rearing the animals in larval condominiums for as much as nine months. Even if Florida lobster larvae could be raised to the puerulus stage,* at which time they can be housed in communal tanks without difficulty, it has been estimated that three to four years would be required to rear them to market size. Whether that is economically feasible is unlikely.[593] After a flurry of interest, and in which some money was made by the snakeoil salesmen who hoodwinked a few unwary investors out of their money, little is heard about Florida lobster culture today.

Crawfish

While there are some 300 species of crawfish in the world—about 200 of which occur in the United States—two species are of particular commercial importance in this country. Those are the white river crawfish and the red swamp crawfish. These species support a large capture fishery in Louisiana and were selected for use by crawfish culturists, first in Louisiana and more recently in Texas.

Crawfish were consumed in any quantity at all almost exclusively in Louisiana prior to the time when Cajun cuisine became fashionable nationally. Even though consumption was limited regionally, not all the consumption was by the local population. Crawfish dishes were very popular with tourists, though unavailable to out-of-staters after they left Louisiana for other climes. By 1970, more than 10 million pounds of crawfish with a value of about $5 million were being harvested annually in Louisiana, with most of that production coming from wild populations in the Atchafalaya River Basin.[594]

Crawfish aquaculture, which may have been initiated by accident, began around 1950, when a rice farmer reflooded a field following harvest so he could provide himself a place for duck hunting. The following spring he found the duck pond teeming with crawfish, which were subsequently harvested.[594] The industry is said to be based on that experience.

Crawfish farming was developed as a means of helping meet demand, particularly during years when the Atchafalaya Basin did not produce bumper crops. The big crops were obtained, on average, about 40% of the time, so during most years crawfish farmers could be expected to have ready markets. By the end of the 1950s some rice fields were being managed for crawfish production to provide a secondary commercial or recreational crop.[595]

Up until the 1970s, little had changed in terms of culture strategy from the approach that was accidentally hit upon by the rice-farming duck hunter turned crawfish culturist:[594]

> Crawfish are currently being farmed in three types of ponds: rice-field ponds, wooded ponds, and open ponds. In rice-field ponds crawfish are rotated with the rice. The general procedure . . . is to remove water from the rice fields about two weeks before harvesting. This permits drying of the field to facilitate harvest. When drying begins, crawfish burrow. Four to six weeks after draining, the ricefield ponds are

---

*This stage occurs two molts before the larvae become juveniles that look like the adults.

reflooded. The second growth of rice and grasses, along with rice straw, provides food for the crawfish.

Wooded areas are also used, but make poor crawfish ponds. Dense growths of trees and shrubs hinder harvest. Wooded ponds usually have poor circulation. . . .

Open ponds are often constructed solely for crawfish farming. The procedure . . . is [to stock] the ponds in late May or June. Brood stock, usually bought from a dealer, is stocked at rated of from 25 to 50 pounds per acre depending on the amount of vegetation and number of native crawfish present. Once stocked, the crawfish burrow. In July the ponds are drained. . . . When young crawfish are found in the burrows, generally in September and October, the ponds are flooded to release them. Once the ponds fill, the young crawfish forage on native aquatic plants such as alligatorweed, water primrose and smart weed. If the winter is mild, crawfish can be harvested the same year.

Harvesting is accomplished with baited traps that are checked daily for periods of several weeks to a few months, so the process is labor-intensive. That situation has not changed significantly from the early days of crawfish farming, though many modifications in harvesting and other aspects of crawfish culture have been recommended as a result of research conducted primarily by Louisiana State University, Southern University, and the University of Southwestern Louisiana. While the beginnings of crawfish culture were somewhat primitive, there was sufficient interest to attract the investment capital required to construct at least a few large farms. One 250-acre farm, for example, was constructed in 1966.[596]

Frogs

Interest in frog culture goes back to about the 1880s, though for decades there was little success achieved. The Bureau of Fisheries apparently attributed failures to lack of experience on the part of those involved in attempts to raise frogs and insufficient capital to maintain frog culture operations long enough for success to be realized.[597] The state of Pennsylvania began work on frog culture in 1903[598] and began distribution the following year as described by the commissioner of fisheries for that state:[599]

In May, 1904, a four-line item sent out by the Associated Press appeared in the Pennsylvania newspapers, announcing that the Department of Fisheries would receive applications for frogs or tadpoles for public planting. In anticipation of this announcement the Department of Fisheries had prepared about 1,000 blank application forms. To the astonishment of the Department, the 1,000 blank application forms were taken up within ten days and it is safe to say that nearly 1,000 letters in addition were received, asking to be supplied with frogs for stocking purposes.

The Pennsylvania commission learned a good deal about frog culture during the early years of the twentieth century, including the fact that at certain stages, the frogs would only accept live foods.[599] Eggs were collected from nature, and during the early years it was not possible to determine which species of frog was being collected. Through experience, culturists learned to identify bullfrog eggs[598] and concentrated their efforts on rearing that species. Early problems associated with bullfrog culture included predation on tadpoles by certain aquatic insect larvae, bird predation, and a propensity for metamorphosing frogs to experience broken appendages.

Tadpole culture turned out to be a relatively simple proposition, but providing live food for young frogs was a vexing problem:[597]

> The young frogs may be fed upon insects by light at night, upon live maggots, and upon flies attracted by various baits, in addition to the food items they secure naturally. Feeding the older frogs, however, is by far the most difficult problem in frog cultivation. Medium and large bullfrogs feed only upon large living objects and are seldom attracted by anything under 1/2 inch in length. Two hundred-watt . . . lamps will attract many June beetles and medium sized moths. Arc lights will attract even larger insects and sometimes in very large quantities. For using a lamp, the early part of the night is best, and the hotter the night the better. Willow trees also attract large numbers of June beetles which later in the night drop to the ground to lay their eggs and are taken by the frogs. Flower beds will attract butterflies and other day flying insects to the vicinity of the [frog] pens. The pond itself will also attract many other insects seeking water, and dragon flies go there to deposit their eggs. While all these things are helpful it does not provide the bulk of food necessary for rapid growth, and some bulky food must be provided.

Topminnows and other small fish were suitable foods for medium-to-large bullfrogs, though the most ideal foods discovered by the 1930s were crayfish, green frogs, and small bullfrogs. Lacking sufficient quantities of live food, the frogs often became cannibalistic.

Frog culture research seems to have languished until Dudley D. Culley, Jr., began his work at Louisiana State University in the late 1960s. Demand for frogs had certainly not diminished over the years, and many would-be entrepreneurs had attempted to rear them commercially. All such ventures failed. According to Culley and his colleague Claude Gravois, writing in 1970, no one had successfully raised bullfrogs in a commercial venture, but the demand was high from the food industry, along with educational and research institutions:[600]

> The demand for bullfrogs and closely related species has become so great that many of our national populations have become depleted and commercial sellers are having to move outside the United States to obtain enough wild frogs to fill their orders.
>
> Unless culture techniques can be worked out in the near future, we can expect a severe shortage of bullfrogs in a few years due to a continued intense harvesting and man's land clearing and draining activities.

Culley and Gravois identified five problems that would have to be solved if bullfrog culture were to become a reality:[600]

- requirement for living food
- susceptibility to a variety of predators
- susceptibility to diseases
- cannibalism
- time to maturity

The early work at Louisiana State addressed each of those problems and made some early discoveries that helped Culley and Gravois reduce or circumvent some of them.* A summary of their early results included the following:[600]

1. Bullfrogs grow quite well under crowded conditions if cages are kept clean and plenty of food is provided.
2. Survival of bullfrogs is better than 98% when raised indoors.
3. The most rapidly growing frogs reach marketable size (8 inches total length or 130 grams) eight months after metamorphosis. Most will be marketable within 12 months. As selective breeding and rearing techniques improve, growth to a marketable size should fall below eight months.
4. Growth rates vary with size. Small frogs tend to double their weight each month for two to three months. The growth rate then gradually declines in succeeding months. . . .
5. Food conversion for frogs up to three-fourths marketable size has averaged about 2.5 (wet weight basis).
6. Rearing facilities are still in the developmental stage. Several types of materials used in various cage designs have been tested. Minor modifications are still being made to facilitate handling and reduce labor costs. Present cages are designed to hold 200 to 300 marketable frogs.
7. It is possible to obtain tadpole stocks from ponds, but disease problems cause high tadpole mortality in the late stages of metamorphosis.
8. Tadpoles held in the laboratory grow quite well on rabbit pellets or commercial trout (feed).
9. The crowding of tadpoles delays growth and metamorphosis, but probably can be corrected by raising tadpoles in a continuous supply of fresh water.

Culley and Gravois predicted that large-scale frog culture would be possible within a few years. We'll take another look at Culley's work in the next chapter to see how things progressed.

DEVELOPMENT OF AQUACULTURE TRAINING PROGRAMS

Many universities had developed fisheries teaching and research programs by 1970. The pace of that development was significantly enhanced beginning in 1960, when the U.S. Fish and Wildlife Service began establishing cooperative laboratories under Public Law 880-686. Cooperative Fisheries Research Units and Cooperative Wildlife Research Units were ultimately created within universities in the majority of the states. In most states a single institution housed both fisheries and wildlife units, though five states had fisheries-only units up until the late 1980s. Combined Cooperative Fish and Wildlife Research Units were developed in the 1980s to streamline the system and reduce expenses. Most of the so-called super units were created by combining existing fisheries units with wildlife units, but others were created from scratch in states that previously had no unit or had one type of unit but not the other.

The Cooperative Fisheries Research Unit at the University of Missouri was prob-

---

*Two years were normally required to produce a marketable frog, not counting the year required to rear the frog through the tadpole stage. To shorten the culture period, the researchers moved the frogs indoors to provide a year-round optimum temperature regime.

ably typical of those existing prior to 1970. I was not directly affiliated with the coop unit when I was at Missouri, but many of my graduate student colleagues and two faculty members were supported by the unit.* Richard Anderson was the coop unit leader, and Dan Coble was his assistant. Both were involved in fisheries management research, but Dick taught a fish culture course (which, incidentally, was not offered during the period of time I was at Missouri; in fact, I've never taken an aquaculture course). The Missouri coop unit was involved in research on the streams and reservoirs of the state. Research was also being conducted in conjunction with a governmental pesticide research laboratory located in Columbia, Missouri. Tom McComish was completing his Ph.D. work in 1967, which involved feeding energetics in bluegill and had some aquaculture ramifications. The coop unit maintained vehicles and equipment that would not have otherwise been available to the University of Missouri fisheries research program. I worked with Bob Campbell, a limnologist. Art Witt was a faculty member involved in fisheries, and there were other faculty members in the Zoology Department who conducted related research. Most of the faculty with fisheries interests collaborated at some level with the Anderson and Coble.

Cornell University was heavily involved in fisheries management research by 1950.[601] Many of their studies involved the use of hatchery fish, and some research was conducted at the Cortland fish hatchery. A training program for U.S. Fish and Wildlife Service employees was established at Cortland in 1946. Opportunities for training non-Service individuals were infrequent, but by 1957 Cortland's program had led to training a few individuals who were subsequently employed by state fisheries agencies. Students from Canada, Mexico, Iceland, Bolivia, Poland, and the Philippines had also attended at least a portion of the training course.[602] Subjects covered included nutrition, freshwater biology, physiology, diseases and their control, water analysis, hatchery management, and fish culture practices.

The two largest programs in the nation continued to be located at Auburn University and the University of Washington. By 1940, five state and federal fisheries agencies had been established in Seattle, and cooperation involving research and training between the University of Washington and those agencies was being developed.[603] In 1941, the Laboratory of Radiation Ecology (LRE) was established under the direction of Lauren Donaldson. That unit was sponsored by the federal government and charged with conducting research on the effects of radiation on fishes. It was a top-secret activity established by the War Department during World War II in conjunction with research associated with development of the atomic bomb. The LRE continued its involvement after the war in conjunction with atmospheric testing of atomic and hydrogen bombs in the South Pacific following the war. Those activities were chronicled in detail by Neal Hines in his book *Prooving Ground*.[344]

In 1950, a new fisheries building was completed on the University of Washington campus. It provided, for the first time, what were then modern hatchery facilities:[603]

---

*Faculty housed in coop units are actually employees of the U.S. Fish and Wildlife Service, but they are provided with faculty rank and are allowed to supervise graduate students. Funding for their research comes from the Service and other agencies. Support in the form of free space and overhead waivers is provided by the university.

Fresh-water biology is provided with a 32-trough hatchery which also accommodates 14 tanks and 3 circular pools for rearing young trout and salmon. Five rectangular and two circular concrete pools are located within a wire enclosure on the lake side of the building. Running freshwater is pumped from Lake Union into the 18,000-gallon tank on top of the building to supply the hatchery and ponds. Supplementary emergency water can be diverted from the city water mains and can be dechlorinated in two large activated-carbon filters. The other laboratories for fresh-water research are equipped with tanks and troughs. One laboratory is for the study of fish diseases and has its own sewage system. The other laboratory has thermo-regulated mixers in the water supply for constant-temperature work.

The School of Fisheries building is located several miles from Puget Sound, so the saltwater system was of the recirculating type and ultimately, until it was decommissioned in about 1990, was used primarily as a public display of local marine life.

As part of the construction project in 1949 and 1950, a fish ladder was constructed behind the new building. Donaldson was interested in establishing fish runs that would home in on his hatchery. He was roundly jeered by those who thought he was a nut, but his idea worked, and in 1953 the first fish climbed the ladder and entered a raceway.[338] It was the first time that fish voluntarily entered the classroom to undergo study. Over the next several years Donaldson established runs of both Pacific salmon and steelhead trout. To provide additional space for returning adults, he decided to add a return pond to the facilities. He obtained private funding for the project which was completed in 1961. The facilities at the University of Washington, now largely outdated, continue to be actively utilized, and while the salmon runs initiated by Donaldson, like many others in the Pacific Northwest, have declined in recent years for a variety of reasons, they continue to be maintained by Bill Hershberger.*

In 1958, Kenneth D. Carlander, a zoology professor at Iowa State College (now Iowa State University), conducted a survey of technical fisheries personnel in the United States and Canada.[604] He ascertained that there were 1,673 technical fisheries biologists in the States and 232 in Canada. Of those, 138 were identified as fish culturists. Among the fish culturists, 79 were employed by federal agencies, 47 by state and provincial agencies, and one by a college. Eleven others were employed in nongovernmental and nonuniversity positions. The training level of the fish culturists was as follows:

High school	77
Bachelor's degree	40
Master's degree	17
Doctoral degree	1

---

*Bill typically works with "Doc" Donaldson looking over his shoulder.

While it is not possible from Carlander's data to identify the fish culturists working in a university of the identity of the lone Ph.D., I suspect that Lauren Donaldson filled both roles.

At about the time Carlander was conducting his survey, H. S. Swingle (an entomologist, not a fish culturist) was writing a short paper on his view of career opportunities in fisheries. He was convinced that the increasing human population and demand for natural resource conservation would provide job opportunities, but that fisheries biologists needed proper training.[605]

> The problem of adequate personnel is increasingly difficult of solution, because of the increasing amounts of education and training needed in these positions. The fisherman, the hatcheryman, the management biologist and the research biologist each must learn to use complicated tools and skills unknown to his predecessors of a generation ago. Usually a college student, while unfamiliar with the various professions, must decide for which field he will prepare himself and what skills he will need.

Swingle indicated that the problem could be solved, in part, if high school and college students had the opportunity to work in the profession on a part-time basis. Such opportunities would help them select the direction in which they wanted to proceed in terms of their academic training.

Auburn developed its International Center in 1970.[606] The center was established to administer international fisheries activities of the university, which were largely contained within the Department of Fisheries and Allied Aquacultures. Long-term support for the International Center was provided by the U.S. Agency for International Development (USAID). Foundation support was garnered over the years to provide for program expansion. The center was involved in the training of international students in the United States and with aquaculture development projects in a number of countries. By 1979, for example, there were 40 professional faculty and 100 graduate students supported by the International Center. Projects were being or had been undertaken in Thailand, the Philippines, Indonesia, Nigeria, Brazil, El Salvador, Panama, Colombia, Honduras, and Jamaica. Other countries were added to the list over the years, and the center prospered until the 1980s, when USAID changed its philosophy, pulling back from its earlier objective of helping to feed people in developing nations. Much of the International Center's activity in aquaculture was designed to meet that objective, so support began to wane. When the Soviet Union collapsed, USAID began turning its attention from the traditionally supported developing nations and began undergoing complete reorganization and restructuring, leaving very little in the way of fisheries and aquaculture programs. Auburn's last USAID-funded aquaculture project (one in which Auburn was collaborating with some other institutions) was based in Rwanda. That project collapsed with the Rwandan civil war of 1994, though it may be resurrected and moved to another location.

While various other fisheries programs developed prior to 1970, one is particularly noteworthy with respect to aquaculture. That program, which was housed in the Cooperative Fisheries Research Laboratory, was initiated at Southern Illinois

University by William M. Lewis.* Bill Lewis was born in Faison, North Carolina, in 1921 and returned to North Carolina to develop a striped bass farm in 1983 upon his retirement from the university, though he continued his involvement with graduate students and a large research program funded by a power company after retirement. He made frequent trips back to Illinois from North Carolina to oversee his projects.

Bill received his B.S. from North Carolina State, after which he served in the U.S. Navy in 1943 and 1944. He returned to school following the war and took his M.S. and Ph.D. degrees from Iowa State University, the latter degree being proffered in 1949. The Cooperative Fisheries Research Laboratory was formed in 1949, the year Bill Lewis joined the faculty at Southern Illinois University. Aquaculture and fisheries management research were conducted under the auspices of the laboratory. Bill was involved with a variety of aquaculture projects and species over the years. The plaque honoring Bill Lewis in the National Fish Culture Hall of Fame in Spearfish, South Dakota, dedicated in 1988, reads, in part:

> His most noteworth[y] achievements include early research on fish pathogens and disease control, pioneering work in the cage culture of fishes, and practical production of hybrid sunfish and largemouth bass by the use of prepared diets. Dr. Lewis also conducted research that was the first to successfully combine hydroponic vegetable production with tank culture of fishes.

Bill Lewis developed a large recirculating system for the production of striped bass and contributed significantly to the development of procedures for rearing the larvae of striped bass and hybrid striped bass. He was a founder of the Illinois chapter of the AFS and served as chapter president in 1979. He was president of the AFS in 1982–1983. During his career, Bill Lewis supervised or served as co-adviser for 71 M.S. and 19 Ph.D. students.

On a personal note, I got to know Bill Lewis in conjunction with the meetings of the AFS and other professional activities. He served on an aquaculture program review team that evaluated the Texas A & M program in 1983 and suggested I apply for the position from which he was retiring at Southern Illinois. I was retained in the position of director of the Cooperative Fisheries Research Laboratory in 1984 and retained that position for 18 months before being lured to the University of Washington. The year and a half in Illinois was a period I fondly remember.

BIRTH OF THE WORLD MARICULTURE SOCIETY

Much of the information upon which this book has relied came from the *Transactions of the American Fisheries Society*, a journal that was one of the few places in which aquaculture information was available outside of governmental publications

---

*Bill Lewis' biographical information was provided by Roy Heidinger. The Cooperative Fisheries Research Laboratory has never been affiliated with the U.S. Fish and Wildlife Service Cooperative Research program, so the name is a bit misleading. Furthermore, a Cooperative Wildlife Research Laboratory was created at Southern Illinois at the same time. Both undertook cooperative research with the Illinois Department of Conservation and received funding from various federal agencies over the years.

until the 1960s. It was during that decade that the commercial industry developed sufficiently to support trade magazines. At the end of the decade, in 1969 to be exact, the seeds for development of what was to become the World Mariculture Society (WMS) were sown. That organization quickly became a primary source of information on aquaculture research and development.

The WMS developed as a follow-up to a small workshop hosted by the Louisiana Wild Life and Fisheries Commission at its Grand Terre Island Marine Laboratory from January 31 to February 2, 1969.[607,608] Attendance was by invitation, with all but one of the 45 attendees coming from 10 Gulf of Mexico and southeastern Atlantic coast states. The exception came from the United Kingdom. Strongest interest seemed to relate to marine shrimp culture and the culture of pompano. It was concluded that another meeting should be held later in the year, and a steering committee was appointed.

Gordon Gunter, a participant in the workshop and director of the Gulf Coast Research Laboratory in Ocean Springs, Mississippi, offered to host the second meeting, and it was eventually scheduled for September 1969. The steering committee met in advance of the general session and drafted a charter and bylaws for presentation to the entire body of 44 people, who met on September 9. Attendance was considered quite good since a hurricane had decimated the Gulf Coast, including the Gulf Coast Research Laboratory, only days before.

Once the idea of forming a new organization was adopted, it needed a name. Gordon Gunter suggested World Mariculture Society, and the name stuck, even though it was never voted on by the membership. G. Robert Lunz, a pioneer in shrimp culture researcher from Bear's Bluff Laboratory in South Carolina, was elected president. (No other officers were elected that year.) James W. Avault from Louisiana State University and Bryant F. Cobb III from Texas A & M were appointed to determine the location of the next meeting.

Avault and Cobb selected the campus of Louisiana State University in Baton Rouge as the venue for the first official workshop of the World Mariculture Society.[607] That meeting was held February 9–10, 1970. Interest in the new organization had spread quickly. The 1970 meeting was attended by about 200 people. Over 20 speakers presented information on a variety of subjects, including oysters, pond management, nutrition, diseases, and larval culture.

The keynote address for the meeting was offered by Paul F. Bente, Jr., president of Marifarms, Inc., the only shrimp farm that seemed to survive for more than a year or two in Florida. Bente felt that the time for mariculture had come, and was convinced that the commercialization of shrimp farming had come at the proper time.[609] Among his reasons for holding that opinion were that natural supplies of shrimp were decreasing, most of the suitable sites in the world for shrimp farming were still available, and domestic cultured shrimp could defray what was already a $150 million annual import deficit in that commodity. Bente went on to indicate that commercial oyster farming was poised for rapid development.

He then cautioned that for mariculture to become successful it would need cooperation from government. Approval of the use of coastal waters for mariculture was seen as a critical need. He pointed out that the state of Florida was the only coastal state that had passed a law and developed guidelines for the conduct of maricultūre operations within state waters. Presentations detailing that law were made later in the meeting.[610–612]

Interestingly enough, the keynote address mentioned the fact that opposition to mariculture had already begun to develop: "There are voices in our country that cry out against mariculture, saying that it means surrendering a heritage, a traditional right of the public to have access to and to use any and all coastal waters."[609] Bente elaborated on that point:[609]

> There are . . . voices which cry out saying that the waters being proposed for mariculture use are the most productive of all waters, or that they are absolutely essential for the propagation of all the fish that the local commercial fishermen catch out at sea. They say the waters are worth as much as $6,172/ha/year.* Exaggeration seems to be the name of the game.

He was convinced that mariculture could be developed in concert with the environment, not in spite of it. He closed by charging the WMS with a fairly prodigious task:[609]

> Because of the highly diversified and interdisciplinary requirements in understanding our shoreline and marine environment, attempts should be made to use the scientific and technological skills available. This new society has, collectively, the expertise that is needed and may very well hold the key to changing the situation in our country for the better of all mankind. This society may be able to promote legislation that is needed to open up this new frontier of mariculture. When such laws have been enacted, this society will then be in a much stronger position to have a meaningful exchange of scientific information which will catalyze the growth of mariculture for the benefit of all mankind.

Presentations of note at the first meeting included a historical overview of oyster culture in Louisiana.[613] Regulations in that state dated back to 1870, and an Oyster Commission was created in 1902. That state was said to have over 500,000 acres of land that could be used for bottom culture of oysters. A leasing program had been established, and there appeared to be sufficient space for numerous small and large enterprises.

On a broader scale, oyster farming in North America was reviewed by William N. Shaw with the U.S. Bureau of Commercial Fisheries.[614] He commented that little advancement in the way oysters being cultured had occurred in the previous century. He discussed off-bottom culture as one means by which additional control could be exerted and higher production rates achieved.

John Finucane discussed progress that was being made in pompano culture,[615] George Klontz updated the group on the status of treating diseases of mariculture species,[616] Cornelius "Corny" Mock and Alice Murphy discussed the rearing of penaeid shrimp larvae,[617] and Frieda Taub from the University of Washington spoke on the subject of culturing algae as food for marine organisms.[618] Robert Stevens, one of the pioneers of striped bass culture in the United States, gave a summary of how that activity had developed.[619] So, the meeting was diverse and it was limited to

*One hectare (ha) is approximately 2.4 acres.

species that spent all or much of their lives in the sea. It was also a U.S. show. The name World Mariculture Society seemed a little pretentious since little was presented on mariculture outside of North America, with the exception of some historical background information. All that would change, or at least attempts would be made to change it, as the society matured.

The proceedings of the first WMS workshop, published by Louisiana State University's Division of Continuing Education, was the first publication of the society.[607] We'll take another look at the WMS, including changes in structure and in the name of the organization, in the next chapter.

CHAPTER 5

The Growth Years Following 1970

PREPARING FOR A GLORIOUS FUTURE

The well-known oceanographer John Ryther predicted in 1969 that the fisheries of the world could be harvested at no more than about 100 million tons a year and that the maximum harvest level would be reached within several years.[3] With projected increases in human population growth and concomitant increased demands for fish and shellfish, the capacity of the ocean to satisfy the markets was being reached. Aquaculture provided the obvious means for meeting increased demands for fish and shellfish in the face of static or declining capture fisheries. World fishery landings actually didn't peak until about 1990, at just about the 100 million tons predicted by Ryther. By that time about 15% of the total came from aquaculture.[620]

By 1970, the scientific and technical foundations that had been laid in the United States were providing the springboard from which U.S. aquaculture was expected to explode. At the meeting of the World Mariculture Society (WMS) in 1971, a call for additional research to solve the many problems still facing commercial aquaculture development was made by Robert B. Abel, then director of the National Sea Grant Program, an activity within the U.S. Department of Commerce. The Sea Grant Program was patterned after the land-grant system, which had been established within the Department of Agriculture more than a century earlier. Bob Abel is generally credited with having been the father of Sea Grant. He expressed the view held by many that aquaculture held potential for feeding the world's hungry, but he did not feel that the tools had been developed to realize that potential:[621]

> I might as well give you my opinion on a key issue right now. It is a regrettable but inescapable fact that the great resources of food fish such as tuna, cod, mackerel, and the like are presently not susceptible to available aquaculture techniques. On the other hand there are seafoods in high demand, which for that reason and others are attractive for aquacultural techniques. As previously mentioned, these include mollusks such as

the clam and oyster, crustaceans such as the crab, shrimp, and lobster, and such prized fish species as the pompano. With some exceptions, however, it is not likely that such products will soon be developed in the huge quantities needed to become a recognized food source for millions of undernourished people. Their annual market value in the United States is, of course, appreciable, aggregating about $50 million, and it will continue to grow rapidly.

What I'm trying to say is that in considering the oceans' living resources it might be possible to feed large masses of starving people or it may be possible to make money but it will not be done in the foreseeable future with the same species.

Those words turned out to be prophetic. In the 1970s many looked to aquaculture as a means to feed the starving peoples of the world. What was often ignored was the profit motive and the fact that with few exceptions, the produce of aquaculture sells at prices beyond the means of the poor. The U.S. Agency for International Development (USAID) pumped in millions of dollars in aquaculture development projects during the 1970s and 1980s to help feed people in third-world nations, but ultimately it all but abandoned that activity. Aquaculture became well established in many of the nations where USAID established projects, but in most of those nations much of the produce of aquaculture was (and is) exported to developed nations and available domestically only in large-city retail outlets that can demand and obtain high prices. Be that as it may, the enthusiasm for a rapidly expanding aquaculture industry was epidemic in the 1970s, and those of us who were actively involved saw a major opportunity for involvement in the process.

Abel's call for additional research was underscored by presentations at the 1971 WMS meeting that outlined priorities for research on mollusks and crustaceans.[622–624] The relationship between the success of aquaculture and the quality of the food being produced was also discussed,[625] as was the need for close cooperation and collaboration between researchers and the private aquaculture sector.[626] The WMS was beginning to define its role as an organization bringing industry and scientists together to identify research needs and communicate the results of research aimed at addressing those needs. We'll look at the WMS and its evolution in more detail at the end of this chapter.

The land-grant colleges, responding to calls for additional aquaculture research, began taking a more active role in providing facilities and hiring personnel with interests in aquaculture after 1970. Some of the people and modest programs had been in place, but their visibility increased dramatically. Some land-grant institutions in which aquaculture programs were developed or greatly expanded are Mississippi State University, Louisiana State University, Clemson University, Texas A & M University, the University of Florida, the University of California at Davis, Oregon State University, the University of Wisconsin, Virginia Polytechnic Institute and State University, the University of Georgia, and Southern Illinois University. Major players like Auburn University (a land-grant university) and the University of Washington (not a part of the land-grant system) continued to be active, and a few less well known institutions, such as the University of Arkansas at Pine Bluff and Skidaway Institute of Oceanography, had active programs. Many other institutions dabbled in aquaculture research from time to time and a few significant programs achieved prominence in the 1980s. North Carolina State University is one example. By 1971,

Sea Grant was supporting aquaculture activities at Oregon State University, the University of Hawaii, the University of Washington, the University of Miami, the University of Rhode Island, San Diego State College, the University of North Carolina, and the University of Georgia system.[621] I was receiving Sea Grant funding by that time to investigate the culture of flounders at the Skidaway Institute of Oceanography. Details of the programs or portions of them at Louisiana State University, Mississippi State University, the University of Wisconsin, and Auburn University have been published within the past few years.[627–631]

The roles of federal and state agencies in public aquaculture continued much as it had in the past, but detailed records of the numbers of fish produced and stocked do not seem to be readily available. For the most part, things had become routine, though some changes had been adopted. The research role of the federal government had already been established in places such as the Marion (Alabama), Stuttgart (Arkansas), Cortland (New York), and Hagerman (Idaho) Fish and Wildlife Service laboratories. Some aquaculture research was also being conducted by the National Marine Fisheries Service in such places as Washington and Hawaii. Studies that provided information useful to aquaculturists was being generated at various other federal and state laboratories as well.

With projections for rapid development and the actual increase in aquaculture activity around the nation, states began developing aquaculture plans. At the federal level the National Oceanic and Atmospheric Administration (NOAA) developed an aquaculture plan that was published in 1977. The roles and responsibilities of the federal government as related in the plan were as follows:[632]

> Federal leadership and guidance should be expressed by a national policy to encourage aquaculture as a means of expanding food production. Federal actions are needed to channel the diverse efforts within and without government into a coordinated program which will provide the scientific and technical information, environmental protection and institutional arrangements required for expansion of aquaculture.
>
> . . . Many of the concepts and techniques which have made private aquaculture possible in the United States have resulted from research and development conducted in government laboratories or sponsored in universities by the federal government. Continuation of federal efforts will be needed to provide an adequate information base for development of aquaculture of additional species and solutions to long range problems of currently farmed fish and shellfish.

Speaking about the NOAA aquaculture plan at a WMS meeting, John Glude, NOAA's aquaculture coordinator indicated that NOAA would provide leadership among the federal agencies in terms of joint planning and coordination of programs.[633] NOAA's plan eventually gathered dust on the bookshelves of libraries and the offices of bureaucrats because the mechanism put in place to coordinate federal programs did not put NOAA in the leadership position.

In 1978, the National Academy of Sciences published a report indicating that U.S. aquaculture would not be a major contributor to the nation's food supply unless research-and-development funds for aquaculture were significantly increased.[634] The report also indicated that absence of a comprehensive government policy and the lack of political power within the aquaculture community were major constraints to aquaculture development. U.S. aquaculture production in 1975 was 65,000 tons, but the report predicted that 250,000 tons could be produced by 1985 and 1 million tons by the year 2000 if governmental support was increased to the required level. The prediction for 1985 wasn't too far off. World aquaculture production that year was 10,587,300 tons,[620] and assuming that U.S. production was about 2% of the world total (as was the case in 1988), it amounted to slightly over 210,000 tons, not far from the projected 250,000 tons. Of some interest is the fact that the dominant group of fishes being cultured in the world are carp (various species).[635] World carp production in 1989 was 4,008,000 tons, more than 10 times higher than the runner-up, milkfish, at 334,000 tons. Tilapia came in third at 325,000 tons. World rainbow trout production was estimated at 245,000 tons, while channel catfish production (almost exclusively produced in the United States) was 184,000 tons.

Perhaps in response to the National Academy study, the U.S. Congress included aquaculture as an appropriate subject for consideration by the Department of Agriculture and put language to that effect in the 1978 Farm Bill. Aquaculture became a recognized word on Capitol Hill and even in the Executive Branch of the government. During the administration of President Jimmy Carter, the Joint Subcommittee on Aquaculture (JSA) was formed. Membership on JSA involved over a dozen federal agencies, though chairmanship of the subcommittee was scheduled to rotate among the agencies with the greatest amount of involvement in aquaculture: the U.S. Department of Agriculture (USDA), the U.S. Department of Commerce (USDC), and the U.S. Department of the Interior (USDI). Those three agencies carved out their particular niches as follows: the USDC took the marine environment, the USDI covered inland recreational species and fishes that spawned in fresh water but migrated to the ocean (anadromous fishes), and the USDA took the inland species of commercial interest. We can assume that ornamentals and baitfish would come under the purview of the USDA, though that agency does not seem to have taken much interests in those aspects of aquaculture.

What was considered to be an important step was taken in 1980 when the National Aquaculture Act was passed by Congress. That act included a charge to the Secretaries of the Departments of Agriculture, Commerce, and Interior to establish a National Aquaculture Development Plan. The plan, in two volumes, was published in 1983.[636,637]

There was some discussion about which agency should take the lead in U.S. aquaculture, but by the time the National Aquaculture Development Plan was published, both the USDC and the USDI were treating aquaculture as a subject in which they were not interested. They maintained their hatchery programs, but expressed a strong lack of interest in commercial aquaculture. Use of the word *aquaculture* in agency publications was, according to people within the agencies, seen as objectionable. In the USDA, on the other hand, the phrase "aquaculture is agriculture" was frequently heard, and an office of aquaculture was established within that agency. The aquaculture office, initially manned by Bille Hougart, was literally a one-man show. There was not even a secretary assigned to that office. Things changed, of course, though not as significantly as one might imagine for *the* aquaculture office within USDA. Today there are three people within the USDA who are assigned to devote their time to aquaculture: Hank Parker, Meryl Broussard, and Gary Jensen. Each, I believe, has access to some level of secretarial assistance.

Money for aquaculture research in the universities has continued to be made available through the USDC's Sea Grant Program, and the USDA has supported aquaculture research through the land-grant system. A major step forward in aquaculture research supported by the USDA was taken in 1987 when the U.S. Congress approved a section in the Farm Bill that established four regional aquaculture centers. Each center was initially funded at $750,000 per year, which increased after the first year's funding was authorized (though never entirely appropriated). Designations of the first four centers and the states in which their administrative offices are located are as follows: the Northeastern Regional Aquaculture Center in Massachusetts, the Southern Regional Aquaculture Center in Mississippi, the Western Aquaculture Regional Center in Washington, and the Tropical Pacific Aquaculture Center in Hawaii. Several states were included within the jurisdiction of each center except the Tropical Pacific Aquaculture Center, which involved the state of Hawaii and the trust territories in the Pacific Ocean. Several states were not included in the first four centers, so in 1988, Congress established a fifth center, designated the North-Central Regional Aquaculture Center. Additional funding came with the creation of the fifth center so that all five received $750,000 per year for a total of $3.75 million. Funding was increased to a total of $4 million in 1992.

The aquaculture research centers have input from industry, researchers, and extension personnel within the regions in which they are located. Research projects supported through the centers are required to be regional in scope and must involve investigators from two or more states. Industry research needs are proposed by the industry committees; the technical committees (made up of researchers and extension workers) determine which of the industry's needs are researchable, and they work with the people who will be involved in the programs to develop solid proposals and ensure that information developed is extended to the industry. Collaborative projects between or among regions are authorized, though few have been proposed to date. Investigators involved with the research and extension projects come from not only land-grant universities, but also other institutions of higher education and interested governmental agencies. Industry is encouraged to assist by providing facilities and, when possible, financial or other types of direct support (such as feed or equipment) to the projects.

REGULATING AQUACULTURE

As the era of rapid commercial-sector aquaculture expansion arrived, the sky seemed to be the limit. Prospective investors heard tales of high, rapid returns on their money from fail-safe aquaculture ventures. While some of those stories were undoubtedly based in fact, there were also a cadre of "snake-oil salesmen" promoting ventures that were bound to fail. Investors who developed their plans carefully and were careful about the people with whom they associated were often able to establish successful ventures. Many of those who were drawn into the business were honestly interested in developing wholesome foods in the aquatic environment under controlled conditions. They were also interested in a profit, of course. In any case, those involved in aquaculture, whether as producers or researchers, would have indicated, if asked, that they considered themselves to be wearing the white hats. Fish and shellfish could only be reared successfully when environmental conditions were optimized. Pollution, from whatever source, was the nemesis of the aquaculturist. Furthermore, it is likely that most aquaculturists would have been proud to wear the label "environmentalist" in an era when that word could be interpreted as pertaining to those interested in the wise use and conservation of natural resources. Feeding people was a noble enterprise, and few involved in aquaculture during the 1970s envisioned the problems that they would face in later years.

I doubt that any of the 1970s aquaculturists were surprised to learn that the government was interested in regulating their activities. By the 1970s the U.S. government was developing environmental policies that required environmental impact statements prior to development of a number of activities. Aquaculture in public waters certainly required permission from appropriate agencies,[638] and regulations on water use on private lands were already in place in most areas. Aquaculturists were not generally put off by regulations, though in many cases the activity in which they wanted to participate was not around at the time many of the regulations had been promulgated, so agencies were often unprepared to deal with the new enterprise.

Florida, as mentioned above, responded to development of interest in shrimp culture by developing a state policy. Hawaii was the first state to produce an aquaculture plan. None of the other states had plans, and few had policies in place that would provide the prospective aquaculturist with a pathway through the regulatory system. As aquaculture expanded, horror stories related to the permitting process became commonplace. Some states seemed to have dozens of agencies from which approval had to be obtained before a commercial aquaculture enterprise could be established. In a few instances, landowners did not own the fish that existed on their land; thus, sale of the produce of aquaculture was technically illegal. In some states, it was illegal to harvest on Sunday or during certain months of the year (for example, harvesting oysters in months lacking the letter "r" has been widely prohibited). Since many fine laws that had been promulgated to protect fish and wildlife resources had been written before the era of commercial aquaculture, they sometimes impeded those who were attempting to go into business. Other examples include restrictions on the duration that groundwater could be removed. In some regions groundwater can only be pumped during a prescribed number of days that coincide with the traditional farming season. Being without water for a hundred or so days a year is not in the best interest of the fish farmer; with water tables already stressed in many parts of the nation, getting the laws changed or obtaining variances can be extremely difficult or impossible. Some states allocate water on the basis of priority uses, with agriculture among the highest priority and industry farther down the list. In the Rio Grande Valley of Texas, for example, water for aquaculture is classified an industrial use, and in agricultural regions there are sometimes so many users already in place that prospective aquaculturists can not obtain an allocation.

Before a fish farmer can begin producing a crop, it is necessary to obtain the necessary permits from the state, and in many cases from the federal government. This may be as simple in some inland locations as paying a small fee and jumping right into business, or it may involve months or years of intense activity with a number of agencies. Difficulties have been commonplace in some states, particularly with respect to aquaculturists who wanted to establish facilities along the coast. The example of a prospective algae culturist in Washington who spent 20 months and $100,000 and had to deal with 23 different agencies in attempting to obtain permits was not atypical.[639] In many instances permits were ultimately refused or even if granted, the prospective culturist had no money left to establish a facility.

Federal Permits

At the federal level, various positions have been taken relative to aquaculture permitting. The Environmental Protection Agency (EPA) has waffled back and forth on whether aquaculture facilities represent sources of pollution. In 1973 the EPA proposed that National Pollution Discharge and Elimination System (NPDES) permits be required of farms where the animals were being reared in raceways or were exotic species.[640] Discharges of pollutants into U.S. waters are regulated under the Clean Water Act (33 U.S.C. 1251 *et seq.*). Section 402 of the act requires that NPDES permits be obtained from the EPA prior to certain discharges. In June 1979, rules governing discharges from certain types of aquaculture facilities were published in the *Federal Register* (45[122]: 111). NPDES permits would be required for facilities rearing the following types of animals:

- cold-water fish species or other aquatic animals (e.g., trout and salmon) in ponds, raceways, or other similar structures that discharge at least 30 days per year but exempt[ed are] facilities which produce less than 9,090 harvest weight kg (about 20,000 lbs) of aquatic animals per year and feed less than 2,272 kg (about 5,000 lbs) during the calendar months of maximum feeding;
- warm-water aquatic animals (e.g. catfish, sunfish, minnows) that discharge at least 30 days per year but exempt[ed are] facilities which produce less than 45,454 harvest weight kg (about 100,000 lbs) of aquatic animals per year and facilities with closed ponds that discharge only during periods of excess runoff.

The above represents minimum standards established by the EPA. States may adopt more stringent regulations if they desire. In addition, the EPA may authorize a state to administer the federal program, as has occurred in Washington.[641]

The other major federal permit is required by the U.S. Army Corps of Engineers in cases where any type of structure is constructed in navigable waters. The basis for Corps permits is the Harbors and Rivers Act of 1899 (33 U.S.C. 403). The courts have liberally defined the term *navigable* to include nearly all flowing waters and all coastal waters within the borders of the United States. If a prospective aquaculturist plans to construct water intake or discharge structures, bulkheads, docks, or net-pen facilities, or if the aquaculturist plans to undertake dredging or other disruptive activities in waters defined as navigable, a permit application must be filed with the Corps. Other agencies, such as the U.S. Fish and Wildlife Service, are typically asked to comment on Corps permit requests to ensure that fish and wildlife resources will not be adversely impacted by the proposed activity.

State Permits

As indicated above, state permitting requirements vary considerably. In the South, the states of Florida, Georgia, Kentucky, Louisiana, Mississippi, Tennessee, Texas, and Virginia have not implemented regulations that are more strict than those imposed by the federal government.[642]

Net-pen salmon culture is currently being undertaken only in Washington and Maine. Controversy surrounding the establishment of net-pen facilities in Puget Sound, Washington, a state with an extensive permitting process in place that begins with County Council Shoreline Hearings Board proceedings, prompted the state to conduct an environmental impact assessment that addressed all the issues. It has been clearly demonstrated that properly designed and managed facilities can be operated in a manner that is nonpolluting and will have minimal impact on navigation, recreation, and other activities.[643,644]

Shrimp are currently being cultured in Hawaii, Texas, and South Carolina, so those states have regulations specifically relating to that type of culture (among others). Delays in obtaining permits have been experienced by prospective shrimp farmers because of potential impacts of pond effluents on coastal ecosystems.[641] Not all of the regulations relate to discharges of waste, of course. There are also land-use issues, use of the water column, leasing issues surrounding the use of submerged lands, the need to protect native species, regulation of bird predation on aquaculture facilities, exotic-species regulations, and regulations involving aquatic animal health. In many states the prospective aquaculturist must have approval of the state antiquities board or some similar body, which can ensure that construction of a facility will not disturb a historically valuable site, such as an Indian midden or burial ground.

Getting through the permitting process can be extremely difficult and frustrating. In most states no systems have been put in place to simplify the process of working through the system. In fact, maps for navigating the system are usually not available. Typically, there has been no designated agency or individual serving as a point of initial contact. The applicant often has to find his or her way through the permitting system with little or no assistance. That is beginning to change now as many states have designated aquaculture coordinators, often within the state department of agriculture. Whether or not a simple telephone call will put a permit applicant in contact with the state coordinator is, of course, another matter. Once contacted, the state coordinator may not be able to simplify the process, but can usually instruct the applicant on the nature of the system and point the applicant in the direction of agencies within the state that have to be approached and what each might require in terms of information.

The permitting process comes up again and again in surveys and in discussions among aquaculturists. The system has become increasingly ponderous as the number of involved agencies has expanded and as an increasingly strict regulatory environment has developed. On the one hand are those agencies (state and federal) that support aquaculture development. On the other are agencies that seem to be doing everything possible to derail the efforts of aquaculturists to conduct business. The frustration level and amount of money expended can be extremely high.

IDENTIFYING AND EXAMINING THE PLAYERS

As we have foraged through aquaculture's past in the pages of this book, we have examined a number of species that have received attention at one time or another. The intense interest in the culture of such species as shad and carp has waned, but new species have appeared on the scene. While those involved with terrestrial animal husbandry seem satisfied with the relatively small array of species they have available to them and have not added to the list in decades (perhaps with the exception of the buffalo, which continues to be an oddity, not a staple in the American diet), aquaculturists seem always to be looking for new species to culture. I'm not sure if the interest in new species comes from boredom within the scientific community and the desire to try something new (or to try the same old things on a different organism), if the commercial aquaculturists honestly feel that there is a need to expand the array of species being produced, or if some other motive or combination of motives prevails. In any case, one frequently hears about the need to develop so-called new species.

In some cases, the public perception of need is behind the notion of developing new species. Currently, for example, interest is being expressed in some regions of the nation with respect to producing marine fish to enhance depleted commercial and recreational fisheries. Recovery of striped bass on the East Coast and red drum in the Gulf of Mexico were both assisted through the stocking of millions of cultured fingerlings. Some interest in culturing such species as cod, lingcod, Pacific halibut, red snapper, seatrout, flounders, and others has been generated. Commercial culture to market size may be economically feasible with respect to some of those and to other marine species.

As we take a final look at the various species that have been the subjects of attention in the United States since 1970, we will see a few new faces (new fins and shells might be more appropriate) and many old ones. We'll also examine one or two species that arrived on the scene and have since departed.*

We'll begin our examination of species of interest and the updating of their status with the most important fishes, and then consider fishes of lesser importance. That is followed by invertebrate culture and a few words on algae. By importance I am

---

*I've sprinkled some information obtained from personal experience and some reflections on people with whom I've interacted in the pages that follow, though most of what is presented is documented in the published literature. Having been actively involved in various organizations that put me in frequent contact with my colleagues on the scientific side of aquaculture, I could certainly introduce you to many more people than you will meet in the rest of this chapter. My selection of those you do meet is somewhat arbitrary, stemming from something that jogged my memory while I was writing one section or another. I trust that neither those of you who are left out or included will be offended by whichever choice was made. I will say that I have made many friends from among the aquaculture community, both in the United States and around the world. It would not be possible to do justice to the accomplishments of all of those people, so I have not made the attempt.

referring to volume and value of the product, not some personal classification scheme. We start with one of my personal favorites.

Channel Catfish

Catfish culture can be found virtually throughout the United States today, but as the industry began to expand in the 1970s, most of the consumption as well as production occurred in the South. A report published in 1971 related catfish consumption with income level, occupation, religion, education, race, and location. At a time when per capita consumption of seafood in the United States was 11.4 pounds,[645] catfish consumption from February 1969 through January 1970 in various regions of the country was as follows:[646]

Region	Consumption (lb)
New England	0.000
Middle Atlantic	0.000
E. North Central	0.132
W. North Central	0.205
South Atlantic	0.153
E. South Central	1.858
W. South Central	0.718
Mountain	0.016
Pacific	0.030

It is obvious that there was little demand outside the deep South. Demand in that region was apparently sufficient to support a considerable industry. Catfish acreage had grown from 400 in 1960[645] to over 40,000 in 1970.[647] Acreages and production levels from 1965 to 1976 were as follows:

Year	Acres of Ponds	Production (lb)
1965	7,000	6,985,000
1966	10,000	10,975,000
1967	15,000	16,463,000
1968	30,000	32,935,000
1969	40,000	43,901,000
1970	40,400	39,910,000
1973	54,577	49,887,000
1975	54,997	61,860,000
1976	55,978	69,841,000

The bulk of the facilities that existed in 1970, a total of 40,000 acres, were located in Alabama, Georgia, Kentucky, Mississippi, North Carolina, Tennessee, and Virginia. With improved feeds and methods of managing catfish ponds, good producers were averaging about 1,500 pounds per acre annually,[645] though if you do the quick math with the figures above, you'll see that overall production levels had improved from just under 1,000 pounds per acre in 1965 to nearly 1,250 pounds per acre in

1976. That assumes, of course, that all the acreage was devoted to foodfish production. In reality, a portion was used for holding broodfish and for the production of fingerlings, so merely dividing the total weight of fish harvested by the acreage of available water provides a low estimate of production per acre. By 1980, production levels in excess of 3,000 pounds per acre were common, and by providing supplemental aeration, up to 12,000 pounds per acre could be produced.[648] A summary of economic evaluations for various years from 1969 to 1977 showed that, depending on assumptions of food conversion, operating cots, and sales price, a catfish farmer could anticipate a return on investment ranging from 11 to 20% annually.[647]

While the majority of the catfish industry was distributed around the South, the bulk of production was centered in Mississippi and nearby Arkansas and Alabama. Large processing plants, mostly in Mississippi because of that state's dominance of the industry, handled the bulk of the fish being harvested. Some farmers processed their own fish or worked with small local processors, who in most cases were also producers, but most sent them to the big processors.

Beginning in 1979, *Aquaculture Magazine* began publishing information on the round weight of catfish processed, inventory levels, and price paid per pound to producers. In 1979, prices paid between January and September ranged from $0.52 and $0.65, with the price rising monthly until the peak was reached in July.[649] During the early years of the industry, the vast majority of the fish was harvested in the fall because the culture process involved stocking ponds with fingerlings in the spring and rearing them to market size during a single growing season. Ponds were drained during harvesting and restocked the following year. Farmers could carry their fish over the winter and harvest them in the spring to take advantage of higher prices, though the fish could be expected to lose some weight during winter, and the feed costs required to get them back to market weight could wipe out the advantage of the higher prices paid by the processors. As the industry matured, the strategy changed to one known as "continuous harvesting." Jim Avault is one among various individuals who have described the technique.[650] Basically, ponds were initially stocked with about 4,000 fingerlings per acre and seined periodically to remove marketable fish. Fingerlings were then restocked so the total population of fish in the pond remained fairly constant. The technique, once adopted by the industry, continues to be in effect. Many ponds have been kept in production for a period of years. Water levels are not reduced during harvesting, but new water is added as necessary to keep the ponds full. One modification in the technique is that it is now common for farmers to stock 5,000–10,000 fingerlings per acre in their growout ponds.[651] The size of the fingerlings stocked is typically split between 5–6-inch fish and 8–10-inch fish. I spoke with some catfish farmers in Alabama during 1994 who indicated that their ponds had been in production for as long as 15 years. A study conducted in 1985 concluded that the practice of not draining ponds did not contribute to a deterioration in water quality, at least over a period of three years.[652] (Fifteen years of production without draining and drying out the pond bottom to oxidize organic matter seems a bit excessive in my estimation.)

The following table summarizes catfish production and price data from 1980 through 1993:[653–665]

Year	Weight Processed (lb)	Price Range ($)
1980	46,464,000	0.65–0.69
1981	61,128,000	0.56–0.70
1982	99,398,000	0.53–0.59
1983	137,250,000	0.55–0.65
1984	154,255,000	0.61–0.74
1985	191,616,000	0.67–0.78
1986	213,756,000	0.61–0.74
1987	280,496,000	0.57–0.69
1988	295,109,000	0.68–0.80
1989	341,900,000	0.64–0.78
1990	360,435,000	0.73–0.79
1991	390,870,000	0.53–0.69
1992	457,367,000	0.53–0.63
1993	461,220,000	0.63–0.73

Note that the price paid to catfish farmers was as low as $0.53 cents per pound during 1991 and 1992. The lows were in December of 1991 and January of the following year. You may recall that H. S. Swingle, in his paper of 1957, considered that a profit could be made if the fish were marketed at $0.50 per pound.[500] Improved feeds, much high stocking rates, water-quality management, disease prevention, and other factors have made catfish farming much more efficient in the nearly four decades that have passed since Swingle's paper was published. The inflation that has occurred during that period can be measured in the hundreds of percent, while even at the highest prices paid by processors historically ($0.80 per pound) the increase over the Swingle value is only 60%.

With the increases in acreage and numbers of farmers producing catfish, the amount of money that farmers received for their product increased rapidly. Farmers received $1,197,000 in 1969. The amount reached $83,723,000 in 1983.[666] By 1993 there were 461 million pounds processed, and if we assume an average price of $0.68, the farmers received over $313 million.

Channel catfish farming was not, of course, without its problems. In addition to consumer acceptance, there was considerable competition between the domestic catfish industry and imported catfish.[645] Catfish from Brazil, which were not closely related to channel catfish, were being captured from the Amazon River system and shipped to the United States, where they were sold at very competitive prices. Imports were frequently mentioned as contributing 40–50% of the catfish in the U.S. market in the 1970s, and there were rumors at one point that plans were afoot to reduce the price of imported catfish to the point that domestic producers would be driven out of business, after which the price would be raised and imports would control the market. That never came to pass because calls for better marketing were heeded by the industry through an advertising campaign asking consumers to eat farm-raised catfish.

A significant problem that hampered development of the catfish industry was off-flavor. The source of the problem, which gives the fish what has been commonly called an earthy-musty flavor—actually, they just taste like mud: they're nasty—is attributable to the presence of blue-green algae blooms. The algae produce a chemical called geosmin, which is the source of the muddy flavor. Catfish store the chemical in their flesh but metabolize it within a few days. Thus, once the algae blooms die out, the fish return to their excellent normal flavor fairly quickly. High stocking rates, which lead to increased nutrient levels, and warm weather combine to promote the algae blooms. No effective means of preventing establishment of the blooms has been found. The problem occurs today, according to discussions I've had with workers in processing plants, in at least 50–60% of the ponds sampled on a given day. Sometimes the percentage is even higher.

Once you've eaten an off-flavor catfish, you may wait a long time before trying catfish again. Many potential repeat customers have been lost because of the problem. The solution that was adopted involves sampling ponds in advance of harvest. A fish is captured at random and is then taken to the processing plant for evaluation. A quality-control person in the plant cuts the tail off the fish, places it in a microwave oven for a couple minutes, and then tastes it. Fish might be sampled once or twice in the days preceding a scheduled harvest, and a final sample is taken when the fish reach the processing plant. If no off-flavor is detected, the fish are allowed to be processed. If off-flavor is detected, the pond is restricted from harvest until the problem is resolved. If left in the ponds the fish may remain off-flavor for several weeks. Alternatively, the fish can be placed in uncontaminated well water for a few days, but few farmers have the ability to take that approach, so most simply wait until the algae bloom collapses and the geosmin in the fish is metabolized.

Before we look in more detail at the development of the Mississippi industry, which dominates in U.S. catfish production, let's take a look at some of the activities that occurred as the industry was growing.

The Demand Feeder Episode. Demand feeders are activated by the fish. They are containers filled with feed and suspended over ponds or raceways. Typically, a metal rod hangs down through a hole in the bottom of the feeder and descends to just below the water surface. When a fish bumps into the rod, a mechanism is activated that allows a few feed pellets to drop into the water. Catfish (and other species) quickly learn how to activate such feeders. Even relatively small fingerlings can be fed with the devices. When I arrived at Stuttgart, Arkansas, in 1969 to work for the summer, Waldon Hastings, Dewey Tackett, and Bill Simco were involved in an experiment designed to examine how catfish in a pond were using a demand feeder. Wald and Dewey were U.S. Fish and Wildlife Service employees who worked at the Stuttgart lab, and Bill Simco (who became, and continues to be, one of my best friends) was on the faculty of Memphis State University (now the University of Memphis). An engineer from the University of Arkansas, Bill Hinson, had worked with Dewey, as I recall, to design and build a mechanism that would allow the researchers to obtain a mark on an event recorder each time the feeder was activated. Basically, the rod that extended from the bottom of the feeder passed through a metal ring that acted as a switch. When the rod was activated, it would touch the ring and close a circuit that activated the event recorder. The device is described in a paper published in 1972.[667] Data collected in 1970 are presented in the paper, which demonstrates that the fish fed most actively between 1800 and 2400 hours. (This is not a good time to feed fish, incidentally, since that's when dissolved oxygen levels are starting to fall, and increased metabolic activity of the fish during and after feeding can accelerate the rate of oxygen consumption.)

You may notice that I saw Wald, Dewey, and Bill working with the device in 1969, but the data used in the paper to which they contributed weren't collected until 1970. That's because there were a few glitches during 1969. It seems that one or more of the following problems had to be solved before reliable data could be collected:

- Sometimes the rod didn't swing far enough to close the circuit.
- Splashing fish often got water on the ring, so feed would pile up, get wet, and clog up the works. As a result, the circuit either would remain closed or couldn't close.
- Various failures in the electronics occurred.
- The wind could move the rod and activate the mechanism.

The problems were addressed throughout the summer of 1969 but weren't completely resolved before the end of the growing season, so the study had to be completed the following year.

Intrigued by the commercial demand feeder, Bill Simco and I decided to design and build our own model and use it over an indoor fingerling raceway. We started with a design that involved inverted glass cider bottles with metal rods suspended from a hole in the cap. That design didn't work very well, so we moved to one fashioned from plywood. The wooden models each had a chain protruding through a hole in its bottom as an activating device. We were pretty proud of ourselves and decided to write a manuscript describing our simple and inexpensive feeders. We were told (I think by Kermit Sneed) that the *idea* of the demand feeder had been patented and that lawsuits were being threatened by the patent holder against commercial manufacturers who had developed competing designs. We kept our feeder design to ourselves and went off doing other things.

Cage and Raceway Culture. The notion of raising catfish in cages, first mentioned in Chapter 4, received additional study after 1970. Charles M. "Bo" Collins (a different Collins than the one who suggested use of cages in the first place) got involved with catfish cage culture in Oklahoma during the early 1970s,[668] and John Kelley conducted research at Texas A & M University during the same period.[669] Various studies involving catfish cage culture were conducted in Arkansas during the 1970s, and some commercial ventures were established.[670] Cages worked pretty well under certain circumstances. For example, if a pond could not be seined for one reason or another, confining fish in cages made sense.

A good deal of research was conducted on the use of cages in power-plant lakes to take advantage of the heated water effluent that could be used to extend the growing season. John Kelley's cages were used in a Texas power plant for just that type of study from about 1973 through 1976, and Herb Simmons and I began some work at another power plant with cages of his design a few years later. The latter cages were then used for holding fish in conjunction with a nutritional study conducted by Bill Yingst in a reservoir at the Texas A & M Aquaculture Research Center.[671] While I was at Southern Illinois University, Phil Moy and I raised some catfish in cages floated in a strip-mine lake.[672]

Also during the 1970s, the state of Arkansas developed a unique program that allowed private fish farmers to lease portions of some of the state lakes for commercial production of fish in cages. The program was administered by the Game and Fish Commission, and one of the operations was established in Mena Lake.[670] I got involved with that activity in late 1978, when the caged fish began developing broken backs. The fish grower called the feed company from whom he was purchasing what he had been told was a feed that contained all the required nutrients, and indicated that the problem was with the feed. A representative from the feed company got my name from somewhere and contacted me. I went up to Arkansas to examine the situation. Sure enough, a good percentage of the fish had deformed or broken backs, a clear sign of vitamin C deficiency. I learned that the feed formulation was about 20 years old and that the feed had been designed for use in lightly stocked ponds, not in highly intensive cage culture. It contained a low protein level and no added vitamins.

In an attempt to keep the problem from developing further, I turned to the nineteenth-century approach and asked the farmer to feed chicken or beef liver, which would provide the necessary vitamin and might be accepted by the fish. (One of the first things that happens when fish get sick is that they quit feeding; I thought they might go for fresh organ meats, however.) I then reformulated the feed by adding a complete vitamin mix and increasing the protein level. I made a few other changes as well, and the feed subsequently was found to be perfectly suitable for use as a complete diet. It cost the feed company a bundle, however.*

The interest in cage culture has abated to some extent, but it has not died completely, nor is it restricted to channel catfish. Other species that have been grown in cages include rainbow trout, striped bass, tilapia, and recently, black bullheads.[673]

After Jim Andrews at the Skidaway Institute of Oceanography demonstrated that catfish could be reared in circular raceways at densities of a few pounds per cubic foot of water, some commercial tank farms that used flow-through water were established, though most were unable to survive when fuel prices rose dramatically in the early 1970s. The major problem involved pumping costs to move water through the tanks. The expense of pumping, which was greatly compounded in instances where the water also had to be heated, made such facilities unable to compete on an economic level with pond operations. The technology for recirculating water systems was being developed, but commercialization was not possible. As we shall see, that situation hasn't changed a great deal.

Linear raceways have also been used for catfish rearing, and in cases where water of the appropriate temperature is available at little or no cost, such systems have been quite successful. Leo Ray's catfish farm near Buhl, Idaho, which was established in the 1970s, is perhaps the best example. I had known Leo for a few years through our interactions at some of the meetings we both attended, but didn't have an opportunity to visit his farm until about 1980. A student of mine was doing an internship at Leo's place that summer. I always tried to visit the students during their internships and jumped at the chance to see Leo's operation. When I arrived, he told me we'd need five days to see everything. I didn't see how we could keep busy for that long, but after he had toured me through his catfish and trout farms, taken me around the Hagerman Valley to see various other trout facilities, showed me processing plants and a feed mill, and conducted a rafting trip down a portion of the Snake River, the time was gone.

Leo had been on vacation from his catfish farming activity in California early in the 1970s when he passed through the Hagerman Valley of Idaho. He learned that, not only was there an abundant supply of cold water available (the trout industry was already established and growing rapidly), but geothermal water could also be had, often without pumping. He found a piece of land that ran down the side of a moderately steep hill. Relatively shallow wells provided Ray with several hundred gallons a minute of artesian 106°F water and the cold water he required to maintain the temperature to which he wanted to expose his fish: 82°F. He constructed a series of concrete raceways down the side of the hill. The catfish raceways are in parallel rows of four. Water exiting each series flows through a couple hundred feet of open channel where it is aerated before entering the next set of raceways. There are four sets of catfish raceways. A fifth series of raceways below the catfish units contains tilapia, which remove a lot of the pariculate matter present in the incoming water. After exiting the tilapia raceways the water flows into a settling pond before being finally discharged into the Snake River. Ray has a sufficient supply of hot water to grow something like a half million pounds of fish per year plus heat his home and swimming pool. He has assisted other farmers in establishing similar facilities where suitable water supplies are available. As one travels around, facilities with Leo's particular design stamp on them are readily identifiable, not only because of their appearance, but because they work.

Mississippi Leads the Way. The Mississippi industry began to dominate in terms of acreage and production beginning in the early 1970s. Acreage in that state has increased every year with the exceptions of 1977 and 1984. Economic conditions during 1982 and 1983 forced some farmers to ultimately quit the business even though prices paid by the processors were pretty good. In general, farmers who owned their land were able to remain in business, while some who had borrowed money at high interest rates were forced out because they could not make their loan payments. In 1984, there were 373 catfish farms in Mississippi. Total water acreage was 64,822, with 7,307 of those acres devoted to fingerlings and the remainder to foodfish production.[674] Nearly 9,000 acres of ponds were constructed in 1985, taking the total to 73,578 acres.[675] Additional expansion in 1986 lifted the acreage to 85,193.[676] In 1991, the 100,000 acre milestone was passed. In May of that year, there were 102,239 acres of catfish ponds in Mississippi.[677]

Other states were also adding catfish acreage. By 1991, as Mississippi was approaching 100,000 acres, the nation's total acreage was placed at 162,000.[678] Arkansas, which was second to Mississippi in acreage, had a total of 20,000 acres devoted to catfish, and continued to lead the nation in baitfish production with 27,200 acres.[679] In 17 states where catfish were grown, there were 1,818 farms as of January 1, 1991.[678] A survey conducted in the northeastern United States indicated that there were 263,000 pounds of catfish produced in that region during 1992.[680] Those fish were produced exclusively in the states of Maryland and Pennsylvania.

Nationally, the numbers of fish involved in the catfish industry is prodigious. Hatcheries were maintaining 1,340,000 broodfish in 1991 and had a January inventory on hand of 553 million fingerlings that would be available for stocking later in the year. Producers held inventories estimated at 239 million marketable fish on January 1, 1991.

Continued growth of the domestic catfish industry is unlikely to be anything like what we have seen in the past. The seemingly limitless water supply in the Mississippi Delta is certainly not infinite, and drawdowns are beginning to occur. Prodigious growth of catfish farming in Mississippi is unlikely, and because of limitations on water and land availability in other states, whether more than modest growth overall is possible remains to be seen. The catfish farming industry is certainly a major success story. Through aggressive marketing and promotion of what is considered by consumers to be an excellent product, the industry has grown rapidly and is strong. People are beginning to look once again at the idea of growing catfish in nations other than the U.S., and there has been some movement to establish farms in places like Latin America. The assumption is that the market for catfish will continue to grow and that domestic production capacity will be exceeded.

Economics of Catfish Culture. A number of studies examining the economics of catfish culture have been conducted in various states. John Waldrop at Mississippi State University has been responsible for several such studies. One of the more recent ones which he co-authored was a 1988 analysis that I have elected to summarize here.[681] Hypothetical catfish farms of three sizes were considered. Basically, those farms had 8, 16, or 32 ponds of 20 land acres (17.6 water acres when the levee area was subtracted). If we look at the medium-sized operation for purposes of discussion (the facility with 16 ponds), we find that basic investment costs are as follows:

Item	Cost ($)
Land	258,400
Pond construction	225,842
Water supply	61,400
Feeding equipment	23,225
Disease and weed control equipment	3,340
Miscellaneous equipment	268,101
TOTAL	840,348
Investment per water surface acre	2,961

The miscellaneous equipment category includes tractors, trucks, a service building, shop equipment, office equipment, a computer with printer, dissolved oxygen meter, aerators, a mower, seines, and so forth. Annual ownership costs are estimated at $125,843. Included in those costs are depreciation of ponds and equipment, interest payments, taxes, and insurance. Annual operating costs are as follows:

Item	Cost ($)
Repairs and maintenance	21,954
Fuel	44,232
Chemicals	30,357
Telephone	2,400
Water-quality test kits	490
Fingerlings	82,386
Feed	312,939
Labor	97,000
Harvesting and hauling	51,092
Liability insurance	3,252
Interest on operating capital	30,070
Interest on fish inventory	3,257
TOTAL	679,429

Given the above, the total annual cost of owning the facility would be $805,272. It is estimated that the ownership cost per pound of fish produced would be $0.65, so any price paid for the fish above that cost would represent profit. Given the prices paid by the processors, it is clear that during some periods in recent years, production costs have exceeded the value of the fish. It is also clear that in only a few instances have profits margins exceeded a few cents a pound. The only way farmers could make a reasonable profit would be by having large farms or having other sources of income. Many catfish farmers have diversified farming operations. Many also belong to feed and processing cooperatives that provide them with financial returns based on the amount of feed sold and fish processed.

Trout

There is much less information available on the commercial trout industry compared with the channel catfish industry. While trout are cultured in nearly every state, the majority of the production, as we have seen, occurs in Idaho, and specifically in the Hagerman Valley. As there are only a handful of major producers in Idaho, data on production, numbers of fish, costs of production, and so forth are not readily available. (The farmers are not very generous with that information.)

Idaho has long been the major player in trout production. Back in 1928, Jack Tingey, who had been with the Utah State Fish and Game Department, recognized that the 58°F water of the Hagerman Valley might be ideal for producing trout and began the Snake River Trout Company.[682] Warren Meader of Pocatello, Idaho, was apparently shipping as many as 60 million eggs a year by 1940. By 1950, there were eight farms producing nearly 1 million pounds of product annually. Production had grown to about 2 million pounds by 1960. The Hagerman Valley farmers were dependent on eggs supplied by producers in Washington, California, and other parts of their own state. For example, Trout Lodge, established around 1950 near Ephrata, Washington, was producing and shipping more than 50 million eggs a year by 1970.[683] That company continues to be a major trout egg supplier today.

Marketing problems have long plagued the trout industry and were responsible, in part, for its relatively slow growth until at least the mid-1960s. While cultured trout were available year-round in various product forms, fish brokers were unfamiliar with the product and could obtain more highly desirable fish like halibut more cheaply.[682]

Things changed dramatically beginning about 1965, and trout farming began to grow at a more rapid rate. Reasons behind the increased demand for trout seemed to have been related to increased demand for catchable fish by recreational fishermen and a higher demand in the United States for seafood, in general. With demand increasing, the trout industry began an intense marketing campaign which worked quite well, thus opportunities for new farmers, as well as for growth of existing farms, were created. Trout farming in Idaho grew at a prodigious rate, and significant amounts of production could also be found in California, Colorado, Montana, Missouri, and in the southern Appalachian states. Fee fishing became popular in many areas, and the demand for fresh fish accelerated.

Advances in feed formulation and manufacture, improved technology that reduced labor costs, and the use of better-trained personnel on the large trout farms were among the things that improved efficiency and made farmed trout competitive with other seafood products.[682] One of the major trout companies in Idaho, Clear Springs, made large investments in a processing plant and refrigerated trucks in the 1970s. That company also began contracting for additional fish with small local farmers, a mechanism that was subsequently adopted by other growers in Idaho and Montana. Some of the large farms established their own research laboratories and hired trained scientists to conduct studies aimed at solving problems associated with diseases and other aspects of culture. About 80% of the rainbow trout being served in the nation's restaurants by 1971 were produced in Idaho.[684]

By about 1980, the trout farming industry seemed to have peaked. In 1981, it was estimated that no more than 35 million pounds of trout were being produced annually. That amounted to less than 2% of U.S. fresh and frozen seafood sales. At the same time, catfish production was about double that of trout[654] and, as we have seen, was growing rapidly.

The handful of major producers in Idaho currently employ concrete raceways and massive volumes of water to maintain rapid turnover rates in those raceways. Smaller Idaho producers, responsible in the mid-1980s for about 8 million pounds of trout a year,[685] generally employ pond culture. Ponds often have concrete sides and gravel or bedrock bottoms, though earthen levees are also sometimes seen. There are also ponds constructed entirely of concrete. The difference between such ponds and raceways relates to flow rate. Pond culturists do not have the large volumes of water seen in the Hagerman Valley, so they rely on slow water exchange rates and a single annual crop rather than the three crops per year being obtained by raceway culturists.

While growth of the industry has occurred outside of Idaho, the large farms have not expanded significantly, nor have new facilities been constructed recently. Total production increase has not, as a result, been very spectacular.

Some people have expressed the view that the large producers were making good money and saw no need for expansion, but there has also been the view that the trout market is limited and perhaps saturated, at least with the products that have traditionally been available.

Pan-sized trout, grown primarily to provide individual portions in restaurants, are, in my opinion, too small, mushy, and far too bony compared with many other types of fish. A large percentage of those fish have been marketed gutted with the heads and fins on. The American consuming public generally prefers filleted fish and not fish that stare back at them while being consumed. In recent years the industry has begun to respond to consumer demand by moving increasingly to filleted and deboned products. Still, the trout industry lags far behind catfish.

Production figures for the period 1986 through 1991 have been collected by the Western Regional Aquaculture Center for the states of Alaska, Arizona, California, Colorado, Idaho, Montana, Nevada, New Mexico, Oregon, Utah, Washington, and Wyoming. That survey also projected 1996 production levels for each state. Among the states listed, only Alaska produces no trout. Annual figures for all states in the region combined from 1986 through 1991, with projections for 1996, are as follows:[686]

Year	Production (lb)
1986	33,155,000
1987	35,961,000
1988	42,918,000
1989	48,507,000
1990	45,725,000
1991	50,433,000
1996 (projected)	54,375,000

Data from 1992 show that 4,721,000 pounds of trout worth $12,862,000 were produced in the northeastern states of Pennsylvania, Massachusetts, and New York.[680]

The value of imported trout during the first six months of 1993 totaled $3.4 million, which was up 77% from the previous year.[687] Most of the imported fresh trout came from Canada, with some also entering the United States from Mexico. Frozen imported trout came predominantly from Chile. At the same time, the United States was also involved in exporting trout. Exports, primarily to Canada, of frozen trout amounted to $855,000 during the first half of 1993, which was a 6% increase over 1992. Fresh-trout exports amounted to $812,000 for the same period in 1993, and there was a small amount of live fish exported. Total exports were just over $2 million, so there was a trade deficit in trout developing in 1993. Exports and imports had nearly balanced in the period from 1990 through 1992.

Salmon

What appears to have been a final attempt to introduce Pacific salmon to the East Coast of the United States occurred beginning in 1969 when the New Hampshire Fish and Game Department stocked coho salmon smolts in an attempt to initiate a sport fishery. Further stockings of fish obtained from Washington and Oregon hatcheries occurred in 1970 and 1971. Returns to the fishery were monitored, and a report was issued in 1974 in which the opinion was expressed that continuation of the program would be dependent on continued stocking.[688] Whether because interest waned, resistance from opponents of exotic species developed, money for the program ran out, or for some other reason or combination of reasons, the program does not seem to have been continued.

Following the development of experimental net-pens,[531] the National Marine Fisheries Service (NMFS) constructed a research facility at its Manchester, Washington, laboratory on an embayment off Puget Sound. The native coho salmon was the first to receive the attention of the NMFS scientists, and good growth was being reported in early work conducted during 1970.[689] Commercial interest developed almost immediately. In 1971, the Washington Department of Fisheries granted a permit for the establishment of a commercial net-pen facility adjacent to the NMFS net-pens at Manchester.[690] The commercial venture intended to produce coho salmon and indicated that the production time in net-pens would be about 14 months.

The commercial net-pen facility at Manchester, called Domsea, changed hands a few times during its 15-year history. For part of the time it was in operation, it was owned by the Campbell Soup Company. Ultimately, the net-pens covered about 5 acres of water. There were 250 individual pens, each about 20 times 40 feet in area and 10–12 feet deep.[691] The pens were surrounded by wooden decking that allowed personnel access to them and were sufficiently large to accommodate small storage sheds and various types of equipment. Pan-sized coho salmon (maximum size of about 1 pound) were produced in the net-pens from 1972 to about 1990. Operators of the facility saw their primary market niche as being the same one targeted by the rainbow trout industry. Therefore, they continued producing small fish and did not try to compete in the fresh salmon market which demanded fish of several pounds. Domsea was virtually alone in the salmon farming industry for a number of years. By 1980, for example, there was only one other net-pen farm in the state of Washington.

Ocean ranching involves the release of juvenile salmon into marine waters where they will grow to maturity, and capture of them once again upon their return from the ocean. That pattern provides the basis upon which salmon have been used to augment commercial and recreational fisheries almost since the first days of the U.S. Fish and Fisheries Commission in the 1870s. The establishment of private ocean-ranching operations in recent years has provided what some would call "a new wrinkle" for commercial culturists, but the idea was tried previously. There was one private ocean-ranching operation on the West Coast in the 1870s, and early in the twentieth century the government had provided tax incentives to salmon processors who operated hatcheries.[692]

The idea of ocean ranching resurfaced in the 1970s. Oregon and Alaska ultimately established procedures by which private companies could obtain permits for salmon ranching. Approval for that activity was obtained from the Oregon legislature in 1971.[693] Commercial salmon ranches began with chum salmon, but in 1973 the industry convinced the legislature to allow the ranching of coho and chinook salmon as well. Pink salmon were added to the list in 1979.

Alaska developed a not-for-profit salmon ranching scheme wherein the operators were compensated by commercial fishermen who captured returning ranched fish in the immediate vicinity of the hatcheries (which were located in areas where natural salmon runs were weak or did not exist). Money collected had to be used for expenses and facilities construction, but the salmon ranchers could not show any after-expenses profit.* Washington considered the concept of salmon ranching but opted to allow salmon production only in systems that contained the fish throughout the rearing period.[691]

Salmon ranching had some inherent problems that needed to be worked out. Both at sea and during spawning runs, ranched fish become intermixed with wild fish and fish released from government hatcheries. All of those fish are available to sport and commercial fisheries, sometimes long before they reach their home waters. Thus, by the time the fish return to the private hatcheries, their populations may have been decimated. In Oregon, indicating that releases of public and private hatchery fish had not increased the supply of adults (private salmon ranchers had released 14 million coho in 1980, and the state had released 18 million), the Fish and Wildlife Commission placed a five-year moratorium on new salmon-ranching operations beginning in 1981.[694] Pressure was being placed on the state by environmentalists and recreational fishermen who held the common opinion that ocean-ranched fish could compete with wild fish for food and that if the two types of fish managed to interbreed, genetic degradation could occur. Commercial fishermen held similar views and were, in addition, afraid that successful salmon ranching could put them out of business.[691]

Twenty salmon-ranching permits were issued in Oregon between 1971 and 1978, but as pressure from fishing and environmental groups increased and regulations became increasingly strict, the number of active salmon ranches began to decrease. By about 1990, the last of them were forced out of business.

When I arrived in Washington in 1985, NMFS biologists were becoming involved with Atlantic salmon spawning. Various strains of Atlantic salmon from the East Coast had been shipped to Washington, where the nation's facilities and expertise in salmon culture were centered. The approach used by NMFS involved captive breeding and, ultimately, the restocking of Atlantic coastal regions. In the meantime, Norwegian salmon farming (exclusively with Atlantic salmon) was developing into a major industry. The Norwegian approach had been to harvest fish at three years of age and several pounds in weight. Far superior to the pan-sized coho salmon being produced in the Pacific Northwest, Atlantic salmon brought much higher prices and, even with the longer growout period, held a much larger profit potential.

Interest in farming Atlantic salmon in Puget Sound developed as NMFS biologists reported good growth with that species in Washington State waters and as fish from Norway began to flow into the United States in large numbers. By the late 1980s, fresh salmon was available in virtually all the major U.S. cities and in many smaller population centers. At the same time, demands for fresh salmon in Japan turned much of the Alaska production away from the canning that had dominated the industry for generations. Americans, most of whom had never tasted fresh salmon,

---

*The incentive comes from the fact that personnel running the facility can profit quite well. Their salaries are negotiated with the state. Having not-for-profit status means that salmon ranching operations are not places where investors can speculate.

discovered a new delicacy and soon found it available in grocery stores and restaurants across the country.

Norwegian firms, looking for areas in which to expand, began seeking permits in Washington. Ultimately, between 10 and 15 net-pen farms were established. The Domsea net-pen facility at Manchester was sold to a Norwegian firm in about 1990. The old net-pens were removed and replaced with a much smaller number of Norwegian-style pens. The new facility has abandoned the idea of producing pan-sized coho salmon and are exclusively growing Atlantic salmon.

The typical net-pen facility in Washington today covers an average of about 2.5 acres of water, and there are currently about a dozen such facilities in operation. Coho salmon are being produced in modest quantities in Idaho. That production is, of course, in fresh water. California utilizes land-based culture systems for the production of salmon and steelhead. A summary of salmon production in Washington and Idaho and for salmon and steelhead from California from 1986 through 1991 is as follows:[686]

Production (lb)

Year	Washington	Idaho	California
1986	2,562,000	150,000	Not available
1987	3,133,000	250,000	Not available
1988	5,224,000	300,000	550,000
1989	4,000,000	500,000	875,000
1990	3,133,000	500,000	600,000
1991	5,550,000	300,000	310,000

Production of pan-sized coho in Idaho is not expected to continue in the future.

In Maine, where the industry is still being promoted and expansion is allowed, production of Atlantic salmon and steelhead reached 13,500,000 pound in 1992. The value of those fish was over $42 million.[680]

In Washington, opposition from a few upland property owners who object to what they call visual pollution in association with net-pens was the basis for much of the controversy that has developed in the Puget Sound region. Since the law does not provide an upland property owner with an unalterable view, the opponents couldn't stop the expansion of the salmon industry on the basis of what they thought was the establishment of eyesores. Thus, the upland landowners, who obtained support from commercial fishermen, began citing excessive noise, odors, pollution, release of antibiotics into the environment, interference with recreational and commercial fishing, interference with navigation, and a number of other reasons why net-pen aquaculture should be curtailed.[643,644,695] Opponents packed public hearings and demanded that the permit applicants provide one study after another to prove that their activities would not cause environmental damage or interfere with various public activities. The permitting process became so long, arduous, and expensive that many applicants either gave up before completing it or had no money left to exercise the permits that they ultimately obtained.

While the expansion of net-pen culture in Puget Sound became bogged down in the quagmire of the permitting process, salmon farming continued to expand in Norway and was initiated in Scotland, Chile, and Canada. The state of Maine, as mentioned, also became involved, and while there was a considerable amount of opposition in that state as well, the industry has become established and continues to grow in a controlled fashion.

Problems not associated with opposition began to occur, particularly in Norway and Canada, during the late 1980s, when it became apparent that the industry had become overcapitalized. Banks began calling in notes from companies that had yet to get their fish to the normal market size. Norway was accused of dumping tons of fish on the market, and everything together caused a major price reduction that made it impossible for many new firms to compete. Canada's industry suffered greatly, with many companies being forced into receivership. Undersized fish were marketed, the price continued to fall, and producers around the world went from excellent profits to marginal profits, and in many cases, considerable losses. In 1992, the U.S. government, which was busy negotiating a free-trade agreement, slapped a tariff on imported salmon from Norway causing that country to further develop its European markets and move into the Asian market.

As the marketplace has reorganized and restructured, salmon prices have risen once again, though they have yet to reach their pre- 1990 levels. Some of the U.S. net-pen farmers have reduced staff, but new farms are not being added. The permitting climate is not conducive to additional development at this time. The industry continues to grow in Maine, and the Canadian industry is recovering. Chile has become a major player in salmon farming and salmon ranching and may soon assume the dominance in the American market that was once held by Norway. During the first half of 1993, imports of Atlantic salmon amounted to 32 million pounds, up 24% from the same period the previous year.[687] The sources of 96% of the imported Atlantic salmon were Canada and Chile.

The overwhelming percentage of salmon imports to the United States are currently in the form of Atlantic salmon.[687] Farmed chinook, and to a lesser extent coho salmon, are also imported. Exports of salmon continue to be dominated by frozen sockeye destined for Japan.

Tilapia

Tilapia was mentioned a few times in Chapter 4 and in conjunction with the catfish farming operation of Leo Ray in Idaho earlier in this chapter. Tilapia had apparently been introduced into the southeastern United States during the 1950s,[696] probably as a weed-control species, and to Hawaii in 1952 as a potential baitfish for skipjack tuna, but it was relatively unknown, even within the aquaculture research community, until the 1970s.[697] Only a bit of research had been conducted on it prior to 1970, and there was no commercial culture of tilapia in the United States at that time. Auburn's research in the 1950s and 1960s laid the groundwork for the introduction of tilapia by that university into several nations in conjunction with its International Center for Aquaculture. Tilapia culture facilities were established in Africa, Asia, and Latin America with support from the U.S. Agency for International Development projects.

Tilapia apparently appear in hieroglyphs in the tombs of the Pharaohs, though actual culture of the fish seems to have been initiated in Kenya in the 1920s.[698] I had never even heard of them prior to accepting a faculty position in the Department of Wildlife and Fisheries Sciences at Texas A & M University in 1975. A power plant at Trinidad, Texas, was built on a reservoir in which research was being conducted by faculty and students in the Department. The reservoir, which was used as a cooling water source for the power plant, had somehow been stocked with tilapia. That stocking incident, which may have involved only a few fish (at least two, we must assume), led to the establishment of a population of 6 million tilapia in 700 acres of water.* Someone mentioned that tilapia were being cultured in the tropics, so I thought it might be interesting to get a few and fool around with them.

I was able to join some of the Texas A & M students on a collecting trip during the summer of 1975, during which we captured about 20 tilapia, which I took back to my laboratory outside of College Station and stocked in a pond. We were soon collecting fry and initiating experiments. Studies at Auburn had already ascertained that tilapia demonstrated excellent food conversion efficiency and exceptional tolerance to degraded water quality, with the exception of temperature. While tilapia tolerate very high environmental temperatures, they become diseased and die quickly during the fall as the water temperature begins to fall into the low 50s Fahrenheit.[696]

To those of us who were becoming interested in tilapia, it became clear that before they could become commercially important in the United States, it would necessary to find ways to efficiently maintain water temperature within the optimum range. It would obviously be possible to put broodstock indoors in warm water during winter, but harvest of marketable fish would then have to take place in the fall, which would mean year-round marketing of fresh fish could not occur. As we have seen, expansion of the catfish industry was hampered by that type of marketing until continuous harvesting was instituted. If the fish were raised in southern Florida or extreme southern Texas, they could be expected to survive except during severe winters.[699] There were also reports that tilapia would survive farther north during mild winters. Tilapia can be reared year-round in the tropics in Hawaii and in U.S. territories (such as Guam and Puerto Rico). Geothermal water and water heated by power plants or industry, used in conjunction with flow-through or closed systems, provide additional options for tilapia production. Some of our research at Texas A & M University involved evaluation of a few types of overwintering systems.[700] Over a period of about a decade we conducted studies on the use of organic fertilization to produce tilapia in ponds, grew them in polyculture with freshwater shrimp, evaluated some aspects of water quality on tilapia growth, looked at the effects of density on growth, and evaluated some of their nutritional requirements, among other things.[701–717] I was fortunate to have the late Jonathan Chervinski, from Israel, spend over a year working with me and my students on various aspects of the research. Jonathan brought a great deal of experience with him and was a dedicated, conscientious researcher who quickly won the hearts of the students. I

*Personal communication from Richard Noble.

was also fortunate to have an opportunity to work closely with two other faculty members, Robert Brick and Edwin Robinson, during portions of the period that I was at Texas A & M. I would be hopelessly remiss in not mentioning the excellent graduate students with whom I associated. Not all of them worked with tilapia— many were involved with channel catfish and a few with freshwater shrimp and other things—but all of them were excellent and have been quite successful in their careers.

The tilapia of culture interest in the United States (Mozambique tilapia, blue tilapia, Nile tilapia, and red hybrid tilapia being the most popular) are all mouth-brooders. In all cases the males dig circular nests in pond bottoms, after which they set about to attract ripe females. Once a female is lured to the nest, eggs are deposited, fertilized by the males, and picked up in the female's mouth. Incubation and yolk-sac absorption occur over a period of a couple weeks in the mouth of the female, after which the fry school around the mother and will dive back in her mouth for a period of another week or so if danger threatens. Thereafter the school of fry and the parent separate.

Females can spawn about once a month, year-round, if conditions are appropriate. Again, that is primarily a function of temperature, though of course the fish must have reached maturity, which occurs within a few months after they are born. (The precise time varies among species but tends to be three to six months.) Since so much of the mother's time is spent in conjunction with reproduction and larval rearing, growth in females becomes very slow once they mature. Several methods of reducing spawning success or producing all-male tilapia for stocking have been developed and are presently in use. Predators can be stocked to pick off tilapia fry, or tilapia can be reared in cages where the spawning success rate is low. All-male stocks can be produced by making hybrid crosses of certain species. Alternatively, once tilapia are large enough that the sexes can be easily determined by examination of the genital openings, they can be hand-sexed and females eliminated from the population that is stocked. Finally, certain hormones can be fed to produce all-male populations. The hormones are provided to the fry for a period of about three weeks beginning when the fish will accept prepared feed (immediately after yolk-sac absorption).[2]

Because of certain problems associated with tilapia production in the United States, including the fact that all tilapia are exotics, there was a significant amount of reluctance to allow the culture of them in some states. By 1980 there were less than 10 sources of tilapia broodstock, fry, or fingerlings in the United States,[718] though interest in culturing tilapia was growing rapidly. One of the questions that remained to be answered was whether people would buy them if the fish were available.

Ethnic markets were found to be excellent outlets for tilapia. People from many Asian countries, in particular, were familiar with the fish and became ready consumers. In Hawaii, where tilapia was being criticized as a nuisance fish by 1982, a marketing program led to development of a significant demand within the food industry.[697] In the Hawaii market, and elsewhere, producers were turning to the red or gold-colored hybrids of tilapia. The dark colors of many species were thought to be unacceptable to some consumers, particularly those used to eating marine fish that tended to have red coloring. Red hybrid tilapia, it was thought, would be as acceptable as marine fish. Different crosses that resulted in gold or red hybrids were first developed in Taiwan. Later, Florida, and Israel also developed red hybrids.

When sold as fillets, the color of the fish is not of particular importance since the meat is the same no matter what the color of the fish from which it comes. Also, the increased price being asked of hybrids pushed some consumers to compare the red fish with those of normal coloration. Those wise consumers found that color had no influence on flavor or texture of the flesh, so some of the attraction for the red hybrids waned. Red hybrids, with the exception of the Israeli type, are highly salt-tolerant, so people interested in rearing tilapia in seawater have maintained their interest in the hybrids. Others still consider the red hybrids to be superior in some way (probably because of the hype used by some advocates of the hybrids) and continue to produce them preferentially in their freshwater culture systems.

Interest in tilapia culture has, as of the 1990s, outstripped domestic production, even though a significant demand has been created. In 1992 about 9 million pounds of tilapia were grown domestically, while over 18 million pounds were imported.[685] The nagging overwintering problem continues to limit locations where successful commercial culture can be practiced in ponds. Many had constructed indoor recirculating water systems* with supplemental heating of the water for tilapia culture, but through 1992 none of those systems had apparently been profitable.[719] Part of the problem can be traced to the expectations of the prospective fish farmers with respect to the price they would obtain for their product. If breakeven costs are, for example, $2.00 per pound, and the fish can be sold in the vicinity of $3.00 per pound as some believe, there is a tidy profit potential. In reality, the price paid for tilapia in the round is unlikely to be more than $0.85 per pound. Fillets could be worth $3.00 per pound, but the dressout percentage is only 40% or even less depending on the size of the fish that are processed, so it takes well over 2 pounds of fish to produce 1 pound of fillets. The economics just don't seem to work out at present. If a recirculating system can obtain free hot water, and if sufficient backup components are in place to allow the facility to operate during power failures, economical production of tilapia in recirculating systems may be possible. (I'll have more to say on this general issue in Chapter 6.)

Those who are making money with tilapia today include people who are growing the fish in ponds and those who use flow-through geothermal water. Successful operations in Idaho exist in areas where winter temperatures can reach $-30°F$. Geothermal water of suitable quality is not easy to find, and prohibitions on dumping thermal effluents (not to mention effluents that contain enhanced levels of nutrients) have removed many suitable sites from consideration. Somewhat surprisingly, tilapia production in the western states is led by Idaho, with additional production by California, Arizona, and Colorado. Production in Colorado began in 1991 and was less than 50,000 pounds. Figures for the other three states from 1986 through 1991 are as follows:[686]

---

*Systems in which most of the water is conserved by treating and returning it to the fish culture units. Some makeup water is required. The amount of new water varies greatly from one system to another.

	Production (lb)		
Year	Idaho	California	Arizona
1986	50,000	Not available	Not available
1987	40,000	Not available	500,000
1988	40,000	283,000	640,000
1989	225,000	250,000	625,000
1990	1,000,000	660,000	500,000
1991	1,000,000	900,000	400,000

Projections for 1996 show doubling of production in California but no increase in Idaho. A return to 500,000 pounds is projected in Arizona, where ponds and cages in irrigation canals are the predominant types of culture facilities. In the Northeast, some production of tilapia has been occurring in Maryland, Massachusetts, and New Jersey. In 1992, 280,000 pounds were marketed, and those fish were valued at $563,000.[680]

Red Drum

Red drum, which occurs from Mexico to Massachusetts in the Atlantic Ocean, has become a species of interest to commercial culturists in the past several years, though earlier interest in some states, particularly along the Gulf of Mexico coast, had been expressed in conjunction with producing red drum for enhancement stocking. Stocks of red drum along the Gulf of Mexico coast were in decline in the 1980s, largely from overfishing, when Cajun Chef Paul Prudholm's recipe for blackened redfish* became the rage around the United States. The depleted stocks were further reduced, leading to bans on fishing in some states. (The ban was imposed in Texas in 1981.)[720] With the ban on commercial fishing and the continued demand, the potential for commercial aquaculture of red drum looked particularly appealing.

Successful spawning was achieved by researchers in 1977,[721] and the Texas Department of Parks and Wildlife began producing fingerlings within a few years.[722] Facilities were constructed for spawning, egg hatching, and larval rearing. Ponds were fertilized and stocked with larvae only a couple of days old. Within a few weeks it was possible to harvest fish of 1 1/4 inches or so, with survivals often reaching 50% of the fish stocked. Small red drum are released into the Gulf of Mexico.[720]

Since red drum will survive in both fresh and salt water,[723] commercial culturists have considered using both types of systems. While spawning and larval rearing were well worked out by the early 1980s, the economics of producing red drum to market size was not well developed even as late as 1985.[720] Prices to producers were ranging at that time from $0.75 to $1.90 per pound depending on season of the year. Judging from the prices of other fishes since that time, it is unlikely that significant changes have occurred. Much of the early commercial culture was centered in Texas, which had at least three farms in place or under development by 1985. Farms have been developed in additional Gulf states more recently.[687]

---

*Red drum is also known as redfish or channel bass, depending on location.

Striped Bass and Hybrid Striped Bass

Interest in commercial striped bass culture developed in much the same manner as what we have seen for red drum. As revealed in Chapter 4, the culture of striped bass, like that of red drum, was initially developed as a means of producing fish to enhance commercial and recreational fisheries. Later, striped bass were introduced into inland waters (which was also true of red drum in Texas). Finally, collapses of the striped bass fishery along the Atlantic coast, which began in 1973, led to bans on commercial fishing by 1988.[724] Recreational fishing was also banned in some regions.

Many of those interested in commercial culture settled upon hybrid striped bass as the culture species of choice. In 1985, an economic analysis and evaluation of the market potential for hybrid striped bass was undertaken in South Carolina, one of the states where interest in commercial aquaculture continued to be high. Given existing technology, and a price to producers at that time of $2.50 per pound or higher, it appeared there was some potential for profit.[725]

The first commercial striped bass farm began operation in 1973 and produced 20,000 pounds before failing after 1974. A second farm, established in 1977, produced 29,000 pounds but failed by 1980.[726] One of the problems facing the industry and causing failures in commercial ventures has been a lack of broodstock. State and federal programs had long depended upon wild broodfish for their activities. Regulations typically prohibited the use of wild spawners by private producers, so a good deal of research was conducted beginning in the 1980s on developing captive broodstock both on the East Coast and in California. The problem continues today, though a number of commercial farms are in operation. Total commercial production from five firms that were in operation at various times between 1973 and 1989 was 2,216,000 pounds, so the market for striped bass and hybrids had obviously not been very thoroughly tested. A limited amount of additional data have been collected more recently from Maryland, Massachusetts, and Pennsylvania. In 1992, production of hybrid striped bass in those states totaled 947,000 pounds. The value of those fish was $2,280,000.[680]

Most of the striped bass and hybrids being reared in the United States by commercial culturists are grown in ponds, though hybrids have also been grown in tanks using geothermal water in California, in power-plant effluents in Pennsylvania, and in net-pens in New York.[727] As with tilapia, there have been attempts to rear striped bass and its hybrids in recirculating water systems, but as of this writing I am unaware of any such attempts that have been economically successful.

Other Fish Species Worthy of a Comment or Two

A number of species that were only produced in state and federal hatcheries in the past are now being produced commercially. In addition, there are a few new species that have been commercially farmed in the United States since 1970. A few words on some of them can be found in the following subsections. After that we'll finish up our look at finfish by examining activities with bait and ornamental species.

Largemouth Bass. Some commercial culture of largemouth bass (along with sunfish) has developed in states where the sale of those species by private individuals is allowed. Even in states such as Texas, where landowners can obtain free fish from the state, fish can also be obtained from commercial fish producers, allowing private pond owners to make their own decisions on stocking rates, species mixes, and so forth. (The states typically provide limited mixtures of species and only at specified stocking rates.)

Somewhat surprising, there was little in the way of advancement in largemouth bass hatchery success between 1930 and 1980. The 70% mean survival reported by state, federal, and private hatcheries in 1980 represented no increase over surveys that had been conducted as much as 50 years earlier.[728] The 1980 survey indicated that the one responding federal facility produced fish for $0.01 each, whereas the average for six state hatcheries was $0.06. A single private producer responded to the survey and indicated that production costs were $0.20 per fingerling. Based on the meager response, the statement was made that states were much more efficient than private producers.

Methodology for the culturing of bass has advanced appreciably since 1980. New technology for spawning and the use of prepared feeds in conjunction with bass rearing have developed in the past 15 years.[729]

Bighead Carp. We have taken a look at grass carp in Chapter 4, and some culture of grass carp continues both in public reservoirs and in ponds. Some catfish farmers have used grass carp for weed control, but more use is being made of them for weed control in farm ponds. With the use of hybrids between grass carp and bighead carp, reproduction of stocked fish is no longer a consideration, and many states have reversed their bans on grass carp if hybrids are used. As a species, bighead carp is not only used to create hybrid grass carp; it is also being stocked, usually in polyculture with catfish.

Bighead carp were introduced to the United States in the early 1970s by a commercial fish farmer.[730] Since bigheads eat primarily zooplankton but will also feed on phytoplankton and detritus, it was thought that they could help improve water quality in catfish ponds. Studies have not confirmed that bigheads actually do carry out that function, but they will provide additional production without interfering with catfish growth.

Successful natural spawning of bighead carp will not occur in ponds, but induced spawning technology has been in place for a number of years. Fingerling production methodology is also firmly in place. Some catfish farmers stock 100 to 150 bigheads per acre. Fingerlings stocked during spring may reach 5 to 6 pounds by fall. By stocking larger fish the farmers will be assured of having marketable fish by fall.

Difficulties encountered include the labor involved in sorting bigheads from catfish during harvesting and the apparent crowding of bigheads around aerators during periods of low dissolved oxygen. In the latter case the bigheads may deprive the catfish of well-oxygenated water. Since the carp have only about half the value of catfish, if one species is to be preserved during an oxygen depletion, it should be the catfish.

Bighead carp are marketed in large-city specialty markets and, in addition, make an excellent canned product. Increased demand will be a key to future expansion of the industry.

Dolphin. We're not talking here about Flipper but about the fish commonly seen on the menu in restaurants under its Hawaiian name, mahi-mahi. Found in tropical and subtropical waters over much of the globe, the dolphin is an excellent foodfish that is also prized for its roe. The first attempts at culture in the mid-1970s showed that juveniles captured from the wild and reared in North Carolina grew rapidly and adapted well to the culture environment (which consisted of pens in an estuary).[731]

Early attempts to spawn dolphin in captivity failed to produce viable fingerlings, but in April 1980, scientists at the Oceanic Institute in Hawaii succeeded.[732] Offspring from their first spawn grew very rapidly, and 15 were grown to maturity in only five months. They were spawned in August of the same year they were hatched. The researchers estimated that 10-pound fish could be produced in six to seven months and that more than 50,000 pounds per acre could be produced in submerged sea cages.[731]

Early feeding trials employed chopped herring and squid,[733] but as has been true with other new species, the research soon turned to study of the animals' nutritional requirements and prepared diets were formulated. Recent studies have shown that dolphin grow well on diets containing high protein and low fat levels. Growth rates have been a phenomenal 13% of body weight daily with food conversions of 0.6![734]

Dolphin continues to be the subject of research in Hawaii, and commercial farming may not be far off. The popularity of the fish, marketed as indicated under the name mahi-mahi, extends throughout the United States, so there is an existing demand.

Paddlefish. The paddlefish is one of the more unusual fishes that occur in the fresh waters of the United States. Originally ranging in the tier of states running from North Dakota to Texas, plus Montana eastward to the Atlantic coast (with the exception of a few parts of New England, South Carolina, Georgia, and Florida),[735] the paddlefish has long been valued for both its flesh and roe. A scaleless cartilaginous fish with a rostrum (the paddle) extending in front by up to a third of the body length, the paddlefish has been extirpated or is in decline in a number of states where it originally occurred. It has been the target of commercial fishermen for many decades and also has appeal to recreational fishermen who capture it by snagging. Once reaching sizes of up to 200 pounds, paddlefish are considered large today if they weigh over 60 pounds.[736]

There has been a long-standing interest in paddlefish culture within the Missouri Department of Conservation. In 1960, personnel within the department first successfully spawned paddlefish in captivity. In the 1970s, commercial culturists were producing paddlefish for sale,[736] but those efforts seem to have failed and there is little or no interest today in producing paddlefish for direct sale into the foodfish market.[735] There continues to be activity associated with reintroducing paddlefish into regions where they have been extirpated and with enhancing their populations in places where populations are declining.

One of the early commercial paddlefish culturists, G. S. Bumpas, who apparently didn't understand a great deal about the concept of salinity tolerance (which tends to keep saltwater fish in the ocean and freshwater fish in lakes and streams), made the following remarks:[736]

> . . . has anyone investigated the possibility of introducing certain schooling fresh-water species, such as the paddlefish, back into the ocean? Such an endeavor could be very helpful in providing the sea with a large, boneless, and especially good-eating fish, with many valuable by-products—which could very well be an improvement over the bony, herring-like fishes which now comprise the bulk of the world's fisheries.
>
> Wouldn't it be possible—if millions of eggs or fingerlings were gradually exposed to different concentrations of salt water—that a few of these would survive and adapt themselves, thereby furnishing mankind with a new source of food from the ocean.

Mr. Bumpas could have got on well with the gentleman who suggested cutting off the spines of catfish and spawning their offspring to establish a breed of polled fish.

Snapper. Marine fish known as snappers have been discussed as potential aquaculture species for a number of years. Of particular interest are species known as red snappers, which appear throughout the United States as a popular item in seafood restaurants. The Gulf of Mexico red snapper, a species of particular interest in this country, was first spawned in captivity in Texas during 1978.[737] That was achieved by control of temperature and photoperiod. Hormonally induced spawning of the same species was achieved in the early 1980s.[738] Red snapper is apparently in significant decline today, so interest in its culture is growing. There have, to my knowledge, been no successful commercial ventures established in the United States to date, but this is another fish which may appear in the marketplace as a culture species in the next few years.

Sturgeon. Various species of sturgeon inhabit U.S. waters, including some that are found in our estuaries. Sturgeon, like paddlefish, are cartilaginous species, so there is no problem with bonyness. Some are of culture interest because they are threatened with extinction (for example, the shortnose sturgeon was considered by 1972 to be an endangered species[739]), while others have been considered for culture as foodfish. Sturgeon are a source of caviar. The value of caviar makes sturgeon culture particularly appealing until one recognizes that the number of years required to rear a sturgeon to maturity (at which time the caviar can be obtained) requires more financial input than can be gained from sale of the caviar and meat combined. The few commercial culture facilities that exist today concentrate their efforts on the production of sturgeon as ornamental or foodfish. The amount of foodfish production available today satisfies a small specialty market since consumer demand does not seem to be particularly strong.

Our old friends Seth Green and Livingston Stone were among those who were involved with early attempts to spawn sturgeon.[740,741] Green, interested in augmenting declining stocks of sturgeons, successfully spawned the Atlantic sturgeon in 1875.[740] He found it necessary to sacrifice females to obtain their eggs, since they could not be stripped. While problems were encountered, about 100,000 fry were produced. Early problems with Atlantic sturgeon culture involved insufficient numbers of males and females that became ripe at the same time and fungal infections on developing eggs.[742]

Lake sturgeon propagation was initiated in the latter years of the nineteenth century. The problems encountered were similar to those obtained with Atlantic sturgeon, though something like 5 million lake sturgeon larvae were produced in 1881 by the Ohio Game Commission and released into the Detroit River.[743] The Pennsylvania Fish Commission became interested in sturgeon production and enlisted the aid of Livingston Stone to assist with their efforts.[744] The first attempts to spawn sturgeon in Pennsylvania were failures. In 1906, some shortnose sturgeon were collected and placed in a yellow perch pond. The following spring two sturgeon appeared to be developing eggs. Additional sturgeon were obtained from the wild to augment the existing pond population. The fish ripened, but the sexes did not mature simultaneously, so they could not be spawned.

In 1909, Pennsylvania culturists succeeded in capturing ripe adults from nature and were able to spawn them. The culturists successfully hatched a few hundred fry. While the work in Pennsylvania did not result in carrying sturgeon through their life cycle, it did demonstrate that eggs could be successfully incubated in the hatchery and that adults would live in ponds, two important requirements of any successful aquaculture operation.

Because the problems of simultaneous maturity of the sexes and heavy egg losses in hatcheries, attempts to enhance sturgeon populations in the United States had been abandoned by 1912.[745] Interest in sturgeon culture within the United States resurfaced during 1979 in the East, when induced spawning of Atlantic sturgeon was achieved through hormone injection,[746,747] and in the West when Wallis Clark and Sergei Doroshov began planning a culture program at the University of California at Davis in 1978.[748] Doroshov, who was considered to be the father of sturgeon culture in Russia, had emigrated to the United States and brought with him the Russian technology he had helped develop over a period of 20 years.[749] It was only natural for Doroshov to continue the pursuit of his interest in sturgeon culture in the United States, particularly since he expressed the viewpoint that ". . . sturgeon is the most highly valued fish in the world for its superior flesh and eggs."[748]

Though dependent on wild fish as broodstock, the researchers on the University of California at Davis campus didn't need many since a single female could produce something like 1 million eggs. In 1980, eggs were obtained from white sturgeon females following the hormone injection of developing females. Those eggs were hatched in McDonald jars, and the fry were fed brine shrimp. Fingerlings of 2–3 inches were produced within six weeks. The research was initially aimed at producing young fish to augment natural populations, though foodfish aquaculture may have also been seen as a possibility.

Those first white sturgeon fingerlings, several thousand in number, were expected to take some 15 years to reach maturity. They were released into the Sacramento River in May 1980, but the researchers were already thinking about the potential of commercial culture.[748] It didn't take long for the potential to be tested by the private sector. At virtually the same time the researchers at the University of California were beginning their work, operators of a catfish hatchery near Sacramento, California, began expressing interest in producing sturgeon.[750] Ken Beer, one of the partners in the catfish farm, indicated that the water temperature at the facility was suitable for sturgeon and was enthusiastic about expanding the operation. Beer might also have been influenced by the fact that he was at the time doing graduate work with sturgeon on the Davis campus. In any case, in 1981 the commercial farmers received a permit from the state of California to obtain wild broodstock. They produced 400,000 larvae during their first year of operation. Under their permit, half of the fish produced were to be returned to the natural environment.

Marketing of young sturgeons was first planned for 1982. By 1983, the commercial hatchery, which by then was also producing striped bass, was selling fingerling sturgeon for $1.50 each to tropical-fish wholesalers.[749] The farm was also exporting some sturgeon to producers who were rearing them to sizes of 2–3 pounds for the marketplace.

California sturgeon production in 1989 exceeded 200,000 pounds, with the producers receiving $3.50–$4.50 per pound.[751] Four California sturgeon farms were projecting production of about 1 million pounds in 1991.[752]

By 1990, sturgeon farmers were able to produce mature males from eggs within four years, compared with 6–10 years in nature. Females had been reared to maturity in less than 10 years compared with 15–20 years in the wild. While some farmers depended on wild females for breeding, one farm was able to spawn captively reared broodstock for the first time in 1990.

The shortnose sturgeon was successfully spawned at the U.S. Fish and Wildlife Service's Orangeburg National Fish Hatchery in South Carolina in 1983.[753] Because of its endangered status, there was renewed interest in culturing the species for enhancement.

Baitfish

Minnow and goldfish culture and the extent of the industry in the United States was reviewed in 1993 by James T. Davis.[566] There are at least 20 species of baitfish being reared, though the golden shiner, fathead minnow, and goldfish predominate. A large percentage of the goldfish sold today are used as feeder fish for predatory ornamentals and as trotline bait. Therefore, I have elected to follow Davis's term *baitfish* to include both minnows and goldfish. Some comments on the rearing of goldfish for the petfish trade are also included later in this section.

In the late 1960s there were about 24,000 acres in baitfish production in the United States, but that acreage grew to 54,000 acres by 1978.[566] Most of the acreage for baitfish is in Arkansas. In 1969, there were 2,550 acres in that state devoted to baitfish culture.[508] That had grown to 29,091 acres by 1972,[754] fell to about 21,000 acres by 1979,[755] increased to 28,000 by 1986,[756] and decreased slightly to 27,000 acres by 1990.[679] In 1984, one Arkansas minnow farmer had 6,200 acres in production, down from the maximum acreage for that farm of 7,200 acres.[757] The total production of baitfish in Arkansas during 1990 was 10,805,000 pounds valued at $26,988,750.[679]

Not all U.S. baitfish production is in Arkansas, of course. Many other states have baitfish producers, though their total acreage probably doesn't approach that of

Arkansas. I can't prove that with data, because they seem to be pretty scarce. For example, in the survey of aquaculture for 1992 conducted by the Northeastern Regional Aquaculture Center, it was determined that baitfish were being reared in the region, but the producers would not provide information on the number of pounds or value of the fish produced.[680] A survey of commercial aquaculture in Virginia in 1978 showed that 15,000 pounds were produced that year.[758]

There are a number of goldfish producers in the United States that target the aquarium trade. Nearly every family, particularly those with children, has or has had a goldfish bowl in the house. It is usually in a child's room out of sight from visitors since it is rarely cleaned, and as a result, often looks and smells bad. The goldfish don't seem to mind, as long as the water doesn't evaporate entirely. Goldfish tanks are also popular in restaurants, nursing homes, physician's and dentist's offices, and in the waiting rooms of various businesses.

While a student at the University of Missouri, I had the opportunity to visit Ozark Fisheries in Stoutland, Missouri. At that time, about 1967, the operation featured over 300 ponds and was shipping goldfish in oxygenated plastic bags, not only throughout the United States but virtually around the world, guaranteeing 24-hour live delivery. In 1972, the operation had about 400 acres of ponds in production and estimated that they were supplying fish to one in five households in the United States.[759] Whereas the rest of the baitfish industry supplies standard run-of-the-mill goldfish, those involved in the petfish industry produce fish with names like comet, fantail, black moor, calico fantail, and Shubunkins, which come in various colors and may feature pop eyes or rather grotesque shapes.

Goldfish were reportedly introduced to the United States in 1874 and were not farmed commercially before about 1900.[760] Apparently, the first farms were located in Maryland where, by 1923, there were some 35 farmers in the business who produced and sold some 20 million goldfish a year, about 80% of the U.S. production. The fish were shipped, in the fashion of the Fish and Fisheries Commission, by rail in 10-gallon cans. Trucks gradually replaced railroads for goldfish transport in the 1930s and 1940s, and goldfish were marketed to households by major dime-store chains like Kresge's and Woolworth's. The use of goldfish as baitfish developed in the 1950s and 1960s.

Ozark Fisheries was one of about 10 large goldfish farms in the United States in the 1960s. Many Arkansas minnow farmers were also producing goldfish, largely for bait and as feeders. The big goldfish producers who supplied the petfish market were selling about 50 million fish per year at that time.

Fancy goldfish are marketed at about 12 weeks of age. They sell for various prices depending on the breed and color. Feeder goldfish were selling for $20 per thousand in 1979, which amounted to $3.25–4.00 per pound depending on the size of the fish (feeders being marketed at a smaller size than fancy goldfish).[760] Trotline goldfish, which are marketed at 3 inches or larger, brought $2.00 per pound in 1979.

Ornamental Fish

For our purposes, the ornamental-fish category includes tropical freshwater and marine fishes destined for the aquarium trade. There has also been, in recent years, some interest the use of native temperate fishes in home aquaria.[761] (Goldfish were discussed in the preceding section as a means of keeping from dividing goldfish up into two categories, so there will be no more mention of them in this section.) Some of the minnows, killifish, sunfishes, and various other species also make suitable aquarium species.

The tropical-fish industry in the United States is centered in Florida and dates back to 1926.[762] The industry had grown sufficiently large by the early 1960s that about 150 producers banded together to form the Florida Tropical Fish Farmers Association to provide a single contact for interaction with governmental agencies and to handle public relations for the industry.[763,764] By 1973, sales of live ornamental fish had reached $200 million per year, with about $160 million (80%) from domestic production and the remainder attributable to imports.[763] Most of the imported fish enter the United States in Florida and are then distributed throughout the country. In the early 1970s it was estimated that tropical fish were being sold in some 150,000 retail outlets. Furthermore, fish were being maintained in 20–26 million households. The aquarium fish hobby was surpassed only by photography.

By the early 1980s, there were about 300 tropical-fish producers in Florida, who were accounting for some 95% of domestic production.[765] Virtually all of that production was associated with freshwater species. A 1983 survey of Florida fish farms obtained responses from 62 ornamental-fish farmers who were producing a total of over 26 million fish valued at $2,678,705, clearly only a fraction of the total production and revenue.[766] Species being produced included guppies, mollies, swordtails, platies, tetras, gouramies, tropical catfish, barbs, cichlids, goldfish, and koi. It has been estimated that there are between 500 and 1,000 species being sold around the world as ornamentals, about 200 of those species being grown in the United States.[767] Total sales of tropical fish in Florida during 1991 were placed $32.8 million, which was down by about $1 million from 1989.[766] The retail value of tropical fish in the United States has been estimated at between $250 million and $700 million.[767]

The competition from freshwater imports had grown to 40% of the industry by 1981,[766] and the total value of imported tropical fish was estimated at about double the value of U.S. production in 1989.[767] Both wild and farmed fish contribute to the imports. Methods of capturing imported fish, both marine and freshwater, came into question during the 1980s when stories about the use of cyanide and other environmentally damaging techniques began to circulate. In addition, of course, are the facts that nontarget fish were being subjected to a toxicant, fish that had been intentionally taken with cyanide often survived only long enough to get them sold to the home aquarist, while intense harvesting of coral reefs and certain freshwater environments in third-world countries were decimating natural stocks. Potential problems arising from the introduction of diseases with imported fish were also being discussed. Declining harvests and urban expansion have reduced the numbers of imported tropical fish in recent years. Farmed production of tropicals occurs primarily in Singapore, Hong Kong, and Thailand.[767]

The taking of native marine tropical fishes was also becoming increasingly restricted to the loss of domestic stocks. Thus, there was increased impetus for increasing domestic production. In addition, virtually all of the ornamental fish being produced in Florida, not to mention those that were imported, were exotics. Many of those were fish that had escaped and in some instances had become established in reproducing populations. Some species, whether they had become established or not, were considered to be potentially harmful. Included were various species of tilapia (which could potentially displace native fish species), walking catfish (once thought to be a scourge capable of eating small dogs), grass carp (which might consume the Everglades down to the last blade of grass), and the dreaded piranha (an overrated predator, schools of which are thought by many able to consume a full-grown cow within minutes). Outright bans were imposed on some of the above-mentioned fishes.

Martin Moe is an acknowledged pioneer in the production of marine tropical fishes. He bred clownfish in captivity in 1973 and neon gobies at about the same time.[768] By 1985, Moe and his colleagues were involved in what was billed as the ". . . world's largest marine fish hatchery."[769] That facility produced clownfish, neon gobies, and sharknosed gobies. Research was underway to develop the technology required to produce a number of other species, but there remain few marine ornamental fishes that can currently be spawned in captivity.

There are ornamental fish being produced in a number of states, often in insulated buildings or greenhouses. In the Northeast, ornamentals are produced in Maryland and Pennsylvania, with ornamentals being the primary aquaculture activity in Maryland.[680] I am personally aware of companies that produce ornamental fish in St. Louis (Missouri), Seattle, and in southern California. There is also some production associated with geothermal water in Idaho (once again in the Hagerman Valley), and there has been at least some interest in establishing similar ventures in Oregon and other states where geothermal water is available. There is probably at least a limited amount of breeding for the retail market occurring in most major cities across the country.

Koi carp are popular with some hobbyists. Bred for centuries by the Japanese, koi come in a variety of colors and patterns and can be valued at thousands or even tens of thousands of dollars each. They are usually stocked in backyard pools in the Japanese tradition. There are a number of koi breeders and suppliers in the United States, but obtaining hard data on the industry is difficult.

The ornamental-fish industry in the United States is augmented by those who grow and sell living aquatic plants that are used in the aquarium trade.* In addition, there is a very large industry in aquaria, aquarium stands, and a multitude of supplies (including plastic plants for those who have problems keeping the real ones alive); medications; and, of course, food. Prepared feeds in the form of flakes and pellets, supposedly designed specifically for aquarium species, are available from several producers at prices so high they would put any foodfish aquaculturist out of business within a few days. There are also frozen, freeze-dried, and living organisms

---

*As wetland protection and renovation activities have increased in the United States in recent years, there has also been a growing demand for aquatic plants for use in natural habitats or those created to mitigate against habitat loss. Although discussion of that activity is beyond the scope of this book, it should not be overlooked entirely.

sold as foods for aquarium species. Total annual sales of ornamental fish and associated products worldwide has been estimated at $4 billion.[767]

Molluscs

Now we'll turn our attention to invertebrates. We'll begin with molluscs and then look at crustaceans. Except for oysters, mollusc culture in the United States had not been developed to any great extent prior to 1970, and it was actually during the late 1970s and 1980s before much activity was seen in conjunction with species such as abalone, clams, and mussels. Because the industries for some species are small, there is not a great deal of information available. This discussion doesn't include all the molluscs that have received attention by aquaculturists, only those that are being cultured commercially. For example, razor clams and geoducks have been produced in state hatcheries in Washington to augment natural stocks but are not being grown commercially.[770] A program associated with a medical school in Houston, Texas, has been involved with the rearing of squid for a number of years, the primary purpose being to supply biomedical researchers with giant nerve fibers. A number of species have been successfully spawned and reared in that program.

Abalone. As with oysters, abalone culture has been developed as an offshoot of a commercial fishery. Depleted stocks of red abalone off the coast of California, caused by heavy fishing pressure (both commercial and recreational) and expansion of the range of abalone-loving sea otters[771] prompted the development of hatcheries during the 1970s to produce small abalone for stocking.[772] Research leading to controlled spawning was undertaken, and by 1981, there were five companies in California producing young abalone, which were sold to local volunteer groups who transplanted them to depleted abalone beds. By 1990, the number of abalone farms had grown to 15, three of which were producing significant numbers of juvenile abalone and rearing animals to markets size.[751] Abalone are being marketed on the East and West Coasts of the United States, in Hawaii, and in Japan. In 1989, about 315,000 food-sized abalone were produced. About 95% of the animals produced in California are red abalone, with the remainder being pink and green abalone.

While some commercial production could certainly be expected to support governmental and private efforts at enhancement, the potential for farming of abalone to market size during the early years was limited because suitable growout areas were not readily available for private use. One California company grew abalone in vertical concrete pipes located in a subtidal area,[773] and another tried rearing the molluscs in offshore cages[774] during the mid-1970s. To avoid high costs of land and facilities for upland pumped-seawater rearing facilities, the notion of rearing abalone offshore was being discussed almost from the onset of culture interest,[772] though no major facilities seem to have resulted.

With improved technology and the development of a workable leasing system, abalone farmers were able to operate in the coastal zone for at least part of the production phase. One of the early California farms, founded in 1972, employed an onshore hatchery and a nearshore growout facility. Abalone, which were first marketed to restaurants by the company in 1981, are grown out in plastic drums suspended from a fixed dock or from longlines.[771]

In the late 1980s an abalone farm was established at Crescent City, Oregon, on an area leased for 40 years.[771] As is typical given the convoluted regulatory structure that exists in many states, the operators of the facility were required to spend three and a half years obtaining the proper permits, but the process was completed for a modest $8,000. The Oregon facility purchases young abalone from a California hatchery (at about $0.30 each) and grows them to about 1 1/2 inches long in modified plastic barrels, after which they are transferred to rectangular cages for growout. One of the permits required for establishment of the farm allows for the collection of kelp from nature, for which the state is paid $1.90 per ton. Labor, fuel, and other expenses run the true cost of harvesting kelp to about $50 per ton, but the expense is necessary since the abalone require kelp as food.

Clams. Robert Winston Menzel, a leading oyster biologist, was a proponent of hard clam culture long before the industry developed. Writing in 1971, he described how hard clams (quahogs) could be cultured.[775,776] The market seemed to be in place, with 15.5 million pounds harvested in 1968 from the wild, primarily from the waters of New York.[776] The biology of quahogs was sufficiently known that the animals could be spawned and reared successfully. As with abalone, onshore hatcheries would be used to support production of marketable animals in leased areas. Menzel's approach was to plant the young clams on suitable leased bottom areas and protect them from predators by covering them with wire cages.

I met Winston Menzel in 1967 when he became my major professor at Florida State University. My impressions of him were included in the obituary I wrote on him:[777]

> He was, at least in the years that I knew him (slightly over 20), among the most humble human beings in my acquaintance. His manner (relaxed, almost plodding) and speech (a slow, down-home Virginia drawl) belied his enthusiasm, keen wit, and outstanding intellectual ability. Winston always spoke fondly of his former students and was intensely interested in their careers. He seemed much less interested in having any glory heaped upon himself, though his research productivity led to some such heaping, which he accepted with humility.
>
> While he was not at all dynamic in the classroom, students who had the good fortune to have him available to them outside the lecture hall found him to be an absolute fountain of information. Winston was knowledgeable in virtually every aspect of marine biology, and in a wide variety of subject areas outside of that broad discipline. In many of his conversations with the students that surrounded him, he always had some tidbit of information that none of us had previously known, and he frequently could drag out a reprint to underwrite his information.

Winston Menzel obtained his Ph.D. at Texas A & M University under Sewell Hopkins, another big name in mollusc biology, in 1954. His only professional position was at Florida State University, where he concentrated on molluscan aquaculture and genetics until his retirement not long before his death in 1989.

Shortly after I arrived with my family in Tallahassee, Florida, late in the summer of 1967, Winston took me down to the Florida State University Marine Laboratory and showed me around. When we went out to the dock, he took out his knife and cut an oyster from a piling, popped the shell, and offered it to me. It was my first experience with a raw oyster (or any oyster for that matter). I had been deprived of that culinary experience for 26 years and have been trying to make up for lost time ever since.

Winston would reportedly eat almost anything captured from the marine environment. There was even a story circulating that on one of the cruises with which Winston was involved, someone captured a living trilobite, but before it could be studied, it disappeared. Winston was supposedly later seen with an odd-looking leg protruding from his mouth. While that story is obviously not factual, it does provide some insight into Winston's love of seafood.*

My interest in research for my dissertation was focused on fish, not shellfish, but Winston had no problem with that. I conducted what I thought was a complete literature review on channel catfish in preparation for writing my dissertation, which, if you'll remember, was on channel catfish nutrition and was conducted at the Skidaway Institute of Oceanography in Savannah, Georgia. When I presented the review to Winston, he read it and indicated that I had missed two important publications. I argued that I felt my review was complete. He went to his reprint collection and pulled out the missing papers, both written by him.[778,779] Needless to say, my face was red.

Menzel's thoughtful discourses on the potential for hard clam culture[774,775] were apparently not immediately received with a great deal of enthusiasm since there was no rush by aquaculturists to jump into the quahog business. However, after some delay, sufficient interest in clam culture developed that by 1981 a few commercial hatcheries were in operation.[769] Those hatcheries produced juvenile clams that were grown in trays until they were about 1 inch in diameter, after which they were planted out on natural beds for growout to harvest size. The aquaculture side of hard clam culture was not very visible until the Atlantic Little Neck Clam Farms of James Island, South Carolina, was established. A detailed description of that facility was published in 1993[780] a time during which the farm was still being developed. Clams were being produced, though it was anticipated that success of the business would not be demonstrated until 1995, when production was projected to be 100 million market-size clams a year. The facility is self-contained in that it has its own hatchery and a variety of rearing systems. The time required for clams to grow from hatching to market size is approximately two years. The company developed markets for relatively small clams, 1–2 inches in diameter and sold, depending on size, as pastaneck, petiteneck, and littleneck clams.

On the West Coast the species of choice has been the Manila clam, which was being produced in hatcheries in California and Washington by 1981.[772] Manila clams continue to be produced in Washington, but were not listed among the molluscan culture species being grown in California in 1990.[751] Young clams are grown on intertidal beds. Predator control devices, such as protective plastic net covers, have been widely used.

---

*He introduced me (and subsequently through me, my family) to some excellent seafood restaurants in Florida, something for which I will always be grateful.

Mussels. Until recent years there has been relatively little interest in mussel culture in the United States. There has not even been much commercial fishing for mussels, even though they are quite abundant in some areas. I used to see them growing all over the place on the East Coast when I was working at the Skidaway Institute of Oceanography, but I never heard of anyone eating them. Locals in Georgia and South Carolina would typically indicate that the mussels were muddy tasting and not worth collecting. For some reason, acceptance of mussels began to increase in the 1970s, and really accelerated in the 1980s as indicated from data collected in Washington, where production from all sources grew from less than 10,000 pounds in 1974 to nearly 200,000 pounds in 1984.[781] Buckets of steamed clams and mussels became common in restaurants, and aquaculturists began to express an interest in the idea of mussel farming.[772]

Mussel farming was becoming popular in Puget Sound, Washington, by the mid-1980s,[781] with culturists concentrating their efforts on off-bottom culture. In California, growout includes thinning and harvest of mussels that are found attached to offshore oil platforms. There is also culture of mussels in bags.[751] Total production in California was about 1.5 million pounds in 1989,[751] a ear during which total U.S. production was over 55 million pounds.[686] Much of the California production is from natural beds, so the Washington culture industry is the largest on the West Coast. Some mussels are also produced in Alaska.[782] On the East Coast, the industry, which marketed $10 million worth of mussels in 1989, also developed beginning in the 1970s.[783] Data from 1985 show that 85% of the mussels harvested that year were from the capture fishery, with only 15% from culture. Rope culture was evaluated in Maine but found to be uneconomical, so bottom culture became the method of choice in that state. None of the other states in the Northeast culture mussels at the present time.[680]

Oysters. The United States is responsible for over half of the oyster consumption of the world,[784] but little attention was paid to modern aquaculture until recently. Oyster culture in the United States was historically based primarily on placement of shell in areas where natural spat production was known to occur; then either the growout occurred in place or spat-laden shell was relocated to other areas for growout. Hatchery techniques were developed by the 1970s, but there was little use of the technology by the commercial industry, which was quite primitive compared with other forms of aquaculture. Gordon Gunter, long-time director of the Gulf Coast Research Laboratory in Ocean Springs, Mississippi and whose name was almost always mentioned in any discussion of marine science activity in the Gulf of Mexico during the middle decades of the twentieth century, summarized (along with a co-author) the history of the U.S. oyster industry up to 1971:[785]

> The Chesapeake Bay area has produced more oysters than any other part of this country, although during 1963 and 1964 the Gulf Coast produced more oysters than Chesapeake Bay. Oyster production on the East Coast has been marked by a huge decline over the years. The peak year for Chesapeake Bay was 1880 when that region produced 117 million pounds of oysters. In 1891 the catch was 108 million pounds, but it declined thereafter to less than 100 million pounds and now runs around 21 million and less a year. Production of New Jersey and Delaware was high in the late

1800's and it has also declined. The greater production of the South Atlantic states has been in the early 1900's with the peaks generally in 1908.

Louisiana oyster production rose greatly following the establishment of the Oyster Commission in 1902 and the annual production attained 12 million pounds in 1911. It was above 10 million in 1908, 1911, 1938–40, 1952, 1956, 1957, 1961–64. During the depression years Louisiana's oyster production fell quite low.

In summary, as the East Coast oyster production fell to about one-eighth of its former size, oyster production around the mouth of the Mississippi River increased. Increases have taken place also in the northwest oyster industry in Washington, Oregon, and southern Canada. Nevertheless, these increases have not compensated for the vast decline of the East Coast industry. As a result, the per capita consumption of oysters has fallen very sharply for the whole nation.

As we have seen, oyster leases were available in some states. During my graduate school years at Florida State, I recall seeing leases in Florida marked with poles and signs to keep out commercial or recreational oyster fishermen. Leases were also available in Louisiana, Virginia, and Washington, to name a few states. By 1980, there were 26 oyster farms on 231 leased acres in Maine.[786] Those who obtained leases were more likely to actively manage the oyster beds than were individuals that had no exclusive right to harvest a particular area. The states often jumped in to manage oyster bottoms in areas that were not leased. Such a program was established in Mississippi in 1960, and led to a fivefold increase in production by 1966.[785] Under the Mississippi program, which was similar to that of other states, new oyster reefs were created and maintained, existing reefs were enlarged, and oysters were relocated from polluted areas to clean locations.

In Washington, the primary source of oysters for planting was from Japan from the time Pacific oyster culture began prior to World War II until well after 1970. Between 1964 and 1970, the supply of young oysters from Japan became limited, and research on how to produce spat domestically increased, while yields declined.[787] Production had declined steadily from over 5 million pounds in 1947 to 1.8 million pounds in 1974. The emphasis in the early 1970s was on trying to find ways of improving natural spat production. Studies on using trays, racks, and ropes to culture oysters were also undertaken.

Off-bottom culture, which had been largely developed in Japan, was also being evaluated in Chesapeake Bay beginning in the late 1960s by the National Marine Fisheries Service.[788] Longlines and string culture from rigid structures and floating rafts were studied. A modification of these techniques, involving submerged racks, was utilized by a commercial oyster culture company in Massachusetts beginning in 1970.[789] Additional research on raft culture was conducted in the mid-1970s in Delaware Bay as well.[790] It was also during the 1970s that the first U.S. research into culturing oysters hung from petroleum production platforms was undertaken. That research was conducted in the Gulf of Mexico off Texas between 1970 and 1975.[791]

While states were beginning to move oysters from areas contaminated by pollution, often involving human sewage, to clean areas, a research project at the Woods Hole Oceanographic Institution in Massachusetts was developed to utilize secondary treated-sewage effluent as a nutrient source for the growing of oysters and other molluscs, as well as polychaetes for bait. (The worms would feed on waste from the molluscs.) Initiated by John Ryther, one of the nation's leading oceanographers and co-author of one of the first modern comprehensive books on aquaculture,[792] the program grew from some small-scale laboratory studies, on using sewage effluent to grow algae, into a program that involved construction of a building wherein a pilot-scale demonstration project was established.[793–796] In addition to John Ryther, whom I have met on several occasions, Bill Dunstan and Ken Tenore were heavily involved in the research. I worked with Bill at Skidaway, where our tenure over-lapped for a couple years, and left Skidaway at about the time Ken joined the program. Ken and I served on the board of directors for a large U.S. Agency for International Development project on fish stock assessment during the late 1980s and early 1990s. John Huguenin, whom I have not met, was the engineer responsi-ble for designing much of the system. The program got a good deal of attention during the 1970s, but the technique has never been adopted commercially.

Commercial oyster hatcheries have been in production in Washington since 1969, but into the early 1980s it was not clear that they were economically viable.[797] A study published in 1980 involving the potential for oyster hatcheries in the Chesapeake Bay area, on the other hand, indicated that a profit could be made, even with interest rates at 12%.[798] The situation in Washington changed significantly during the 1980s when economically viable oyster hatcheries were developed. Inter-est in oyster hatcheries has also been seen in Louisiana since the late 1980s, but problems have retarded their development.[799] One major player in the Washington oyster industry, the Coast Oyster Company, produced sufficient numbers of spat to provide for its own needs with enough left over to market to other growers in Washington and elsewhere. That company produced 3 billion oysters in 1981[800] and has produced tens of billions since. Oyster spat are shipped in baseball-sized lumps wrapped with fine-mesh plankton netting. Several million spat are contained in each ball of oysters. Total production from Washington in 1991 was nine million pounds, 80% of Pacific coast production for that year.[686]

Diseases of the American oyster along the Atlantic seaboard and in the Gulf of Mexico became a major problem beginning in the 1950s.[801] Compounded by peri-ods of lower than desired salinity associated with flooding, and predation from starfish and oyster drills, the American oyster industry found itself in dire straits by the 1980s. Ken Chew, one of the most renowned oyster biologists living today, current Director of the Western Regional Aquaculture Center, and colleague of mine at the University of Washington, told me during the late 1980s that if you were to purchase an oyster in the Chesapeake Bay area the chances were one in two that it was a Pacific oyster grown in the state of Washington. Yet, even though interest in oyster culture was increasing, oyster consumption in the United States was not,[783] so aquaculture was apparently just helping maintain the status quo with respect to overall production.

California was one of the leading states in oyster production between 1880 and 1910, but as water quality deteriorated in the major oyster-growing area (San Fran-cisco Bay), the industry waned.[751] The industry was revitalized in the 1950s when young oysters were imported from Japan and planted in suitable areas within the state. With the development of hatcheries, oyster farmers in California began pur-

chasing spat from Washington, a practice that continues. In 1988, California produced over 1.6 million pounds of oyster meats.

U.S. oyster landings in 1989 from farms and the wild harvest amounted to 36
million pounds of meats.[686] From 1966 to 1985, Louisiana production averaged
10.4 million pounds,[800] and production reached 11.9 million pounds with a value
of $25 million by 1991.[802] In the northeastern Untied States, the industry produced
$63.4 million worth of oysters in 1992. Producing states were Connecticut, New
York, and Massachusetts.[680]

Oyster production has traditionally been highest during the winter months, with
few fresh oysters appearing on the market during the summer. It has long been
traditional in oyster production areas to restrict the sale of oysters during months
that did not contain the letter "r" (May, June, July, and August). While many believe
that oysters are not sold during those months because they are not healthful, in
reality it is because the quality of the oysters is poor: their sugar reserves have been
converted to gametes because the oysters are preparing for and involved in spawning. The poor, milky quality of oysters can be avoided if the oysters are sterile, that is,
do not develop eggs and sperm. Sterility occurs in instances where oysters contain
an extra set of chromosomes (known as triploid oysters). The condition of triploidy,
can be induced by exposing oysters to a certain chemical. The resulting individuals
have been called sexless oysters. They remain in excellent condition throughout the
year and can thus be marketed at any time once they reach suitable size. Triploidy
was first accomplished with American oysters in 1981,[803] and was subsequently
applied to Pacific oysters in 1983.[804] The technique has been utilized by commercial
hatcheries in Washington.

Crustaceans

The bulk of the attention by researchers with respect to crustacean culture has
centered on crawfish (or crayfish) and shrimp (both freshwater and marine). There
has been some interest in developing crab culture, but cannibalism and other problems associated with rearing that group have kept the enthusiasm level low. Once in
a while there is a small flurry of interest, such as occurred with the Caribbean king
crab a few years ago, but the odds are that crab culture in the United States is
unlikely to develop in the foreseeable future, if ever. While lobsters are still seen to
have commercial potential in some circles and continue to be the subjects of research,[805] successful commercial culture remains elusive. Insofar as I can determine,
all the attention on lobsters continues to be directed toward the American lobster,
since problems associated with rearing the larval stages of the Florida or spiny
lobster remain to be solved. Crawfish and shrimp are species upon which commercial culture operations have been based, so the discussion that follows is restricted to
those animals.

Crawfish. The basic approach to crawfish culture, which was developed prior to
1970 (and discussed in Chapter 4), continues to be used today. The bulk of the
industry is found in Louisiana; however, while I was in Texas (1975–1984), several
thousand acres of crawfish ponds were developed in the southeastern part of that

state. Larry de la Bretonne, Jr., who was affiliated with the Louisiana Extension Service, collaborated with Jim Davis at the Texas Agricultural Extension Service to advise and assist the Texas crawfish farmers during the formative years of their industry. Those farmers were shown the latest techniques, which they quickly adopted. Interestingly, according to Jim, Larry had been trying without much success to get the Louisiana farmers to change their ways of managing crawfish ponds. Research had uncovered ways in which production could be increased without much added expense, but the Louisiana farmers seemed to be entrenched in doing business as usual. The Texas farmers, who took a more intensive approach to crawfish culture than their counterparts in Louisiana, were quickly outproducing their neighbors to the east on a pounds-per-acre basis. Harvest rates approached 2,000 pounds per acre in Texas,[806] as compared with a few hundred pounds per acre in Louisiana. Still, in terms of total volume of crawfish produced, the Texas industry paled into near insignificance relative to Louisiana. To achieve their high rates or production, Texas crawfish farmers employed supplemental prepared feeds, aeration, and modified pond designs.*

I don't recall when I first met Larry de la Bretonne, but it must have been sometime in about the mid-1970s. Larry was certainly not a person you would forget once you had seen him. A true gentle giant, he was not particularly tall but had enormous bone structure, which made him an imposing figure in any crowd. I don't think I ever saw him without a smile on his face, so his prodigious size was never threatening. Following his premature death in 1991, the American Fisheries Society enshrined Larry de la Bretonne in the National Fish Culture Hall of Fame. The plaque commemorating his short life (he was born in 1943) reads, in part:

> As an international leader, he was known worldwide as an expert in crawfish culture. He was also a leader in extension work relating to catfish and alligator culture.
> Larry de la Bretonne had a deep appreciation for all aquaculturists and an uncanny ability to communicate highly technical knowledge and present it to all levels, from administrators to pond bank workers. He was a person who eagerly shared his knowledge. He was a natural teacher, able to communicate the importance of fisheries to any audience from school children to elected officials.

While new technology may have seen slow adoption in Louisiana, it was investigators in that state who continued to lead the way in research and development. Jim Avault, who has been on the faculty at Louisiana State University for about 30 years, is a world-renowned leader in crawfish research. He worked closely with, and was involved in the training of, such other well-known crawfish biologists as Larry de la Bretonne and Jay Huner. After obtaining his doctoral degree, Huner took his crawfish interest and experience from Louisiana State University to Southern University, where he became a faculty member. In the past few years Huner has directed a

---

*All of this is not to say that Louisiana farmers continue to maintain their allegiance to the old ways. It just took them longer to adopt new methods, whereas the Texas farmers, with no experience, were willing to accept the advice of those who were willing to assist them.

crawfish research program at the University of Southwestern Louisiana. Huner and Avault have written volumes of material on crawfish culture. As one among several examples of the type of research that has been undertaken by investigators at Louisiana State University, new designs for crawfish ponds were developed that allow for the circulation of water to all portions, thereby facilitating supplemental aeration during periods of low oxygen.[807]

By 1985, Louisiana had over 105,000 acres of crawfish ponds, compared with over 8,000 in Texas, some 600 in South Carolina, and something over 500 in Arkansas and Mississippi.[808] One estimate was that there were 135,000 acres in Louisiana in 1989,[809] while a 1991 estimate was 115,000 acres.[686] In any case, Louisiana crawfish acreage is far and away the highest in the United States and continues to grow. The Texas industry ultimately grew to about 18,000 acres but then fell by 1990 to about 4,800 acres.[810] In South Carolina, where interest in aquaculture and aquaculture research has been strong for years but where the commercial industry has never been fully developed, a survey in the early 1980s showed that the public was receptive to locally produced crawfish.[811] The economics of crawfish production in South Carolina was also the focus of a study in the 1980s that resulted in the drawing up of balance sheets to provide prospective culturists with an indication of production costs and potential profits.[812]

Most of the crawfish culture is in the South, but is not totally confined to that region. In addition to the southern states already mentioned, crawfish are produced in North Carolina, Florida, Georgia, and California.[813] In the Northeast, there is some production in Delaware and Maryland.[680] There has also been interest expressed, and may be a modest amount of production, in a number of other states, including California. In some cases, crawfish are being produced incidental to the production of minnows and other types of fish.[808]

Interest in producing soft-shell crawfish began in 1970.[814] Hormone injections were first used to induce molting, but short-term high-density culture associated with supplemental feeding using high-quality trout feed can be effective and avoids the use of chemicals.[815] The original intent was to meet the demand by the bait industry for soft-shell crawfish in the midwestern states, and it was in that region that interest first developed. By the mid-1980s the impetus to produce soft-shell crawfish increased when they became a gourmet item.[808] At that point, Louisiana culturists began to develop soft-shell crawfish for the restaurant trade. Immature crawfish, which molt much more frequently than adults, are used. They are maintained in trays at densities of 20–30 animals per square foot, and the trays are checked every two or three hours during the day.[816] The same type of technology has recently been used to produce soft-shell crawfish in South Carolina,[817] and an economic evaluation has demonstrated that soft-shell crawfish can also be profitably produced in Mississippi.[818]

Crawfish have traditionally been captured by trapping, and that approach continues to be used. There have been a number of new types of traps developed, and specialized boats and other motorized vehicles have been designed to facilitate the activities of those involved in running traplines.[819] Some research has also been conducted on seining crawfish rather than trapping them.[820] Bait has usually been in the form of fish having high oil content, shad being one of most popular.[821] In recent years, commercial baits have been developed, though fresh fish continues to be widely used.

The crawfish of commercial interest in the United States rarely weigh more than an ounce or two, of which only about 20% is edible meat.* Crawfish much larger than those native to the United States can be found in other parts of the world. One Tasmanian species is supposed to reach 8 pounds,[820] and another in Madagascar is reported capable of reaching similar size.[806] Getting crawfish to such large sizes is a time-consuming process, but research has shown that some of the exotic species can reach up to several ounces within a typical southern U.S. growing season. Research with Australian crawfish was initiated in the early 1970s,[822] but no real interest in their commercialization in the United States developed until the mid-1980s, when ads in trade publications lauding the superiority of Australian crawfish began to appear. There were undoubtedly a number of people who obtained Australian crawfish after responding to such ads, but I can find no evidence that a thriving business resulted.

Research to evaluate the true potential of Australian species was initiated at Auburn University in the late 1980s under the direction of one of my former students, David Rouse. He has been involved in studies with crawfish having such unique, and in one case literally colorful names as marron, yabbie, and red claw.[823,824] Because they are exotic species, the introduction of Australian crawfish into commercial aquaculture in the United States may be restricted or prohibited if it cannot be shown that escape can be prevented, diseases are not introduced, and that they will not compete adversely with native species if they escape.[823] Much of the attention in Alabama has focused on the red claw, which has also been grown in Missouri. Average harvest weights of the red claw have ranged from 1/2 ounce to slightly over 2 ounces, which is somewhat larger than the average for red swamp crawfish. Rouse has indicated that under proper conditions the red claw can reach nearly 3 ounces in six months. One of the advantages of the red claw and marron are that they have tail meat percentages exceeding 30%.

Freshwater Shrimp. The term *prawn* actually refers to any large shrimp. Many aquaculturists have adopted a convention by which they use *prawn* in reference to freshwater shrimp and reserve the word *shrimp* for marine shrimp. Not everyone accepts that convention—including me—nor does it eliminate confusion. For example, one of the marine species grown by the Japanese and others is known as the Kuruma prawn, which is a marine species. I prefer to use the terms *freshwater shrimp* and *marine shrimp,* thereby eliminating confusion. My convention is used in this book.

Freshwater shrimp can be found throughout much of the world. Many species are too small to be of much culture interest, but others can be quickly grown to sizes of several ounces, and if one is patient, shrimp as large as 1 pound can be produced. There are at least two native species that grow to sizes that would be readily marketable, but the two I'm familiar with are highly aggressive (cannibalistic in

---

*Unless you suck the heads, which is an issue, like abortion and weapons bans, that has people polarized and one that I refuse to discuss—though I can usually spot a head sucker when I see one.

addition to just plain nasty), and have not been seriously considered as culture species.

To demonstrate the point about aggression, a short story would seem to be in order. When I was responsible for operation of the Aquaculture Research Center at Texas A & M University, my right-hand man was W. A. Isbell. A farmer, outdoorsman, and most importantly, native Texan, W. A. was responsible for overall maintenance of our facility. On one occasion, Bob Brick, another faculty member, and one or two of his graduate students showed up at the lab with a number of native freshwater shrimp that they had captured somewhere near the Rio Grande. The shrimp were placed in an aluminum raceway. W. A. went over to inspect them and stuck a hand in the water to get them to move around. One immediately grabbed his thumb. He put his other hand in the water to try and extricate his digit, but the second thumb was also grabbed by the same shrimp. When he pulled his hands from the water, he was handcuffed. (Once those of us who witnessed the episode got over our convulsive laughter, we assisted in obtaining W. A.'s release.)

Nearly all the attention of U.S. researchers and commercial producers has focused on a freshwater shrimp from Southeast Asia known as the Malaysian giant prawn.* This is the species almost always referred to when aquaculturists mention freshwater shrimp. The work that led up to aquaculture of freshwater shrimp began in the 1950s in Southeast Asia when a scientist with the United Nations Food and Agriculture Organization, S.-W. Ling, first attempted to rear the species though its life cycle.[825] Ling's story provides two valuable lessons that should be learned by all aquaculturists. First, learn about the natural history of each species before you attempt to rear it, and second, keep trying until you find something that works. The second process can be expedited if the first precedes it, as I'm sure Ling would admit were he still alive to be queried on the subject.

Female freshwater shrimp carry their developing eggs on their abdomens, where the embryos are well protected from predators. Once hatched, freshwater shrimp larvae go through fewer stages than is true of saltwater shrimp, and therefore become juveniles more rapidly. Ling found that freshwater shrimp would readily mate in captivity and that there was no problem in obtaining fertile eggs. Once the eggs hatched, however, the larvae would begin to die after a couple of days. Over a period of a few more days, total mortality would occur. From his studies, Ling determined that starvation was not the problem, and concluded that something was missing from the environment in which the young shrimp were being maintained. He placed small numbers of larvae into dishes of water and added various chemicals from the shelves of his laboratory to see if anything provided the missing ingredient.

After some time, Ling had gone through his chemical inventory but still hadn't found the missing substance. The story goes—and Ling was often heard to tell it— that his wife brought him lunch in the laboratory each day. Not too surprisingly, since the Lings were Chinese, the lunches were composed of the culinary delights of that country. One day, while Ling was eating his lunch and contemplating the small

---

*The name is a bit of a misnomer since the species occurs throughout much of Southeast Asia. Those that were initially imported to the United States apparently came from the Malay Peninsula, which accounts for the adoption of this common name.

dishes of dying freshwater shrimp larvae, he poured some soy sauce into one of the dishes. You guessed it—the larvae exposed to soy sauce survived. (Thus, to this day, every freshwater shrimp hatchery keeps several gallons of soy sauce on hand. No, just kidding.)

Ling, being no fool, figured that there must be something in soy sauce (other than soybeans) that was not a substance he had previously tried but was required by the young shrimp. He concluded, rightly, that the missing ingredient was off-the-shelf table salt, or sodium chloride. When he substituted salt for the soy sauce, he obtained the same result, thereby confirming his conclusion. He was subsequently able to rear freshwater shrimp to juvenile size, and learned that the salt concentration could be reduced over time until the shrimp would survive quite well in fresh water.

The life history of the Malaysian giant prawn is now known to involve migration of egg-bearing females to the lower reaches of rivers where water salinity approaches about one-third that of seawater. It is in those estuarine areas that the eggs hatch. The larvae move upstream into fresher and fresher water as they go through their larval stages. All that is necessary to accomplish successful larval rearing is to mimic the environment that the young shrimp would seek out in nature. Had Ling known about the life cycle of the freshwater shrimp, his efforts at cracking the larval problem would have been much less frustrating, but a good story would also never have evolved.

The major bottleneck associated with marine shrimp was the difficulty associated with the induction of spawning, but there were also significant problems associated with larval rearing. Only the latter problem existed with freshwater shrimp, and that problem was solved once the salt requirement of the larvae was discovered. The rapid growth to market and the potential of producing jumbo shrimp within a reasonable period of time were seen as additional benefits of freshwater shrimp.

However, there was a major drawback associated with freshwater shrimp. As is true of tilapia, a limitation on freshwater shrimp culture is the fact that they cannot tolerate low water temperatures. Thus, it is not surprising that interest in commercial culture initially developed in Hawaii.

In 1966, Takuji Fujimura, an employee of the Hawaiian Division of Fish and Game, began a pilot study to evaluate the potential of freshwater shrimp for commercial culture.[826] Modest success was achieved during the first two years of the project, and some shrimp were marketed in Honolulu at $3.00 per pound. In 1968, a commercial venture was launched in Florida. By 1971 a commercial farm in Hawaii had eight small ponds in production and had developed plans to develop a total of 25 acres.[827]

Other states that became involved in research on freshwater shrimp during the 1970s with the idea of developing commercial industries included South Carolina and Florida.[828] The initial studies utilized domestic species, but investigators in both states soon turned their attention to the Malaysian prawn. Scientists with the South Carolina Marine Resources Institute in Charleston began their studies in 1973. Their early work demonstrated that survival rates of postlarval shrimp stocked in ponds could exceed 81%, that male shrimp larger than 3 ounces could be produced in five months, and that pond production levels exceeding 2,000 pounds per acre were achievable.[829] Research in Florida was initiated by the Department of Natural Resources. Both small operations and those run by large corporations were established in South Carolina and Florida.[828]

In late 1975 or early 1976, Bob Brick joined the faculty of Texas A & M University to develop a freshwater shrimp culture research program. There had been a couple of students already in the program who were working on freshwater shrimp, but there were really no facilities to speak of other than some very small (perhaps 3 × 5 foot), shallow (about 1 foot deep) ponds that had been dug by hand at a field station of the university. That field station was located at Bryan Air Force Base, which had been closed by the government and was taken over by Texas A & M University. The runway was used for vehicle crash testing, and several of the buildings were being used for various activities. The Department of Wildlife and Fisheries Sciences, with which Bob Brick and I were affiliated, had control over the enlisted personnel's swimming facility. Included in the fenced-in site was, naturally, a swimming pool (which we never used) and a bathhouse. Brick turned one dressing room in the bathhouse into a functional freshwater shrimp hatchery and used the other dressing room for research. The basket checkout area was used as a chemistry laboratory.

For pond studies, Brick utilized some of the facilities at the Aquaculture Research Center, which was my base of operations. About 15 miles away from the swimming pool facility and 10 miles from campus, the Aquaculture Research Center had 12 ponds of approximately 0.25 acres in surface area and another 12 of 0.5 acres, an 8-acre reservoir, and a support building that incorporated a garage area that we used as a wet laboratory. The building also had a functional chemistry laboratory and offices. Within a few years we added a 3,200-square-foot dedicated wet-laboratory building. A third building with more offices and laboratories was constructed in conjunction with a project conducted by Mark Chittenden, another faculty member who was involved in marine fisheries research. When Chittenden's project ended, that building reverted to the aquaculture program.

One of the projects that we undertook at the Aquaculture Research Center involved rearing freshwater shrimp together with tilapia. David Rouse, mentioned earlier in conjunction with his activities with Australian crawfish, conducted his Ph.D. research on that particular subject and found that significant production of both species could be obtained within a growing season.[711]

Bob Brick was approached by a private aquaculture firm about 1980 and asked to evaluate that firm's shrimp growing activity. It turned out that the firm was trying to raise marine shrimp in Chicago! We'll come back to this particular activity when we discuss saltwater shrimp, but for the time being you might be interested in knowing that Brick left Texas A & M shortly after his visit to Chicago.

The economics of freshwater shrimp culture in Hawaii continued to look bright through the 1970s. One economic study by University of Hawaii researchers indicated that money could be made if 3,000 pounds per acre were reared and sold at $3.00 per pound. Facilities of 10 acres or more were considered economically viable, though there was an opinion that if an operation were run as a family venture, smaller farms might be profitable.[830]

In 1980 locally produced freshwater shrimp were first test-marketed in South Carolina. Consumers and owners of retail outlets were generally pleased with the product, which sold for $8.70–11.01 per pound.[831] Subsequently, a 1982 study of production costs in South Carolina showed that a profit of $157.91 per acre was possible.[832]

During the 1980s the technology was developed for producing marine shrimp economically, and much of the interest that had swirled around freshwater shrimp evaporated. A number of enterprises came and went during the 1980s, and a few hangers-on continued in business for several years. One commercial farm in south Texas[833,834] was established in the 1970s and remained in business for several years. Mississippi had about 5 acres of freshwater shrimp ponds in production during 1985.[674] At the peak of the South Carolina industry in the early 1980s, there were nine farms in operation with a total of only 3.8 acres.

Even though early economic studies indicated good consumer acceptance, the fact was that the marketplace favored the marine shrimp. Complaints about freshwater shrimp included the fact that females with eggs often appeared in the marketplace and were not accepted by the public (perhaps because people thought the berried females were deformed),* and that freshwater shrimp had poor keeping qualities when frozen. There were also complaints that the flavor was not up to that of marine shrimp. Producers felt that in order to be profitable, freshwater shrimp had to be grown to very large sizes, which meant a longer-than-desirable growing period and increased production costs.

It was not only the United States that saw diminishing interest in freshwater shrimp. World production tumbled during the 1980s as marine shrimp production rose, and there are now few countries producing the freshwater crustaceans. Aquaculturists in the states in which the feasibility of freshwater shrimp culture was demonstrated were not able to make sufficient profits to remain in business. There is an occasional mention of resurrecting a freshwater shrimp culture industry in the United States, but it does not seem likely that will happen. Raising freshwater shrimp for the aquarium trade may provide a niche market for a few enterprising aquaculturists. There are native species that are much smaller than the Malaysian prawn, and it is the smaller freshwater shrimp that would probably be of most interest in the aquarium trade.

Marine Shrimp. In the early 1970s the focus of U.S. shrimp research was on native species. Gulf of Mexico shrimp were being studied at the National Marine Fisheries Service (NMFS) laboratory in Galveston, Texas, and by extension specialists at Texas A & M University. Research was also being conducted by the Texas Department of Parks and Wildlife, the Louisiana Wild Life and Fisheries Commission, Nicholls State College, Louisiana State University, and the University of Miami.[835] Researchers at the University of Arizona were concentrating their work on a brown shrimp from the Gulf of California[836] (which is not the same brown shrimp that occurs in the Gulf of Mexico). A species native to Hawaii was receiving the attention of investigators at the Hawaii Institute of Marine Biology.[837]

---

*By the end of the 1980s there were at least anecdotal reports that berried females were actually in demand, as gourmets decided those animals represented a special treat. Apparently, berried females brought a premium price in some markets.

Shrimp nutrition studies were already underway at Florida State University when I arrived in Tallahassee, Florida, to enroll in that institution during 1968, though the program was terminated within a couple of years. Jim Andrews visited Florida State University and met me during his information-gathering trip to learn how to hold shrimp for studies he was planning at the Skidaway Institute of Oceanography. After I went up to Skidaway to begin my Ph.D. research, Andrews and some of his colleagues embarked on shrimp nutrition research. In the meantime, a good deal of information was being developed in Asia and elsewhere. As time passed, shrimp nutrition research was conducted in Hawaii, Texas, and a few other states. Reviews of the shrimp nutrition literature by Michael New in 1976 and 1980 (which included studies on freshwater shrimp) demonstrated the intensity of activity around the world.[838,839] Several hundred papers were listed in those reviews.

With the development of better feeds came improved survival in ponds. Of course, studies were also being conducted on stocking rates and other management techniques that also contributed to improved survival.

Economic feasibility studies on mariculture showed that only marine shrimp and lobsters could provide a return on investment, and while commercial shrimp farming had become a reality in Asia due to successes obtained by Japanese researchers, there were no profit-making enterprises in operation within the United States. The reason there were no successful shrimp farms in the States was viewed as being simple:[837]

> Although the market value of shrimp is high and the biology is sufficiently well known to begin culture, the labor-intensive, land-extensive, Asian pastoral technique of shrimp farming is neither economically exportable to the United States, nor does it have any economy of scale when expanded from the small family pond of Southeast Asia to the huge diked-off marshes of the southeastern USA.

Richard Neal, then director of the NMFS laboratory in Galveston, felt in 1971 that shrimp culture could succeed if three problems could be overcome. The first involved high mortality of juvenile shrimp in ponds; the second (which was thought to have a close relationship to the first) was inadequate knowledge of shrimp nutrition; and the third was the inability of researchers to induce maturation in captive shrimp. Diseases, which had not been a problem up until that time, were predicted to become a problem as the industry developed.[840]

The spawning problem remained a hurdle that was difficult to jump. Shrimp spawning was achieved by a process known as sourcing. This involved experimental fishing during the spawning season to capture mature females that were already carrying a spermatophore—a packet of sperm that was transferred from the male to the female and was used to fertilize the eggs when they were extruded. When found, mated females could usually be depended upon to expel fertilized eggs within 24 hours under the proper environmental conditions. Hundreds of thousands to a few million eggs could be obtained from a single female, but the culturists had to take what they could get. Sometimes their efforts to collect mature shrimp failed, spermatophores fell off, shrimp died, eggs did not develop properly, and/or the larvae did not survive.

Once fertilized eggs of good quality were obtained, the process of hatching them and rearing the larvae was fairly well developed. Cornelius (Corny) Mock at the NMFS laboratory in Galveston took credit for developing much of the system that became known as the Galveston technique. (Another successful larval-rearing procedure was developed in Taiwan, and both approaches had their champions.) Eggs were hatched and the larval stages reared in conical bottomed tanks in which a small amount of water continuously entered at the bottom and kept the eggs and larvae in suspension without pasting them against the screens that covered the drains. Algae were mass-produced to provide feed to the brine shrimp, which in turn were hatched to provide feed for the shrimp larvae. By the mid-1970s, a large greenhouse had been constructed; it covered two concrete tanks in which studies were conducted on postlarval shrimp.

I first ran into Corny Mock during my Florida State University days. As a part of our training, each student in the Department of Oceanography was expected to spend several days at sea. My trip to satisfy that requirement involved an eight-day cruise on the research vessel *Alaminos,* which was operated by Texas A & M University. A group of students from Florida State University, including me, drove to Galveston, where we met our professor (who had flown to Texas from Florida) and prepared to board the vessel. Lacking some piece of equipment or another, it was suggested that a couple of us should go to the NMFS lab to borrow the item. We were told that a person named Corny Mock could help us.

When we arrived at the NMFS lab, we looked around for someone to assist us. Spotting a very thin man in a tennis cap, we approached him and asked, "Do you know where we can find Mr. Mock?"

"I'm that sorry son of a bitch."

That introduction to Corny Mock was a bit startling but, as I was to learn, was nothing out of the ordinary. Mock is, to say the very least, a colorful character. He and I have crossed paths on many occasions over the years. I once took an aquaculture class to Galveston on a field trip. I had not warned the students about Mock's colorful language, and they had quite an experience. He not only used his repertoire of four-letter words without regard to the fact that we had both males and females in the class; he also punctuated each major point in his canned talk by sneaking up on one of my graduate students, Meryl Broussard, and whacking Meryl on the side of the head with a tennis hat (which was probably not the same one Mock had been wearing several years before when I first encountered him). Broussard would sneak off to the side to avoid the abuse, but Mock would always manage to maneuver into position at the appropriate time and clobber the graduate student again. I don't think the two had every met, so Mock wasn't being vindictive—it was just his style.

Beginning in the 1970s, Mock began trotting the globe as a representative of the U.S. government, advising foreign nations on shrimp culture. After some apparently bloody battles with the administration of the Galveston NMFS laboratory, Corny was reassigned to the State Department, where he became a roving shrimp culture ambassador.

Corny Mock, though now retired from the federal government, continues to be active as a consultant and outspoken opponent of the use of exotic shrimp species in the United States. He has developed an extensive mailing list and sends copies of his correspondence on the exotic-introduction issue, particularly with respect to the spread of shrimp diseases, to people on that list several times a year.

Regardless of the problems associated with rearing native shrimp species, a number of commercial attempts were made, beginning (as we saw in Chapter 4) during the 1960s. Marifarms, the largest U.S. shrimp farming operation and the longest-lasting, was located in Panama City, Florida, and was already in operation when I arrived at Florida State University in 1968. By 1972, when their first harvest occurred, the firm controlled something like 3,000 acres of salt water and had diked off an area of about a half square mile in which the company had developed two ponds.[841] Scientists and technicians who had experience in successful commercial shrimp culture were sent to Florida from Japan to assist with the program. The sourcing technique was used to obtain brood shrimp for spawning. Marifarms continued in their attempts to culture native shrimp, but ultimately they turned to exotics species, as did everyone else.

In 1973, a team of scientists with the Texas Agricultural Extension Service harvested 1,800 pounds of shrimp from demonstration ponds located on the Gulf of Mexico coast not far from Galveston.[842] Jack Parker, who left the university in 1980 to become involved in commercial shrimp farming, headed up the work. Parker and Hoyt Holcomb began their work in 1969 using both native white and brown shrimp.[843]

The problem of inducing captive marine shrimp to spawn was resolved, at least to some extent, when it was learned that ovarian maturation in females could be induced through a technique known as eyestalk ablation. That approach was reported on in conjunction with Gulf of Mexico pink shrimp in 1972.[844] Eyestalk ablation is a polite term for cutting off one or both of the eyes (which are at the ends of stalks) of an adult female shrimp. (Sometimes males are also subjected to the procedure.) Another technique is called enuculation, whereby the eye capsule is slit open and the contents are squeezed out. It sounds a bit disgusting, but it works. You see, control of certain hormones in shrimp is located in endocrine tissues found on the eyestalks. That apparently relates to the fact that light is involved in the timing of hormone production. Removing the eyestalk (or squeezing out the contents of the eye capsule) stops production of a hormone that blocks maturation. Thus, the adult shrimp can be forced to mature, after which mating will occur in captivity.

The technique means that control of the life cycle is possible in captivity; however, since a shrimp has only two eyes, it is not possible to spawn the same individual more than once or twice. It is intuitively obvious that if culturists could mimic conditions that the shrimp experience in nature (which presumably would involve little more than adjusting the temperature and photoperiod appropriately), then maturation could be induced without resorting to surgery. However, that simple-minded solution doesn't work very well and eyestalk ablation continues to be employed. One of my students, Bill Wurts, thought the problem might be related to excessive light intensities in hatcheries. He looked at the depths at which marine shrimp spawn in the Gulf of Mexico and figured out what the light intensity at that depth range was likely to be. From that exercise, he felt that virtually all background light levels in hatcheries were too intense for induction of maturation. We published a little paper on the subject,[845] and subsequently, George Chamberlain, during his student days at Texas A & M University, was able to induce maturation in an exotic species on the Texas A & M campus by controlling temperature and maintaining an extremely low light level.* One problem is that the light level must be maintained so low that people can't see what they're doing.

Even after finding a method for inducing spawning, U.S. shrimp researchers continued to be frustrated in their attempts to improve the methodology for rearing native species, commercial culture continued to grow in Asia and was under development in Latin America. It became apparent that some of the species being reared in other parts of the world could be more easily cultured than the native ones, so there was a shift in focus. Some attention was given to Asian species, but the bulk of U.S. research eventually centered on two species from Latin America. It was those species that were being reared in Panama, Costa Rica, Honduras, Guatemala, Columbia, and most importantly from the standpoint of level of production, Ecuador.[846] Ecuador quickly became the biggest player in Western Hemisphere shrimp production. By 1979, there were over 2,000 acres of shrimp ponds in that country, and that was only the beginning. Expansion continued throughout the 1980s, when tens of thousands of acres of ponds were constructed. As a result, Ecuador became a major exporter of shrimp to the United States.

Captive reproduction of the Latin American species was less difficult than for native marine shrimp. Hatchery systems had been developed to the point that they could be operated efficiently and with high levels of output, and pond management techniques were well established. The exotic species seemed suitable for rearing in such places as California, South Carolina, Florida, southern Texas, Hawaii, and Puerto Rico. Marifarms in Florida had made the transition from native shrimp to one of the Latin American species by 1980 and were once again optimistic about their chances for profitability.[846] (They folded their tent in the United States, sold the Florida facility, and relocated their operation in Latin America not long after expressing that optimism.)

Durwood Dugger, who had long been involved with freshwater shrimp culture in southern Texas, also turned his attention to exotic marine species but continued his emphasis on the use of closed and semiclosed systems for shrimp production. Other companies were also promoting high-intensity shrimp culture systems, including the firm that Bob Brick went to visit in Chicago. We'll get back to his adventures shortly.

In California, Solar Aquafarms, a pioneer in the use of greenhouses for intensive culture, was also working with exotic marine shrimp. In 1980, Steve Serfling, who was the moving force behind the development of Solar Aquafarms, was predicting production rates of 20,000 to 30,000 pounds per acre annually from the facility.[846] Ultimately, Solar Aquafarms turned exclusively to tilapia. The operation was eventually sold to the Chiquita Corporation (of banana fame) and remains in operation as a tilapia production facility.

Corny Mock had left the NMFS laboratory in Galveston by 1980, and shrimp work at that NMFS laboratory was under the direction of James McVey. McVey was soon to move from Galveston to assume a position with the Sea Grant Program in

*George Chamberlain, personal communication.

Washington, D.C., where he continues as of June, 1995, his involvement in aquaculture through that agency. The demonstration project developed by Jack Parker in Texas had been terminated, but the university had established a research facility near Corpus Christi. Jack was in charge of pond facilities.[846] Addison Lawrence, formerly a researcher at the University of Houston, had been employed by Texas A & M to head up the shrimp maturation program. He later became increasingly involved with shrimp nutrition as well. Lawrence and George Chamberlain evaluated the effects of light intensity on the maturation of two exotic shrimp species and found that the optimum level was different between the two species (which may or may not relate to the Wurts theory[845] on optimum light level). Furthermore, Lawrence and Chamberlain found that ablating one eyestalk of males increased gonad size and frequency of mating.[847] Texas A & M University economists got into the act around 1980 and began developing bioeconomic models and budgets to assist commercial shrimp culturists. Publications on that work appeared beginning in 1981.[848–852]

During the 1980s the Oceanic Institute in Hawaii and the University of Hawaii became increasingly involved shrimp culture research. The University of Arizona program was still ongoing, and was receiving support from the Coca-Cola Corporation among others.[846]

There was still a lot of talk by 1980 about the potential for shrimp culture in the United States, but in reality the industry was becoming established and growing at a rapid rate in developing nations where land and labor were cheap, there were few if any regulations to contend with, and optimum year-round climates could be found. Universities were turning out trained shrimp culturists who could not find jobs in the United States because of the lack of a viable industry. They were, however, able to find work as consultants or employees of offshore shrimp farming companies. In 1980, Jim McVey summarized the potential for shrimp culture development in the United States as follows:[846]

> The general feeling of those involved in the industry is one of guarded optimism. Many problems still confront the industry, and very little commercial development has actually taken place. New technological developments, i.e. improved diets, improved intensive culture systems, control of captive reproduction, better species, have had a positive effect on the development of the industry. Other factors favoring development of a shrimp aquaculture industry in the United States are the escalating cost of diesel fuel, the declining shrimp yield per boat effort, the high cost of importing shrimp, and the reduced volume of shrimp for the shrimp packing industry. As a result, many individuals and major companies are seriously considering investments in shrimp farming.
>
> Most industry representatives feel that United States shrimp aquaculture could replace foreign imports and provide for the increased demand projected for the future, assuming that the domestic shrimp catch would remain at present levels.

The optimism expressed by McVey never did materialize. Most of the activity continued to be located outside of the United States.

Bob Brick's trip to Chicago revealed that the venture was housed in a large warehouse located in an industrial park. The facility consisted of long, plastic, segmented raceways mounted on wooden frames; they were stacked several high from the floor almost to the ceiling, which was at least 20 feet high. Rows of the raceways took up a good amount of the space, but even with more space occupied by a hatchery, offices, and a processing area, there was plenty of room for expansion. Postlarvae from the hatchery were introduced into the first section of one or more of the raceways. As the animals grew, they were moved to a second, larger section of the raceway, and the empty section was restocked. After going through several moves to increasingly larger segments as they grew, the shrimp were removed from the system and taken to the processing area. Salinity in the system was controlled by mixing fresh water with various salts to produce a solution that mimicked seawater. All the water was retained in the system and recirculated through a treatment facility to maintain quality within defined limits. Brick said the company employed about 25 people. Funding for the enterprise, called the King James Shrimp Company, was from a group of building contractors who also supplied much of the material used in constructing the raceway and hatchery systems.

Brick said he had been asked to look at the King James operation and provide his advice on what he saw. After examining the situation, he told management that essentially the idea was nuts. It just couldn't work. The response was that the managers continued to believe in the system and wanted to hire Bob to take charge of the facility. Brick would be in a position to either make it work or prove that management was, indeed, nuts. As is often the case when contracts are being negotiated, Brick was provided assurances that the investors didn't expect a return on their money for a few years since they recognized that it was a research-and-development program. He was also told that he would be paid handsomely for his work if he would sign on with King James.

When he got back to Texas, Brick had a long discussion with me. He said that at first he had dismissed the job offer as being ludicrous insofar as his career was concerned. However, upon additional thought on the plane back to Texas, he became intrigued with the notion. Ultimately, he decided to take a flyer and accept the position.

A year or so after Bob Brick left College Station, Texas, for Chicago I had the opportunity to stop off and visit him.* By that time, the situation at King James had changed considerably. The investors, it seems, found themselves in extreme financial straits because there was a recession in the United States and the construction industry had taken a major hit. It suddenly became very important to stop the "hemorrhaging" associated with cash being poured into the King James shrimp operation. By the time the finances began to collapse, production had reached about 6 pounds per week, a bit low to support 25 employees. At the time of my tour, Brick was the only employee left, and he was in charge of shutting down the system and selling off whatever he could.

Even though the operation was on the rocks, there had been a major change in

---

*This trip was made in conjunction with one of my trips to the Philippines, where I was involved with developing a tilapia facility for the Bureau of Fisheries and Aquatic Resources, a project supported by USAID. It was during the winter—February, I think—when I had the chance to visit King James, since my flight between the United States and Japan, a fueling stop on the way to Manila, left from Chicago.

Brick's view of the facility. Having immersed himself in the venture, he had become convinced that marine shrimp could be produced in a closed system and, further, that it could be done profitably.

Bob Brick's career has taken him on long- and short-term shrimp consultancies in various countries. Among the places where he has spent considerable amounts of time are Pakistan and Indonesia. He also became involved with another closed-system shrimp culture venture, Texas Mariculture, Inc., located back in College Station, Texas.[853] At the time of this writing, that venture continues to be in operation. While not as distant from salt water as Chicago, College Station is about 150 miles from the Gulf of Mexico, so the need for artificial seawater still exists.

A prodigious amount of research information on shrimp culture was developed during the 1980s and continues in the 1990s. Meetings of the World Mariculture Society in the 1980s seemed to be dominated by information on shrimp as exhibited by a summary of the 1984 meeting published in *Aquaculture Magazine*.[854] While enthusiasm for commercial shrimp farming continued to run high, there was still little in the way of domestic shrimp production. By 1988, there was one commercial shrimp farm in Hawaii that was producing over 500,000 pounds of exotic marine shrimp and 50,000 pounds of freshwater shrimp annually. A few smaller firms were in operation but had not marketed any product at that time.[855] Marifarms sold the Florida facility to Continental Fisheries Ltd. in 1982.[828] The latter company operated a couple of the original ponds for two years but ultimately concentrated its efforts on the hatchery side of the business. Exotic postlarvae produced in their hatchery were sold to shrimp farmers in South Carolina, Texas, and Latin America.

A commercial shrimp hatchery was established in Summerland Key, Florida, in 1971. That operation apparently went in and out of business over a period of several years and was being operated as Shrimp Culture, Inc., in 1988. At that time the company was trying to obtain permits for expansion. They produced postlarval shrimp for shipment to growout facilities in Honduras.

Palmetto Aquaculture was established in South Carolina in 1981 and began making a small profit from its shrimp sales in 1987.[828] Plantation Seafarms was established in 1983. It later became the Edisto Shrimp Company and was the largest operation in the state by the late 1980s. Seven new shrimp farms were established in South Carolina during 1988.

Jack Parker founded Laguna Madre Shrimp Farms near Bayview, Texas, in 1980:[853]

> This farm began [operating] in 1981 with a 3-phase, [67 acre], semi-intensive pond system and a pilot-scale greenhouse-covered hatchery. A second [67 acre] pond module was added in 1982. The pilot-scale . . . hatchery . . . began producing sufficient post-larvae for stocking all the grow-out ponds . . . in 1982.

In 1983, the farm was expanded to over 425 acres, and a large permanent hatchery was constructed. Production reached nearly 3,000 pounds per acre annually. In the mid-1980s, postlarvae became available from hatcheries in other states, and one of the bottlenecks to development of shrimp farming in Texas seemed to have been overcome. As a result, several additional commercial ventures were initiated: Ocean Ventures in 1985, Guffey Seafood Farms in 1987, Port Lavaca Shrimp Farms in 1988, and Bowers Shrimp Farm in 1989.[853] An additional two farms were in the planning stages in 1990.

In Puerto Rico, two shrimp farms were in operation by 1991. One had a total of 60 acres and had developed plans to more than double that acreage. The other was on about 140 acres leased from the territory.[856]

Pressure in the last couple of years from the Texas commercial shrimp fishing industry, in the wake of the capture of a few exotic shrimp in their nets, has caused an uproar in the Texas shrimp production industry, and that controversy has spilled over to other states as well. The future of shrimp culture in the United States remains uncertain, but those optimists who once predicted the development of thousands of acres of domestic shrimp ponds are suddenly difficult to find. We've listed quite a number of farms, which might lead one to believe that a burgeoning industry is developing. When compared with the enormous quantities of shrimp being grown in Thailand, China, Ecuador, the Philippines, and a few other nations, the U.S. industry pales into insignificance. Even with the modest increase in domestic shrimp farm production, imports of cultured shrimp continue to increase.[687]

Amphibian and Reptile Aquaculture

Frogs and diamondback terrapins have been mentioned in one or more of the preceding chapters. While there doesn't currently seem to be much interest in terrapin culture, some work has been conducted on the culture of sea turtles, and frogs continue to receive at least some attention. Alligator culture, which has not been mentioned before, has matured over the past two decades and deserves some consideration. All of these topics are discussed in the following subsections.

Frogs. Interest in frog culture never seems to disappear, though it has never taken center stage either. Primary interest has settled on bullfrogs since their legs are a valuable commodity in the gourmet food industry. Bullfrogs, as well as leopard frogs, are used in research and for teaching anatomy. The industry in the United States was described by Priddy and Culley as it was in 1972:[857]

> Much of the biomolecular and medical research in this country utilizes bullfrogs. However, wild-caught frogs can no longer be utilized for the sophisticated level of research now being undertaken. Unless well-defined strains of frogs are developed, much of this research will be terminated. An estimated 50% of all frogs used in education in the United States are now imported due to the increase in demand and dwindling of native stocks. Only by mass production of frogs can we hope to fill the demand and reduce exploitation of our native stocks. The increased pressure on biological supply houses for more frogs has only accelerated the depletion of native populations and suppliers now state that they will be unable to obtain sufficient frogs within the next 10 to 15 years to justify collection and marketing.
>
> Most current attempts at frog culture in the United States still employ outdoor facilities, mainly ponds. Success has been minimal, and the only economical market has been through the sale of these pond-reared frogs for pond stocking purposes or breed stock. The demand for frogs for educational and research purposes cannot be met by these operations, nor can the frogs be produced cheaply enough to realize a profit by selling to biological supply houses.

In the 1970s and 1980s Dudley Culley, at Louisiana State University, was continuing the research on frogs that he had initiated in the 1960s. Other colonies of frogs that were being maintained for research by 1970 were located at the University of Michigan and in Japan.[858] Research problems that had been identified by the early 1970s included the need for developing commercial feeds, genetically defined frog strains, simplified breeding techniques, disease control methods, and better rearing systems.[857] Research was being conducted on the effects of crowding on the growth and survival of bullfrogs, and data were available that indicated males could mature four months, and females six months, after metamorphosis.[859]

An example of a commercial frog farm was the Southern Frog Company, which was established in Dumas, Arkansas, in 1970.[857,858] Frogs were stocked at densities of up to 5,000 in 20-foot-diameter concrete tanks and thinned as they grew to a final density of about 1,000 frogs per tank. The frogs were fed crickets, worms, fish, and tadpoles. Crickets and earthworms were cultured for the frogs. The minnows were purchased. It appeared as though marketing frogs to research institutions would be more profitable than producing them for restaurants.[858]

Successful frog culture remained elusive, with companies coming and going. As of 1981, not one economically successful commercial pond-rearing operation could be found.[860] Dudley Culley felt that chances of success improved when indoor intensive facilities were used. Culley was able to obtain support from the National Institutes of Health in 1970 and from the Sea Grant Program in 1975 to continue his research into developing an economical culture system. By 1981 he was able to control frog reproduction in captivity through the use of hormones, thereby providing a year-round supply of eggs. Tadpole production techniques were also well established, the water-quality requirements of the young frogs had been determined, and management techniques for the rearing of juveniles had been developed. Prepared feeds had been developed for tadpoles, though fish were still being used to feed the metamorphosed juvenile frogs. Attempts to develop acceptable pelleted feed for juveniles had been unsuccessful. Frogs of sizes sufficient to satisfy the research market could be produced within a few months, but those large enough for the food industry required considerably longer, so the economics still favored producing frogs for sale to the research community.

In 1986, Culley characterized intensive bullfrog culture as ". . . one of the most complicated forms of aquaculture."[861] An intensive frog culture system required a breeding area, a hatchery, a tadpole growing facility, feed preparation areas for tadpoles and metamorphosed frogs (including facilities for producing living food for juvenile frogs), a growout facility for metamorphosed frogs, a processing area, a disease laboratory, a shop, and storage areas (some of which would be refrigerated. Commercial culture was still considered to be a high-risk activity, and Culley was still not able to find a domestic operation that was economically successful. He listed problems facing the industry as follows:[861]

(1) predation by mammals, birds, reptiles and small aquatic insects,
(2) cannibalism,
(3) diseases,

(4) an inadequate supply of living food,
(5) poor water quality,
(6) sporadic egg production,
(7) seasonal temperatures,
(8) poor sanitation, and
(9) season[al] availability of frogs.

Items 3 and 4 were identified as being major deterrents to successful commercial culture.

In 1988, a report was published that indicated success had been achieved by a private frog farmer in training juvenile frogs to accept nonliving food. The technique, which reportedly is the culmination of 14 years of work, is as follows:[862]

> Special feeders were developed that train juvenile frogs to eat an inert food until they reach sexual maturity. Crawfish heads are then used in other types of feeders until the mature frogs reach marketable size. The juvenile frog training feeders and the associated food for the successful training of the post-metamorphic frogs appears to be one of the keys to inducing frogs to eat inert food.

Tadpoles are fed a mixture of fish meal, rice bran, crawfish meal, yeast, soy protein, bone flour, whey, alginate, vitamins, and lipid. Once they metamorphose, the frogs are trained, through the use of a mechanical device, to continue taking prepared feed:[862]

> This device simulates living food by pulling a piece of food on a thread past the frogs. The juvenile frogs attempt to capture this moving food. When they are successful, the food stays attached to the thread until the frog can pull it off. The resistance from the thread simulates a live creature struggling to escape and increases the frogs feeding instinct.
>
> The food is a mixture of the following ingredients:
>
> 4.0 lbs. Crawfish Meal
> 3.6 lbs. Fish Meal
> 4.3 ozs. Bone Meal
> 2.5 ozs. Alginate
> 2.5 ozs. Vitamin Premix
>
> 2.5 lbs. of the above mixture is combined with 1 oz. of horse hair (from the mane or tail) and 2.75 lbs. of water. 15 mg. of a sodium metaphosphate is dissolved in the water to promote solidification. This dough is extruded through 3 mm circular holes (an ordinary meat grinder works well) and sprayed with a 5:1 mixture of $H_2O:CaCl_2$. After drying, the long strings of food are broken into . . . pieces. The hair protrudes from the food and helps stimulate the frog's reflex to swallow. The food is similar in appearance to an insect.
>
> After the small frogs have been trained to eat the food dragged from threads, they will take the food when it is dropped near them while feeding. An improved dropping feeder has been designed to dispense the food in the feeding area.

An alternative method involves throwing floating catfish feed into a tank containing mosquito fish. The fish nip at the feed pellets, causing the food to move around:

"This attracts the frogs attention and the food is eaten. The food is 32% protein. If the frogs are feeding well it is readily eaten."[862] Not too surprisingly, the amount of labor involved in training juvenile frogs to accept prepared feed is seen as a major roadblock to commercialization.

That's seems to be where we currently stand with respect to frog culture in the United States. Interest will undoubtedly continue, and perhaps a viable commercial industry will be developed. In the meantime, investing in culture frog futures may not be a very good idea.

Turtles. There are a number of sea turtles that occur in the marine waters off the United States. Some of the species have historically nested on beaches along the southeastern U.S. and Gulf of Mexico coasts, but with the expansion of the human population, many of those nesting sites have been abandoned. The activities of people, including the proliferation of streetlights along historical nesting beaches, has almost certainly had a negative impact on the night-nesting turtles. So too has the propensity of humans to dig up turtle nests. Until the nests became protected under federal law, it was not uncommon for fun-loving folks to have turtle egg fights.

Sea turtles often swim hundreds or even thousands of miles from their feeding areas to nesting beaches. Mated females drag themselves out of the sea, dig nests, and deposit their eggs. After covering the eggs, the females make their way back to the ocean. If not disturbed by humans or other predators, and if environmental conditions are appropriate, the eggs will hatch after several weeks, and the young turtles leave the nest to scamper to the sea. They cannot dive for several weeks after entering the water, so they are subjected to heavy predation from birds, marine mammals, and fish. As juveniles they are capable of sounding and so become less susceptible to predation, but they still may be captured in shrimp trawls. To reduce the latter source of mortality, turtle exclusion devices (TEDs) have been developed, and their utilization, which was generated a great deal of controversy, is now required in the U.S. commercial shrimping industry.

Loss of nesting habitat and high rates of mortality have pushed some species of sea turtles toward the brink of extinction. The green sea turtle, which had once nested in large numbers in Florida, was the subject of early attempts at restoration. Successful nesting had become rare in Florida by the late 1960s since the nesting beaches were so heavily used by humans. The threat to the green sea turtle was exacerbated by the fact that its meat was in high demand, particularly in Florida, where it was a popular item in many restaurants.

Recovery attempts with the green sea turtle were led by Ross Witham, who developed a program using volunteers to establish hatcheries in Florida. I met Witham in about 1972 while I was working at the Skidaway Institute of Oceanography. He was employed by the state of Florida and was involved in research on lobsters in addition to his work with sea turtles. He came up to Georgia to look for lobster larvae out in the Gulf Stream, and I went with him on his collecting trip. The subject of turtles came up, and I later went down to see his turtle hatchery, after which arrangements were made for me to obtain about 100 green sea turtle and 100 loggerhead turtle hatchlings for research. Two of my technicians, Dave White and

Dan Perlmutter, worked with me to set up a holding facility for the young turtles and to conduct feeding studies.[863] We soon learned that the turtles were very susceptible to diseases and that if very good water-quality conditions were not maintained, the disease problems were compounded. We were able to maintain a few turtles of each species for several months and ultimately tagged and released them off Savannah Beach, Georgia. None of the tagged turtles was ever recovered as far as I know.

In about 1980, the NMFS Galveston laboratory began a program to headstart Kemp's ridley turtles. Headstarting involved incubating turtle eggs and maintaining the young turtles in captivity until they reached sufficient size that the chances of being subjected to predation were reduced. That program was apparently less than a roaring success (since they experienced many of the same problems we had encountered with greens and loggerheads), and it has, I believe, been terminated.

Commercial aquaculture with sea turtles has never been developed in the United States and is not likely to be developed in the future since all the species that occur in our waters are protected by federal law. The mere possession of a sea turtle or parts thereof (like a shell) is a federal offense. If you were to purchase a turtle shell overseas—in, for example, the Philippines, where they are readily available in the markets—it would be confiscated by customs agents when you enter the States, and you might be facing federal charges if you happened not to declare the fact that you had the item in possession.

The only commercial culture activity that has ever been established in association with green sea turtles occurred close to, but outside of the United States. Though discussion of that one operation that persisted for a number of years is a bit outside the scope of this book, it is appropriate because the target market for the turtle meat was almost exclusively America. The venture to which I refer was Mariculture Ltd. (The company changed hands and became the Cayman Turtle Farm Ltd. in the mid-1970s.) Founded in 1968[864,865] it remained in operation as a private enterprise until 1983, when it was taken over by the government of the Cayman Islands.[865] The firm began harvesting turtles in 1973 and was soon processing between 12,000 and 15,000 animals annually. That represented a total of 1 million pounds of live weight in sea turtles.[865] Initially, all of the eggs obtained for hatching and rearing on the farm came from the wild, and the company had an agreement to release a certain percentage of the young turtles that were reared. That practice was objectionable to some groups who were attempting to conserve natural populations, and a captive breeding population of green turtles was eventually established.[865,866] The company became independent of the need to obtain wild turtle eggs in 1978.[865,867]

To be profitable, virtually the entire turtle had to be sold. Much of the meat, as I said, was targeted for the United States where it was used for turtle steak and stew meat. Green turtle soup was produced in Europe. Leather and jewelry made from the skin, and shell was sold in Europe and Japan, while Mexico and Japan purchased the fat, which was rendered into oil and used in cosmetics and folk medicines.[865] The firm offered tours and had a gift shop in place to augment its cash flow.

Things changed dramatically for the turtle farm after the United States passed a ban in 1978 on the taking of green sea turtles in U.S. waters, along with a ban on the import, export, or transshipment of either the turtles or products made from them through this country. Cayman Turtle Farm asked for an exception to the ban but lost its appeal in 1980.[867] When the ban was first proposed in 1975, there was an exemption for mariculture products, but opposition from environmental groups caused that exemption to be removed from the final legislation.[865]

With the U.S. market closed, the farm turned to Japan as a market for turtle meat. However, the company was soon caught up in the Convention on International Trade in Endangered Species (CITES). While CITES provided for the shipment of sea turtles that were farm-bred, the fact that half of the broodstock of the Cayman Turtle Farm had been taken from nature muddied the water with respect to whether their product fit the CITES criteria.[865] While CITES continued wrestling with interpretation of their own rules, the turtle farm turned to marketing turtle products, including meat, only in the Cayman Islands.

In 1980, a breeding colony of Kemp's ridley sea turtles was established at the Cayman Island farm and a headstart program was initiated. The farm added crocodiles to its inventory of cultured species. Since becoming a government entity, has depended primarily on tourist dollars as a revenue source. Some 100,000 people a year visit the farm to see the various species, sample the food produced therefrom, and browse in the gift shop.*[865]

Alligators. The farming of alligators has now become a significant enterprise, particularly in Louisiana and Florida.[765,868] Research on alligator culture began in the 1950s,[868] though alligators were being held, sometimes in large numbers, as tourist attractions and for the marketing of their hides prior to that time. (I recall my father saying he had visited what he called an alligator farm in Florida when he was a youngster, which would have been in the 1920s.) Serious culture of alligators was not undertaken until the late 1970s.

What little bit of alligator farming did exist previously was curtailed in 1969 when the American alligator was declared an endangered species.[869] The designation was made even though the Louisiana population had apparently recovered by that time.[868] Recovery of alligators throughout their range was much more rapid than most people would have predicted, and by 1978, the classification of alligators had been changed from endangered to threatened.[869] In 1981, Florida became the 48th state to remove the ban on local sale of alligator products.

Since the status of alligators relative to the Endangered Species Act changed so that they could be farmed legally, at least 150 alligator farms have been established in Florida and Louisiana (some of which have subsequently failed).[870] Alligator farms were holding an inventory of about 250,000 animals in 1993. In 1989, about 16,000 alligators were harvested from farms in Florida, while the Louisiana harvest was placed at 67,000.[871] The industry faces competition from the wild-harvest (about 25,000 animals in Louisiana during 1989), and imports of wild and farmed croco-

---

*Prior to the ban, I tried green sea turtle steak at a restaurant in southern Florida that was famous for that so-called delicacy. I thought it was quite similar to chicken-fried steak and was nothing exceptional, but I suppose one should rave about how wonderful it was now that the opportunity to consume it has passed—unless, of course, you visit the Cayman Islands.

diles. Hides sell for as much as $32 per linear foot. The meat of alligators has now become an important product as well.*

Feeding protocols for alligators harken back to the early days of fish culture. While some pelleted feed is often provided, alligators also typically receive fresh fish, beef, chicken, horse meat, and nutria. As with other aquaculture ventures, feed costs amount to about half of the expense involved with alligator production.[871]

Water quality in alligator farms is often poor, and its release into receiving waters is becoming a significant problem for the reptile farmers as effluent standards are tightened. Treating the water prior to release is expensive, and meeting the requirements on effluent standards could seriously impact profitability.

Algae Culture

In this section, we'll take a look at seaweed culture as an end product.† The culture of algae as food for aquacultured animals is a routine activity in many facilities, and there has been some interest in culturing certain types of microscopic algae by the pharmaceutical and health-food industries, but those topics are not the focus of our discussion. We will stick strictly to macroscopic marine algae.

Products from seaweeds are a part of everyday life in the modern world. Substances extracted from marine algae are found in everything from automobile tires to toothpaste. It is extracts from algae that allow dentists to make impressions of your teeth, lets ice cream maintain its shape in a bowl so it doesn't form a puddle when it warms up, and is present in all types of cosmetics and pharmaceutical products. In their native form, seaweeds are used directly as food, particularly in Japan. Sushi wrappers are made from sheets of seaweed.

The possibility of rearing one type of seaweed, nori, in Puget Sound, Washington, was raised at a workshop on the subject held in 1982. Interest was significant, since over 100 people attended the event.[872] Research was soon initiated to determine which of the native species would produce an acceptable product, and to estimate potential production levels.[873] At least one commercial farm was put into operation.

Interest in growing the California giant kelp as a source of biomass has been discussed for several years. The idea, along with the design of a proposed kelp farm, was detailed in 1983, with the purpose of the conceptual facility being to produce massive amounts of kelp that would be converted to methane gas or methanol and used as fuel.[874] The concept has been discussed relative to other types of marine algae[875] and some freshwater species, but little in the way of actual development has occurred.

---

*I've tried alligator at some of the World Aquaculture Society get-togethers where farmers provided aquaculture products. My view is similar to the one expressed about sea turtles: it's okay, but I wasn't so excited that I'd want to run out and buy it every week. My wife, on the other hand, thought it was very good. Take your pick.
†Yes, for the purists, we have turned phylogeny on its ear by taking up algae at the end instead of the beginning of our survey. The reason for that is that while algae culture represents a significant percentage of total aquaculture production on a worldwide basis, it has received very little attention in the United States.

The commercial nori farmer in Puget Sound attempted to obtain permits for expansion to several hundred acres a few years ago. That effort prompted heavy resistance from some segments of the public that foiled that expansion. Because of the large areas required to produce significant quantities of algae, development of the industry may have to occur offshore. Some of the conceptual designs for algae farms would cover large offshore areas. Problems associated with working in the offshore environment, including the costs involved in establishing such facilities, have precluded development of the industry to date.

POTPOURRI

This chapter concludes with discussion of a few, often unrelated topics that have been important to aquaculture development since 1970 but that haven't been discussed in detail in the previous sections. Included is information on chemicals that are available to aquaculturists, a discussion of how the World Mariculture Society developed and became the World Aquaculture Society, and a few other topics.

Aquaculture Development

The 1970s was a period of optimism by aquaculturists, both those who were actually producing products and those who were conducting research. Agency personnel involved in the production of fishes for stocking felt that their activities were having a positive influence on recreational and, in some instances, commercial fisheries. Commercial culturists were developing new species and expanding the culture of those that had already been established. Researchers were finding ways of improving efficiency, developing feeds specifically designed for specific types of animals, finding ways of diagnosing and treating diseases, and developing technology to make the determination of water-quality variables simple and inexpensive. Producing fish for recreational fisherman and the hobbyist, augmenting commercial fisheries, growing fish for bait, and producing aquatic organisms as human food were all activities designed to serve humankind. Aquaculturists typically saw themselves as environmentalists. After all, if a high-quality environment wasn't maintained for the animals that were being cultured, it would not be possible to optimize production. Producers and researchers were the cowboys in the white hats.

While there was enthusiasm and optimism in the 1970s, there was also a realization that commercial aquaculture is a risky enterprise. Snake-oil salesmen were still out there touting the financial gains available to those who invested in aquaculture. (I once heard a shrimp culture entrepreneur make a presentation in which he indicated that his worst-case scenario was that there would be a 200% return on investment the first year.*) However, it was becoming apparent to those actually involved in the business that a good deal of thought should precede establishment of a facility.

*My view is that the worst-case scenario involves complete crop loss, which typically means that money will be lost by everyone except the person who cooked up the scheme and talked investors into wasting their money.

One of the most important decisions that needs to be made prior to the establishment of an aquaculture facility involves species selection. It seems as though aquaculturists are always looking for a new species to culture. That's something that is particularly exciting to researchers because they have the opportunity to solve some interesting problems and develop a new technology. It goes against traditional agricultural practice, however, where the emphasis has been on continuous improvement of just a few standard animals that have been grown in captivity for thousands of years. A diversity of domesticated breeds represents the norm in terrestrial aquaculture, while on the aquatic side a diversity in species using virtually wild animals is standard practice.

Pleas by those involved in aquaculture development for the production of domesticated stocks of fish and other types of aquatic animals have been made for many years, and while selective breeding has been practiced for some time, commercially produced fish and shellfish are usually indistinguishable from their wild counterparts. Wild turkeys and chickens, on the other hand, bear little resemblance to their cousins being produced in poultry houses.

Harold Webber and his colleague Pauline Riordan were heavily involved in aquaculture development during the 1970s. They took up the subject of candidate species for aquaculture in 1976 and compiled a list of attributes that a successful candidate should possess.[876] Among them, not surprisingly, was sufficient genetic variability to allow for the development of a domesticated breed. Other attributes included the ability to withstand and grow well under crowded conditions and the potential for economic culture; that is, the costs of production must be less than the value of the crop. Finally, and perhaps as important as any other factor, is that the product must be marketable. It often happens that economic production of an aquaculture species looks very good on paper, but aquaculturists sometimes (too frequently) make the mistake of assuming there will be a ready market within reach of the producer. That is sometimes not in fact the case, and a crop of an excellent-quality product may have to be buried if it cannot be sold. Many prospective culturists also assume that if the price of a particular species is several dollars in the supermarket, the farmer will obtain a price close to that obtained from the retailer. In reality, the price obtained by the culturist is generally less than 25% of the retail price.

Several candidate marine species were proposed by Webber and Riordan.[876] Among them were abalone, various species of clams, three types of crabs, lemon sole, English sole, Dover sole, Pacific halibut, various species of flounder, pompano, and a few members of the drum family (including seatrout, white sea bass, and corvina). Commercial culture has been developed in the United States for abalone, clams, and red drum. (The red drum was not specifically mentioned by Webber and Riordan.) The Atlantic halibut is being commercially cultured in Europe and has been the subject of research, as has the Pacific halibut, in North America. There has been some research on a few of the other candidate species listed, though no commercial culture of them is occurring in the United States at the present time.

Getting into the aquaculture business requires a good deal of advance planning. Prospective aquaculturists must not only find appropriate species and sites, they must also develop business plans and find financing. As we have already seen, finding a path through the regulatory quagmire that exists in many states is another major undertaking. Beginning with what was the first, and most intensive effort by any state to develop a plan for aquaculture development, Hawaii produced a state aquaculture plan in 1978.[877] The planning process undertaken in Hawaii was described in 1979 to demonstrate how the process was developed and implemented.[878] Other plans soon followed. I worked with Jim Davis at Texas A & M to develop one such plan for the state of Texas, which was produced in 1981.[879] Missouri developed a plan in 1982, and Florida began working on its plan the same year.[880] Ultimately, many states developed plans, none of which was as comprehensive as the Hawaii plan, but all of which provide help for prospective aquaculturists. The plans can help people identify suitable locations for aquaculture and provide them with assistance with respect to the permitting process.

While U.S. commercial aquaculture was growing rapidly during the 1970s and 1980s, there were still those who invested and lost large amounts of time and money. The situation got particularly difficult for new entrants into the field during the early 1980s when interest rates and land prices skyrocketed. Many who bought in at that time were unable to make sufficient profits to pay off their debts, while those with established businesses and little in the way of long-term debt were able to make reasonable profits. A paper published in 1984 listed six ways in which money could be lost in aquaculture ventures:[881]

1. Pinning hopes on as-yet-undeveloped technology.
2. Failure to realize that aquaculture is, first and foremost, a farming business.
3. Undercapitalization.
4. Aiming at the wrong market.
5. Failure to realize that the product is a live animal.
6. . . . lack of business experience.

Environmental Considerations

I took a course in ecology at the University of Nebraska in about 1965, several years before television commentators became aware of the topic. As my interest and opportunities to become involved in aquaculture developed, I always felt I was involved with an activity that was environmentally sound. As I've indicated previously, the rearing of aquatic animals at high densities requires maintenance of good water quality. Marketing of products that meet the demands of consumers (which means they must be nutritious, flavorful, attractive, and perceived as being healthful) is paramount. If anyone was wearing the white hats, it had to be the aquaculturists.

When criticism of aquaculture began to develop in the 1980s, it was at first dismissed, within the aquaculture community at least, as absurd. Some objections, such as those lodged by upland property owners in the Puget Sound region of Washington, who described salmon net-pens as visual pollution, were viewed as emanating from people with special interests. Failing to make a case for the right to unalterable views, the opponents began to decry the use of antibiotics, point to destruction of the benthos community, and find a number of other activities by net-pen culturists offensive and threatening to the environment.[643,644] In an attempt to counter the criticism, an environmental impact statement was prepared,[882] which demonstrated that there were, indeed, some environmental impacts, but that they could be avoided by proper siting of the facilities. Still, opposition was so strong that development of the industry was halted with about a dozen small farms in place.

While the net-pen issues in Washington were heating up (there was also a similar controversy brewing in Maine), criticism was beginning to be heard in other circles. Destruction of mangrove swamps in developing nations associated with the creation of shrimp ponds was one activity that was being roundly attacked. Given discussion overheard at professional meetings, I am convinced that the scientists associated with aquaculture were sensitive to that issue and opposed to further destruction of mangrove swamps, that they recognized the importance of those areas as nursery grounds and for the protection of coastal areas during storms. A case can certainly be made that the greed of politicians and commercial culturists in developing countries for cash flow overcame any interest in protecting sensitive environments. This was certainly a case where some of the aquaculturists were wearing black hats. The severity of that problem and others associated with the rapid development of aquaculture in the third world has been documented in an excellent publication of the International Center for Living Aquatic Resources Management in Manila, Philippines.[883]

Back in the United States, criticism of aquaculture began to increase and come from new quarters in the late 1980s. Much of the criticism was voiced in the press and electronic media—there was not much documented in the professional literature. In any case, some of the backlash to aquaculture involved perceptions such as the following:

- Effluents from trout hatcheries in Idaho were polluting the Snake River and causing increased growth of noxious vegetation.
- State hatcheries in Michigan and elsewhere were dumping large amounts of nutrients, particularly phosphate, in their effluent waters causing eutrophication* of receiving waters.
- Catfish farmers were rarely draining their ponds, and when they did, they dumped extremely nutrient-rich water, causing degradation of water quality in the receiving waters.

The initial response of the aquaculturists was disbelief. When criticisms such as those mentioned above first appeared, it was common for the scientists gathered at World Mariculture Society meetings and meetings of other professional organizations to scoff and dismiss them as insignificant. There wasn't much information in the scientific literature on the subject, but at least one study on the effects of trout

---

*Eutrophication is the natural process of aging that occurs in lakes. Highly eutrophic lakes are characterized by high levels of nutrients, dominated by rough fish, and may have heavy concentrations of aquatic vegetation of various kinds. Taken to its extreme, eutrophication will lead to the formation of marshes and, eventually, dry land. The natural rate of eutrophication may take thousands of years or longer in areas where nutrient inputs are low, for example, the Great Lakes region. Human activities have accelerated the process. Examples of lakes that became highly eutrophic, but were returned to a more pristine condition when nutrient inputs were reduced and the process was reversed, include Lake Washington (Seattle, Washington) and Lake Erie.

hatchery effluents on receiving waters had been conducted in Georgia, North Carolina, and South Carolina.[884] That study revealed no detrimental impact.

Government took the criticisms seriously and began more strict implementation of regulations, such as the requirement for NPDES permits, on the aquaculture industry. Aquaculturists in the United States, being strong advocates of a healthy environment for all, began to examine their practices, altering them to reduce the negative impacts. Trout farmers and others have constructed settling ponds to reduce the levels of solids being released. Solids can even be settled out in trout raceways.[885]

Aquaculturists in the United States and elsewhere have taken up the challenge to reduce the potential impacts their activities might have on receiving water bodies. The technology to treat effluent water exists, but it can be so expensive as to be an economic disaster for the aquaculturist who has to purchase the equipment. There is a good deal of research currently under way to produce feeds from which the phosphorus is better utilized by the animals being fed, thereby reducing the amount of that element in the feces and ultimately in the water. Jim Avault recommended the following ways to conserve pond water.[886]

(1) Reduce loss of water from seepage by constructing ponds on soils with a high clay content.
(2) Store rainfall whenever possible.
(3) Don't drain ponds to harvest.
(4) In a system of levee ponds, one pond can be pumped (drained) into an adjacent empty pond.
(5) Practice wise use of make-up water.

Other suggestions for reducing environmental impacts are to raise aquatic animals in recirculating water systems or move facilities offshore. Those topics are discussed in more detail in Chapter 6.

Not all of the issues relate to the impacts that aquaculture might be having on the natural environment. There are also issues surrounding how nature can negatively impact aquaculture. Marine mammals, which are protected by federal law, have been known to cut through the mesh of net-pens and allow fish to escape. Barrier nets of material that the mammals cannot chew through easily have been used to alleviate that problem.

Algae blooms have caused serious problems for net-pen salmon farmers. Various species of algae produce toxins that may directly affect salmon,[887,888] while others may clog gills and suffocate the fish.[889] The most widely recognized and most devastating to the industry is, however, bird predation.

Birds consume fish and aquatic invertebrates that are being cultured and also serve as intermediate hosts for a number of parasites. The most significant of the two problems is direct predation. In a survey of the problem in Arkansas, it was determined that entire minnow ponds were sometimes wiped out by birds.[890] The major culprits were great blue herons, snowy egrets, great egrets, double-crested cormorants, green-backed herons, little blue herons, black-crowned and yellow-crowned night herons, gulls, terns, diving ducks, grebes, and common grackles. Aquaculturists in Arkansas have estimated their annual losses to birds from as little as $200 to as much as $100,000.[891] The different birds have preferences for different aquatic species, though great blue herons, little blue herons, great egrets, and snowy egrets all seem to prefer golden shiners over other species grown in Arkansas. Herons, egrets, and ibises are heavy predators on crawfish in Louisiana.[890] The double-crested cormorant is the most serious bird predator on catfish in Mississippi.[892] Kingfishers are also often mentioned as fish predators.

Various methods have been devised for deterring birds. Among the common ones are noise cannons (which the birds tend to ignore after a few days), bird netting (which is prohibitively expensive for a large pond operation), stringing wires across raceways, and constant patrols by humans and/or dogs. It is possible to obtain federal permits to kill small numbers of some predatory birds, but because of the numbers of birds involved, that means of control is usually ineffective. Many of the species that prey on fish are strictly protected under federal law. One hatchery manager in Washington told me that if he could keep gulls from invading his raceways for three days, they would get so hungry that they would leave in search of other feeding grounds. He used dogs and human patrols to keep the birds away from the raceways.

Pelicans can also cause significant problems for fish farmers. Someone told me that about 400 pelicans showed up in Idaho in the early 1990s and decimated some of the small trout farms over a period of just a few days. How the pelicans got wind of the easy pickings is not known.

Registered Chemicals

The number of drugs chemicals that have been approved for use on humans and livestock in the United States since 1970 must number in the thousands, but the number of approved chemicals for use on aquatic animals has changed little during the same period. With respect to aquatic animals, there are two categories: food and nonfood. There are more chemicals available for nonfood animals than for those being produced as human food. If a chemical is not registered for use on foodfish, it cannot be used on those animals at any time during their life. Withdrawal periods between the time of use of a registered chemical and the time at which the animals can be harvested are also sometimes specified. Interestingly, fish that may be caught by recreational fishermen are placed in the nonfood category even though they might have been treated just before they were caught with a chemical not allowed for use on animals in the food category. All the chemicals approved for use on food animals can also be used on nonfood species. Compounds that were registered for use on food and nonfood aquatic animals during 1976, 1979, and 1989 are as follows[893–895] (with U indicating an unregistered chemical or one not mentioned):

Chemical	1976	1979	1989
2,4-D	Food	Food[a]	Food
Acetic acid	U	Food[b]	Food[b]
Aluminum sulfate	U	U	Food[b]
Amitrole	U	U	Nonfood
Antimycin	Nonfood	Nonfood	Food
Aquashade	U	U	Food

Chemical	1976	1979	1989
Calcium hypochlorite	U	Food	Food
Carbon dioxide	U	Food[b]	Food[b]
Casoron	Nonfood	Nonfood	Nonfood
Copper	U	Food	Food[b]
Copper sulfate	Food	Food	Food[b]
Dichlone	U	Nonfood	Nonfood
Diquat	Food	Food	Food
Endothall	Food	Food	Food[c]
Fluorescein sodium	U	Food[b]	Food[b]
Fluoridone	U	U	Food[d]
Formalin	U	U	Food
Furanace	Food	Nonfood	Unavailable
Lime	Food	Food[b]	Food[b]
Masoten	Nonfood	Nonfood	Nonfood
MS-222	Nonfood	Food[c]	Food
Potassium Permanganate	U	Food[b]	Food[b]
Potassium ricinoleate	U	U	Food[c]
Povidone-iodine compounds	U	U	Food[b]
Quaternary ammonium compounds	U	U	Food[b]
Rhodamine B	U	Food[b]	Food[b]
Rodeo	U	U	Food
Romet	U	U	Food[c]
Rotenone	Nonfood	Nonfood	Food
Sanaqua	U	U	Food
Silvex	U	Nonfood	U
Simazine	Food	Food	Food
Sodium bicarbonate	U	Food[b]	Food[b]
Sodium chloride	Food	Food[b]	Food[c]
Sulfamerazine	Food	Food[e]	Unavailable
Terramycin	Food	Food	Food[c]
Xylene	U	Nonfood	Food[c]

[a]Restricted to use by federal, state, or local public agencies.
[b]Listed as GRAS "generally regarded as safe" and exempt from registration.
[c]Withdrawal period specified.
[d]Not to be used in marine or brackish water or where crawfish are being raised.
[e]Approved only for salmon and trout.

In 1992 the Food and Drug Administration (FDA) reviewed the list of chemicals used by aquaculturists and considered the following drugs to be of low regulatory priority: acetic acid, carbon dioxide gas, calcium chloride, povidone iodine, sodium chloride, and sodium sulfite.[896] In 1993, the FDA decided that sodium chloride was a drug and planned to regulate its use in aquaculture. The argument that marine fish are exposed to salt all the time didn't seem to have much impact initially, but the FDA finally relented and decided to allow table salt to be used by aquaculturists.

The situation with respect to obtaining approval for the utilization of chemicals in aquaculture is not improving; it may even be growing worse, as the incident with respect to sodium chloride indicates. The costs involved with clearing a chemical are enormous, hysteria about the use of chemicals is growing, and the FDA is increasing its stranglehold. When aquaculturists look in their bags of tricks to pull out chemicals with which to treat aquatic animals, they find those bags to be nearly empty.

The World Mariculture Society/World Aquaculture Society

The World Mariculture Society (WMS) has become the leading organization in which aquaculture scientists gather each year to discuss the results of their research; learn about new technology through the trade show that is a part of each annual meeting; and interact with friends, colleagues, and practicing aquatic farmers. As the organization grew from the handful of people who attended the first couple of meetings to over 2,000 members today, many changes occurred. Although the majority of the members in the WMS were from the United States, there was dedication on the part of the society to maintain its international focus, so a decision was made after a few years to hold one annual meeting every three years outside of the States. The WMS board of directors attempted to maintain breadth within the society not only by meeting outside of the United States but also by having the officers and board members be represented by individuals from around the world. That initiative has been successful, particularly with respect to the annual meeting locations. The following table lists the presidents of the WMS through 1994 and the meeting location each year:[607]

Year	President and Country	Annual Meeting Location
1969	G. Robert Lunz (USA)	
1970	Theodore Ford (USA)	Baton Rouge, Louisiana
1971	Samuel Munroe (USA)	Galveston, Texas
1972	Terrance Leary (USA)	St. Petersburg, Florida
1973	Gordon Gunter (USA)	Monterrey, Mexico
1974	Wallace Klussman (USA)	Charleston, South Carolina
1975	James Avault, Jr. (USA)	Seattle, Washington
1976	Harold Webber (USA)	San Diego, California
1977	Kenneth Chew (USA)	San Jose, Costa Rica
1978	John Glude (USA)	Atlanta, Georgia
1979	Paul Sandifer (USA)	Honolulu, Hawaii
1980	Carl Sindermann (USA)	New Orleans, Louisiana
1981	W. Guthrie Perry, Jr. (USA)	Seattle, Washington and Venice, Italy
1982	George Lockwood (USA)	Charleston, South Carolina
1983	Fred Conte (USA)	Washington, D.C.
1984	Gary Pruder (USA)	Vancouver, Canada
1985	Guido Persoone (Belgium)	Orlando, Florida
1986	S. Ken Johnson (USA)	Reno, Nevada
1987	David Aiken (Canada)	Guayaquil, Ecuador
1988	John Manzi (USA)	Honolulu, Hawaii
1989	John Castell (Canada)	Los Angeles, California
1990	William Hershberger (USA)	Halifax, Canada
1991	Robert Stickney (USA)	San Juan, Puerto Rico

Year	President and Country	Annual Meeting Location
1992	Louis D'Abramo (USA)	Orlando, Florida
1993	Susan Waddy (Canada)	Torremolinos, Spain
1994	LeRoy Cresswell (USA)	New Orleans, Louisiana
1995	George Chamberlain (USA)	San Diego, California
1996	Meryl Broussard (USA)	Bangkok, Thailand

In 1978, the society met in Atlanta, Georgia, in conjunction with the Catfish Farmers of America and the Fish Culture Section of the AFS. The collaborative effort was successful and was repeated in 1980 in New Orleans, Louisiana. Thereafter, what came to be known as the triennial meetings were scheduled on a routine basis. Meetings were held in 1983 (Washington, D.C.), 1986 (Reno, Nevada), 1989 (Los Angeles, California), 1992 (Orlando, Florida), and 1995 (San Diego, California). There have been changes in participants over the years (such as the U.S. Trout Farmers Association and the Shellfish Institute of North America), though the WMS and the Fish Culture Section have participated in each of the triennial meetings. The third current participant, and one that has been actively involved in most of the meetings, is the National Shellfisheries Association.[607]

The Washington, D. C., triennial meeting in 1983 was developed with the idea of highlighting aquaculture to the U.S. Congress. Secretary of Agriculture John R. Block was the keynote speaker. Senator Thad Cochran of Mississippi and Representative Daniel K. Akaka of Hawaii were other featured speakers.[897] There were great expectations associated with the reception that was held in the ballroom of the conference hotel. Members of Congress were invited, and the reception was set up so that if they showed up, they would be treated to a variety of aquacultured delicacies. There was only one problem. Congress was not in session, so the members were back home, not in Washington, D.C. A number of aides did show up, but there didn't seem to be a great deal of positive impact on Congress generated by the meeting.

While the WMS had been initially dedicated to the culture of marine organisms, it was, with the exception of the Fish Culture Section of the AFS, the only scientific organization at which U.S. aquaculture scientists could gather on an annual basis. Thus, somewhat by default, the number of presentations made at annual meetings and the interest of the membership in freshwater aquaculture increased as the organization grew and matured. A survey was conducted in 1985 to determine whether the name should be changed to the World Aquaculture Society (WAS) to better reflect the mission of the organization. That survey won the support of the membership, and the name change became official in 1986.

During my years of service on the WAS board and as its president-elect, president, and past president, I frequently reflected on the notion of having a truly global society. No successful scientific societies of that nature jump to mind, so in some ways the WAS has plowed new ground. Many in the international community viewed the WMS, and later the WAS, as a U.S. organization with the audacity to use the word *world* in its title. Similar organizations were established in other regions or individual countries. The WAS established an affiliation policy designed to assist development of those organizations by providing full membership benefits, including at a very modest rate the WAS newsletter (which became *World Aquaculture* magazine in 1988, but excluded a subscription to the *Journal of the World Aquaculture Society*). The European Aquaculture Society, the Aquaculture Association of Canada, and the Caribbean Aquaculture Association became affiliates and sent representatives to the WAS board meetings for several years. The Asian Fisheries Society became an associate of the WAS in 1989, which provided members of that society with reduced registration rates at WAS meetings and member prices on literature produced by the WAS. Chapters, which were defined as national or regional subgroups of the parent society, were also approved. The first one, the U.S. Chapter, was formed in 1989. The Japan Chapter was formed two years later.[607]

In 1993, the board decided that the three affiliates had matured to the point that they could stand on their own. The affiliate category was deleted, and the three organizations that had held that status were offered associate status instead. Each associate was told it could negotiate its own arrangements with the WAS with respect to member benefits. That change led to some hard feelings, particularly with respect to the European Aquaculture Society, and negotiations continue as of this writing.

The WAS began publishing annual meeting proceedings after its 1970 meeting in Baton Rouge, Louisiana. In 1981, the proceedings became the *Journal of the World Aquaculture Society*. The proceedings and journal were published at Louisiana State University through 1985, after which a commercial publisher was retained. With creation of the journal, papers presented at the annual meeting were not automatically published. Instead, papers had to be submitted to the journal editor, after which they were subjected to peer review. The papers published in the journal in some cases have been presented at an annual meeting, but more often that is not the case. The quality of the publication has improved greatly over the years, and it is now one of the leading aquaculture journals in the world. The WAS also launched a book series in the 1980s.

The WAS home office has been located on the campus of Louisiana State University ever since it was established in 1979. Prior to that time the contact point for members was the treasurer of the society since that was where annual dues were sent. Sam Meyers, a professor at Louisiana State University, held the position of secretary from 1977 through 1983, and his home was in reality the WAS office during much of that time. The current home office is efficiently managed by Juliette Massey, who had dedicated herself, body and soul, to the society.

CHAPTER 6

Projecting into the Twenty-First Century

In this final, short chapter, some thoughts on the future of U.S. aquaculture are presented. As I think you will agree has been made clear in Chapter 5, the optimism surrounding continued rapid expansion of aquaculture in this nation has largely dissipated. That does not mean that aquaculture will not grow, for certainly it will, at least in some sectors. However, we have to face several facts with respect to what aquaculture might contribute to the nation's overall food production in terms of commercial culture and to the maintenance and restoration of recreational and commercial fisheries.

While various figures are thrown about, it is a generally recognized fact that the bulk of the U.S. population lives within 50 miles or so of the sea. We could also include the Great Lakes, which are inland freshwater seas in the minds of many, in that figure. Pressure to develop coastal lands, on the one hand, and to protect them, on the other, led to high tension levels in those regions where more and more people want to live. The coastal regions are also the best locations for certain forms of aquaculture, but I think most people who critically examine the situation will recognize that aquaculture cannot compete economically with the construction of resorts, condominiums, marinas, upland-view property developments, coastal industry, and general urban development. Coastline areas hostile to aquaculture, such as the one along the Laguna Madre in Texas are not difficult to find but have severe limitations, such as no fresh water with which to control salinity when evaporation causes the salt content of the ponds to increase rapidly.

Beachfront property that was virtually worthless along some of our coasts a century ago now sells for up to thousands of dollars a front foot. Aquaculture can be profitable, but no one is going to make sufficient money growing fish, molluscs, or crustaceans in the face of such high land costs. Moving aquaculture into bays and estuaries by growing fish in net-pens has provided a simple, effective solution to the problem of high-priced coastal land, but it has resulted in outrage from at least some of the population, who feel that aquaculture intrusion into the commons should not be allowed. Interference with recreation and commercial fishing have already been mentioned. Then there is just the gut feeling that nobody should make a profit in regions that are in the public domain.* Residents in coastal areas may, in fact, feel that aquaculture is an appropriate use of their waters, though the NIMBY (not in my backyard) syndrome operates effectively when proposals to actually begin establishment of facilities are generated. Typically, strong opposition comes from a very small minority of the people within any given region, but it only takes a few vocal people willing to attend public meetings and express their outrage to derail a proposed aquaculture facility.

Inland, places like the Mississippi Delta, with its once seemingly unlimited water supply, and the Hagerman Valley, Idaho, where the Snake River Canyon walls exude water by the millions of gallons per hour, are rare indeed. In these two cases, aquaculture development has reached the point where the water resource is being almost completely utilized or where expansion is being constrained by those concerned about environmental impacts, so opportunities for growth are extremely limited.

Geothermal resources exist in many regions that could provide significant quantities of warm water even in high temperate climates, thereby making it possible, at least in theory, to produce warmwater fishes. Much of the geothermal resource cannot be used directly because of high levels of sulfur and other water-quality problems.

Obtaining water for aquaculture can be a considerable problem. In many parts of the nation the available water has been entirely allocated to users who were in place before aquaculturists came along. In places where crops are irrigated with ground or even surface water, there is no opportunity for expansion because there is no additional water available. Withdrawal of water from rivers and streams is strictly controlled in many parts of the country, and there may be no opportunity to drill additional wells because the water table cannot tolerate additional withdrawal. In many regions, not only is the amount of water that can be removed from the ground controlled, but the number of days per year that wells can be pumped is regulated. A fish farmer who can purchase land and obtain a water allocation for what was previously irrigated farmland may be in a position where water is only available for six to eight months a year.

Public-sector aquaculture provides the foundation upon which all existing activity has been built. Initially aimed at augmenting declining commercial fish stocks, public-sector stocking programs concentrate much more on recreational fisheries today. The most visible remnant from the original charge of the Fish and Fisheries Commission can be found today with respect to salmon stocking programs in the Pacific Northwest. No emphasis is being placed on aquaculture as a solution to declines in cod in New England, nor can regulators bring themselves to taking the

---

*Persons holding that view sometimes forget that public lands are being leased for oil and gas production (or they may decry the fact that such leasing was ever authorized by Congress) and that commerce, including commercial fishing, utilizes the commons without much criticism, except when an oil tanker runs aground.

obviously necessary action required to allow the fisheries to rebuild. That action has only been taken when recreational and commercial fishing interests go head to head in confrontation. This has happened with respect to striped bass along the East Coast and red drum in the Gulf of Mexico. In both instances, at least some commercial fisheries were closed, perhaps for all time, and recreational fishing was put on hold until stock rebuilding—with the help of aquaculture—took place. Recovery tends to be rather quick, or at least it has been in the two cases mentioned. The same could be true with respect to cod in New England, though experts tell me that no one really has a good prediction as to how long the fishery would have to be closed before it would recover sufficiently to allow even moderate exploitation once again.

Salmon is a very hot topic in the Pacific Northwest at the present time, with much of the hue and cry being directed at stocks that spawn in the Columbia River Basin. The issues are international in scope, the 1994 salmon war between the United States and Canada being a good example. The result of poor returns of fish to the coast of Washington in 1994 resulted in a virtually complete closure of both the commercial and sport fisheries. Everyone was outraged. Calls were renewed for removal of the hydroelectric dams on the Columbia River, and other draconian and economically (as well as politically) untenable solutions were proposed. While little is being discussed in the media about the future of salmon fishing along the northern California, Oregon, and Washington coast, past experience predicts that commercial fishing will give way to recreational interests, for it is the recreational fishing industry that involves not only more people (many of whom are very vocal), but also more money. In Washington there is the added dimension of the American Indian tribal fishermen, who have, under federal law, the right to take half of the fish allocated to the fishery.

Hatchery bashing has become popular in some fisheries circles over the past few years, particularly by fish geneticists who, in some instances, feel that hatchery fish are leading to a decrease in genetic diversity (whatever that means). We have already seen, in Chapter 4, how the situation with respect to dams and hatcheries on the Columbia River system developed. My view has been, and continues to be, that hatcheries should continue to be a part of the overall salmon recovery approach.[427] Basically, I believe that hatchery practices can be modified to the extent that the fish produced are behaviorally and genetically indistinguishable from their wild counterparts. This will mean some significant changes in the way hatcheries are operated; it will also require expensive retrofitting of existing facilities or construction of new ones in order to emphasize fish of high quality with good survival chances rather than high numbers with little chance of living long enough to return as adults.

Restoring habitat, improving the means of getting fish around and over dams, and improving the survival of natural populations of salmon through other methods are all laudable, but they have their limitations. There is only so much habitat that can realistically be improved, and some dams just don't lend themselves to modification of the type needed. In the meantime, the human population will continue to grow and pressure on the resource will increase, not be ameliorated. The question can be legitimately asked: *can we afford to protect individual stocks of fish?* As written, the Endangered Species Act (ESA) is being interpreted very narrowly. Rather than protecting species, we are being asked to protect salmon stocks. The media decry what they assume to be a fact that several species of salmon are in danger of extinction. In reality, there are five species of salmon native to the West Coast of the United States, and none of them is in any danger of extinction. Not one of those species is even threatened. There are a number of stocks that are present in very low numbers, but those fish are not separate species. They can be identified as distinct through DNA and electrophoretic fingerprinting, but they are still members of one of only five species.

One wonders how much evolution could have occurred since the last ice age about 10,000 years ago. Prior to the retreat of the glaciers there were no salmon in the Pacific Northwest, because there wasn't any place for them to live. The fish invaded once conditions were appropriate. Some of the invaders have been extirpated over the years from various streams due to landslides and other acts of nature. In addition, new habitats have been created by the same means. So, are these fish really sufficiently distinct in the genetic sense to make the effort and expense to protect them as individual "species" under the ESA worthwhile? There is a danger that once the costs of protecting isolated stocks of fish become known, there will be a backlash by the public and Congress. One result could be the gutting of the ESA. That could, in the long run, have more serious consequences than losing a few fish stocks.

These issues are, of course, highly charged emotionally. Large amounts of time and money are being, and have been, expended in trying to resolve them. I anticipate that those efforts are only just beginning and cannot predict the outcome with any certainty. It seems to me (and this is going to be viewed with derision in some circles) that humankind is also part of the equation. Many within our community would like to see a return to those pristine times of the early nineteenth century when the forests were largely intact, there was little pollution from the burning of fossil fuels, cities did not sprawl across some of the most beautiful regions of the continent, and graffiti did not cover every visible rock. We cannot go back. It is far more likely that unless we solve the real problem, we will begin to appear more like a developing nation than one that is the most highly developed in the world. The root of the problem is, of course, too many human beings in our population. We do little more than pay lip service to that problem, but it is the one that underlies many of the problems that we face today, and until it is solved, there is little hope that the tension level with respect to environmental issues, including those associated with fisheries and aquaculture, will do anything but increase.

Most of the remainder of this chapter relates to the future of commercial aquaculture. I think there will be some changes in public aquaculture, but for the most part the existing situation should continue largely in its present form. The next section deals with the public sector, after which the commercial sector and its future are discussed.

THE FUTURE OF PUBLIC-SECTOR AQUACULTURE

State and federal agencies can be expected to continue operating hatcheries to provide fish for stocking recreational waters. There will be continued emphasis on reducing the levels of nutrients released from both public and private facilities, and that topic of research can be expected to occupy the time of government and university scientists for some years to come. The issue is sufficiently important that it has become a major factor in the current situation, in which there are very few new public facilities being constructed in the United States. The problem can be overcome with modifications in feed and feeding practices, and through the development and utilization of new technology that will be used to treat effluent water.

Two significant changes can be anticipated in the way fish are produced for release into public waters. One is that the private sector can be expected to take on an expanded role and the second involves extension of current activities into some areas that were abandoned many decades ago—that is, the development of marine fish enhancement programs. Such programs have already demonstrated their effectiveness with respect to striped bass and red drum recovery programs. The maintenance of reasonably good fishing in inland waters is achieved through well-run stocking programs and regulations. In the marine environment, pressure on certain species has reduced stocks to extremely low levels. Bans of fishing, such as those put in place with respect to striped bass, red drum, and salmon in some locations and years, are examples of how fishing pressure can lead to drastic remedial measures. In Washington, the nearly total ban on salmon fishing in 1994 did not dissuade avid fishermen from pursuing their sport. Pressure on inland trout waters increased dramatically, and those who are determined to fish in the marine environment have been increasing the pressure on already depleted nonsalmonid fishes in Puget Sound and are filling charter boats that target species such as halibut and tuna.

Recognition that nonsalmonid stocks in the Pacific Northwest had already been hammered by heavy fishing pressure and that further limitations in salmon fishing were inevitable led to the convening of a meeting in Seattle, Washington, on the subject of marine fish enhancement in October 1993.[898] Speakers at the meeting documented the declines in marine fishes in the Pacific Northwest and examined the potential for developing an enhancement program. Examples of how such activities had been developed in other regions and nations were presented. Included were presentations on red drum,[899] Atlantic halibut,[900] Pacific halibut,[901] and cod.[902] A list of potential species that might be used to enhance populations in Puget Sound was developed. Legislative funding to initiate a marine fish enhancement program was provided in the form of a $10 increase in the fee charged for salmon stamps. The fact that salmon fishing was almost closed down in 1994, which translates to few sales of salmon stamps, will obviously have a severe impact on development of the program. However, the groundwork has been laid, and preliminary research plans have been developed. The basic idea will almost certainly receive attention in other parts of the country where similar declines in native marine fish populations have occurred.

Commercial aquaculturists are already playing an increased role in the production of recreational fishes. Most states provide, and for many years have provided, fish free of charge for stocking private waters (such as farm ponds). Private fish farmers have the ability to sell their fish for stocking private waters in many states. (There may be restrictions on the sale of exotics species, of course.) There has been at least some discussion as to the appropriateness of providing free fish for stocking private waters because it is public funds that provide this benefit, which is not generally available to the population at large. While it is true that in many states, fishing access to private waters is approved with permission of the landowner, the access to most private waters by the fishing public is restricted. Changing the process to require persons wishing to stock private waters to obtain their fish from commercial fish farmers makes a good deal of sense.

Public hatchery personnel frequently complain that commercial fish farmers have inferior fish and are more prone to spreading diseases around than are state and federal facilities. That view is hotly contested by many private culturists who also profess to produce more fish much more economically than their public-sector counterparts. No matter where the truth lies, it should not be difficult to develop a process by which a disease certification and hatchery inspection program for private culture facilities could be put in place that would ensure the maintenance of quality.

Marine fish enhancement programs could also benefit from public-sector input. State and federal programs involved with the hatching and stocking of most of the marine fish under consideration either have never developed or were abandoned long ago, so the public-sector aquaculturists are not in a better position to move into marine fish enhancement than are private producers. The interest in the private sector became apparent at the Seattle meeting in October 1993.[898] The lack of facilities for (and in some cases interest in) rearing nontraditional marine fishes by biologists working in state hatcheries also became apparent at that meeting. Public funding for such programs will be required whether production comes from the private sector, the public sector, or both. User fees seem to be the most appropriate mechanism for obtaining funding for enhancement. The future of commercial aquaculture is the subject of the next section, but it is clear that private enterprise has a role to play in what have been areas dominated by the public sector.

Before we leave the subject of enhancement, it is important to mention that capture fisheries seem to have peaked and that a number of species that have contributed significantly to the harvest of animals used in the human food supply are in serious decline.[903] Aquaculture will not be able to make up for the declines in cod, pollock, and similar species because those fishes cannot be reared to market size profitably. There may be instances where cod and similar species can be produced for release as fingerlings to enhance recreational fisheries, but the notion of producing them in the quantities required to restore them to commercially harvestable levels is wishful thinking. Regulation (including closure of fisheries, perhaps for several years when necessary), limiting entry, and imposing strict controls on harvest once fishing is resumed together provide a much better approach.

VISIONS OF THE FUTURE FOR THE PRIVATE SECTOR

To set the scene for what may be forthcoming in U.S. aquaculture, I would like to look back to some views expressed by others over the past couple of decades. First, let's look at what Harold Webber had to say about the risks facing aquaculturists in the 1970s.[904] At that time the risks were associated with constraints on biological knowledge, the physical environment, and economics. Webber saw the most critical biological risks as emanating from diseases, production of juveniles for stocking, predation and competition, inventory control (that is, developing the ability to accurately determine the number of animals within a culture system at any given time), nutrition, and managing the culture system to enhance the production of natural food organisms. Physical factors that were seen as posing significant risks included water-quality degradation and storm damage. Many of the social and economic risks discussed by Webber were of particular importance outside of the United States though such things as the availability of trained labor, poaching, the availability of suitable land, and regulations had, and continue to have, significant implications domestically. All the emphasis at the time Webber was writing was focused on constraints to rapid development of the aquaculture enterprise. There was little or no recognition at the time that aquaculture might have negative impacts on the environment.

Jim Avault has long been an outspoken supporter of aquaculture development and has played an important role in the growth of the industry in Louisiana. In his recent writings, he has begun stressing the need for sustainability* in aquaculture,[903] which represents a change in focus. In 1986, Avault argued that federal government support of aquaculture was in the national interest since it would reduce the importation of fishery products and reduce the deficit in that area.[905] By 1990, the issue of waste production from aquaculture had come to center stage, and Avault's column in *Aquaculture Magazine* focused on that issue. His outlook for the expansion of aquaculture with respect to many species was not optimistic. He expressed the following points of view:[906]

Trout: Increased production will come primarily from existing facilities rather than expansion of new lands.

Salmon: We will always have social/legal problems as long as public waters are used. Even if we clean up all fish wastes from pen culture, the public still perceives floating pens in pristine waters as an eye sore.

Crawfish: Two major constraints impede expansion, costly harvesting methods and a volatile market. . . . [If those problems are solved,] the current acreage of 135,000 could double in 5 years time.

Shrimp: Legal constraints . . . will impede shrimp farming from becoming a major industry in the United States.

Bait fish: Some farmers are having problems with bird predation.

Oysters, Clams, Other Mollusks: We must clean up pollution in our waters if acreage is to expand.

Exotics: Bighead carp, grass carp and tilapia have great potential in polyculture and with ethnic markets if legal constraints can be overcome. . . .

New Species, New Industries: Hybrid striped bass have excellent potential. Markets await this product on the east coast. Red fish has potential too but will not develop as quickly as the hybrids. Alligator farming will continue to burgeon, producing both hides and meats. Bowfin cavier [sic], soft-shell crawfish and soft-shell prawns are new industries waiting in the wings.

---

*Sustainability is a term that has been thrown around for the past few years. It developed as a buzzword with respect to agriculture. While there are many definitions, the most simple is that sustainable agriculture or aquaculture involves year-to-year production without depleting the natural resources.

Avault obviously felt that aquaculture on private land would be successful and could expand, while the use of public water (used extensively for the production of trout and salmon) represented a severe constraint, though he later indicated that consumption of those products would become more commonplace (perhaps through the use of imports?). He presented his overall view of aquaculture's future in the United States as follows:[906]

> I see aquaculture as one of the most challenging, dynamic, exciting, growth industries in the United States. I see per capita consumption of fish in the U.S. doubling from 15 to 30 pounds. Aquaculture will furnish 25% of all seafoods consumed. In the future, . . . it wouldn't surprise me if [per capita consumption of catfish] reached 5 or even 10 pounds. I see trout and salmon as regular table fare in many homes across the country.

Another stab at predicting the relatively near-term future of U.S. aquaculture was taken by David Harvey in 1994.[907] His predictions included the following:

- . . . the commercial availability of major fresh water fish such as striped bass, walleye, and yellow perch and shellfish species will have shifted almost entirely to aquaculture production.
- With relatively large amounts of suitable land and water supplies, plus a large domestic market, the United States will likely remain a major supplier of cold and temperate fresh water species.
- . . . [there will be] a growing body of knowledge in the areas of reproduction, growout strategies, disease prevention, and nutrition.
- The United States' position as a major grains producer . . . will be a stimulus to aquaculture research.
- . . . gains in aquaculture production efficiency will lower real production costs.

I would argue that suitable land and water supplies are extremely limited (with the exception of offshore waters, which are discussed later), so the second item is not one to which I, or Avault it seems, would subscribe.

Also contrary to Avault's position,[906] Harvey predicted a continued decline in per capita seafood consumption.[907] The trend in per capita consumption was up between 1982 (12.5 pounds) and 1987 (slightly over 16 pounds), after which it declined to 14.8 pounds in 1992. Harvey's position was that while per capita consumption would not increase, the overall demand for seafood would increase because of a growing human population. Not everyone agrees with that assessment. While no one can doubt that the human population will increase, a recent article showed that per capita consumption in 1993 was up slightly (to 15 pounds) from 1992.[908] In any case, per capita seafood consumption in the United States is low compared with many other nations; even with calls for healthy eating by incorporating more seafoods in the diet, per capita consumption of red meat appears to have bottomed out and is on the rise. Fish and shellfish, for all their laudable characteristics, are not competitive with respect to cost with poultry and will continue to be compete with red meat for a place on the American dinner table.

Before presenting my own views on where U.S. aquaculture might be going, let's take a look at information from a recent summary of its status. *Aquaculture Magazine* annually publishes a status report on aquaculture in the world, and the following information comes from their 1993 report:[909]

The value of 1991 fish farm production exceeded 760 million dollars; it is estimated that the 1992 value reached one billion dollars.

. . . As the aquaculture industry grows, it is coming into direct competition with commercial fisheries and imports.

. . . demand for catfish in 1993 led to a price increase at the pond [bank] from 63 cents in January to 73 cents in August and remained firm for the balance of the year. Estimated total production at the pond site is expected to be about 460 million pounds.

. . . Trout production for food was 60.9 million pounds for the . . . period September 1, 1992 through August 31, 1993. Although the weight was down 5 percent from the previous year, the sales value was $54.3 million, up 7 percent. . . .

. . . The leading producers of food size trout are Idaho, North Carolina, California, Utah and Pennsylvania in that order. There are 630 commercial trout producers in the United States concentrated in approximately 15 states.

. . . Live weight production [of tilapia] in 1992 was 9 million pounds and the estimate for 1993 is 12 million pounds.

. . . 1992 shrimp farm production was 4.4 million pounds live weight. . . . the United States does not appear to be destined to be a major producer.

Salmon production in the United States continues to grow in spite of serious fluctuations in the market and several periods of over production. It is estimated that the 1992 figures for salmon production in the United States are a total of 20 million pounds of live weight. . . .

. . . Louisiana produced 60 million pounds of crawfish with a value of $33 million for the year 1992.

. . . Ornamental fish . . . had $33 million in farm sales in 1991.

. . . Baitfish had a total production of over 15 million pounds and a value of $45 million for the year 1992. The state of Arkansas as the major producer accounted for 14,225,000 pounds and $40,143,750 income.

. . . Washington . . . [produced] an estimated 52 million pounds live weight [of oysters in] 1991. California produced 9,975,000 and Oregon 1,298,000 pounds. Maryland reports its oysters by individual count. . . . Its estimated production for 1993 is 3,364,500 oysters. Florida reported oyster sales for 1991 at $1.4 million.

. . . Mussel production for the west coast of the United States in 1991 totaled 676,000 pounds live weight. . . .

. . . Currently there are nine cultured redfish operations along the coast of Texas.

. . . Alligator exports of raw skins had a spectacular growth from 1977 to 1990, from 2,000 to 125,000. Record prices for a good quality six foot skin was $250. Louisiana and Florida has 150 farmers and over 250,000 alligators. Alligator sales for . . . Florida in 1991 were $4,393,000.

No summary of constraints and opportunities for U.S. aquaculture would be complete without at least mentioning fish in space* (the rearing of fish to provide food for humans living and working in space). The National Aeronautics and Space Administration (NASA) has had some interest in raising algae and various types of animals (including tilapia) in space for a number of years and has funded some

---

*When I use that phrase I think about the "Pigs in Space" skits on television's Saturday Night Live.

research in that area. The Biosphere 2 program in Arizona, which was designed, in part, to demonstrate how humans might survive in space, also incorporated tilapia culture. Whether we'll ever see aquatic animals reared in space is questionable, but using algae and other types of aquatic plants to assist with cleaning up wastewater, which will have to be recycled, and for producing oxygen from carbon dioxide exhaled by humans and perhaps other animals, would seem to have merit.

I'll not speculate further on fish and other aquatic organisms in space, but will look at what I think are two most likely directions that U.S. aquaculture will take in the future: closed systems and offshore systems, which are quite distinct. Before discussing them, I would like to make the point that I don't believe that U.S. aquaculture will develop in the absence of more traditional approaches. Usable water and suitable land are becoming more difficult to find, but they do exist, and aquaculturists will continue to search them out and put them into production. It is not likely that we will see anything like the level of production that has been achieved with trout in Idaho, catfish in Mississippi, or crawfish in Louisiana, since the resources that allowed for the development of those industries are no longer available. What I envision is modest expansion in both the traditional approaches to aquaculture and in the closed and offshore systems.

Closed Water Systems

Closed, or recirculating, water systems are those in which the water that exits the culture tanks is treated and reused. We've discussed a few such systems previously, and I have pointed out that no system is totally closed. At least some water must be routinely added to replace evaporation and other losses. It may be possible to reach something like 95% recirculation, and at that level we can consider a system to be closed. In the paragraphs that follow I'll use the word *recirculation* in conjunction with systems that add more than 5% of the system volume in new water daily.

While I have yet to see a closed system producing foodfish that is economically viable, that day is not far off. In fact, I have recently seen some proposals for facilities that just might be successful. New biofilter technology, the incorporation of ozone into recirculating systems,* and other technological advances bode better for the potential success of closed systems each day. It is possible at the present time to operate systems that recirculate at least 60–75% of their water daily and to do it economically if (and this is a big if) the water is available at little or no cost and does not have to undergo temperature alteration. It may be possible to recirculate economically at higher rates, and the time might be upon us when a truly closed system can be operated at a profit.

Recirculating systems such as the King James Shrimp Company in Chicago (mentioned in Chapter 5) were forced to utilize fossil fuels to maintain water temperature within the optimum range for the shrimp. Keeping shrimp warm in Chicago during the winter was a highly fuel-consumptive process. Many who have attempted to rear tilapia in recirculating or closed systems have employed groundwater that had to be

---

*Ozonation following biofiltration kills disease organisms, oxygenates the water, and oxidizes organic matter.

heated before it could be exposed to the fish. Some firms that never lasted very long, attempted to heat water used in flow-through systems.

It may be possible to employ water purchased from municipalities in systems that approach the status of being truly closed since makeup water requirements are not very high (no more than 5%, as previously indicated). A 500,000-gallon closed system would require, at most, 25,000 gallons a day of new water. At something like $1.00 per thousand gallons for domestic water, that is $25.00 a day, or a little over $9,000 a year. Using a more realistic recirculation rate of 60% requires 200,000 gallons per day, or $73,000 in annual water costs. The producer who has a source of artesian water available will have a clear financial advantage, particularly if that water is at the proper temperature.

The costs of heating or cooling water are generally prohibitive if fossil fuels or electricity (which may be derived from hydroelectric dams, fossil fuels, wind, solar energy, or nuclear power) are used. Heating water for aquaculture is more common than cooling, and in most cases the need of aquaculturists is to provide heat, not cooling, so this discussion centers on increasing rather than decreasing the temperature of incoming water. To get around the costs associated with water heating, aquaculturists using recirculating systems have employed geothermal water in conjunction with heat exchangers* or have planned their facilities as cogeneration operations. Examples of cogeneration facilities are those established adjacent to power plants or industries that produce large quantities of waste heat. Heated water from such facilities is usually not suitable for exposure to fish or invertebrate aquaculture animals directly, but it can be passed through heat exchangers from which heat is transferred to the water used in conjunction with the aquatic species. Since the waste heat is usually a liability for the company that generates it, reducing the problem of one industry by utilizing the heat for another (aquaculture) is generally seen as a beneficial activity. The aquaculturist will also require a suitable supply of good-quality water for the rearing of the aquatic animal crop, of course, but that requirement is reduced as the extent of recirculation increases.

Pumping costs are another significant expense in closed and recirculating water systems. It is necessary to move water through the system. Typically, turnover rates in fish tanks are at least several times per day, so for a 500,000-gallon system, it will be necessary to pump a few million gallons daily, and that requires a good deal of electricity over and above what would be required for pond filling and maintenance of appropriate flow-through raceways receiving artesian or spring water.

Disposal of water from aquaculture has become a problem in recent years, and that problem continues to grow. Recirculating systems produce continuous effluents, and even closed systems have to be emptied periodically. There is the need to empty settling basins on a frequent basis, and the effluent from those units is highly concentrated in terms of solids and nutrients. I recently saw one recirculating system that had piled up sewage bills of about $10,000 per month! More common than using a sanitary sewage system for disposal of effluent from highly intensive systems

---

*If geothermal water of appropriate quality and quantity can be found, recirculation is not necessary and flow-through systems can be employed. Several such systems are currently in operation. The one established by Leo Ray in Idaho, for example, was discussed in Chapter 5.

is the notion of using it in hydroponics or in irrigation. There is still the problem of solids to deal with. Collecting and drying the sediments from settling chambers can provide a secondary product in the form of fertilizer. Little use is being made of that material at present, but the technology to collect and dry it is readily available.

On balance, I believe that given low-cost or free water and low-cost or free heat, closed or nearly closed water systems can be economically viable. I envision relatively small facilities: a 500,000-gallon facility (which represents the amount of water in the culture chambers—the total volume of the system would be somewhat higher) can produce $1/2$ million pounds of fish a year if the density in the culture chambers is 5 pounds per cubic foot (7.5 gallons) and 1.5 crops per year. Such a facility will not make anyone wealthy, but it could be profitable. Two or three people could operate such a facility and could conceivably run one that is two to three times that size if everything works efficiently, thereby increasing the profit potential. That would not include the labor involved in processing the crop if that aspect of the business is included in the operation.

I continue looking for a profitable closed system other than one I know of that produces tropical fish. I honestly believe I'm going to see one that produces food animals in the near future. There has been a long string of failures, yet there continues to be people who are constructing systems and who believe that they will not solve all the problems but that they will be able to make money. I know one of those ventures will succeed if it avoids utilizing excessive amounts of energy to pump water and heat it.

I believe the technology and expertise exists today to produce a successful operation. In the past those two components don't seem to have been properly put together. The failures of some systems have been related to the fact that they were designed by biologists without benefit of engineering input; thus, only part of the expertise was available and the proper technology was not employed. I see design of an effective system as requiring not only the input from biologists and engineers. It is also necessary that those people understand each other's needs and work very closely together until the system has been constructed and thoroughly tested.

Offshore Systems

To date, all of the successful marine aquaculture in the United States, and in the world for that matter, occurs within a mile or two of shore.[910] In the states that means that marine aquaculture is being conducted within the territorial waters of the states. Paul Sandifer, who is well known within the United States and international community of aquaculture scientists, conducted a survey of coastal state aquaculture coordinators to determine the level of activity that is currently in place with respect to coastal aquaculture.[911] All 24 states that were polled responded to Sandifer's survey. The results, published in 1994, showed that only two states, Pennsylvania (with almost no coastline) and Oregon (because of legal restrictions) have no coastal aquaculture in place.* Only one state, Massachusetts, reported that its industry was

*You may recall that ocean ranching was practiced in Oregon for several years but is no longer practical because of the regulatory environment that exists in that state.

growing rapidly. Of the remaining states, 15 (69.2%) reported either no growth or very slow growth. Technological problems and the availability of properly trained aquaculturists were not seen as major problems. Lack of adequate financing, the regulatory environment, and the presence of competing users in the coastal region were seen as significant problems.

In 1992, the National Research Council published results of a study on the potential for marine aquaculture in the United States.[912] Among the conclusions reached in that study were that the federal government should take a more active role in assisting with development of the mariculture industry and that offshore aquaculture will need to be developed as a means of avoiding the nearshore conflicts. Incidentally, the study also acknowledged the potential role of private aquaculturists in producing animals for enhancement stocking in the marine environment.

Moving offshore means extending aquaculture toward the 3-mile statutory limit of the coastal states* and beyond. When the 3-mile limit is passed, jurisdiction falls to the federal government, which controls the Exclusive Economic Zone (EEZ) that extends out to 200 miles. It is doubtful, though I suppose possible, that aquaculture will be developed beyond the 200-mile limit. If it does, international law will govern such enterprises.

The idea of moving aquaculture offshore has a certain amount of appeal. Some advantages are:

- The activity can be placed out of site of upland landowners, who would not be able to claim visual pollution.
- Leaseholds of virtually any size can be developed.
- If facilities are properly located, natural circulation patterns will ensure a constant and dependable supply of high-quality water while carrying away waste products.
- Dilution factors will be so high that no measurable changes in water quality due to the aquaculture activity will occur.
- Offshore facilities will not conflict with *most* users of the coastal zone.
- Pollution problems that can occur in nearshore waters will be avoided.

The advantages are pretty much self-explanatory and require no additional embellishment.

Offshore aquaculture is not, of course, without its problems, and it is the problems that require additional comment. Among the more important ones are the following:

- lack of federal regulations in the EEZ
- lack of appropriate technology
- logistics of operating offshore facilities
- possible interference with shipping and commercial fishing

---

*A couple states have broader territorial waters for historical reasons, but for the most part 3 miles is considered to be within state jurisdiction.

It was only in about 1992 that the U.S. government became aware of the fact that anyone might even be considering the establishment of an aquaculture venture in the EEZ. At that time, Congress was approached with a request for permission to establish a salmon farm more than 30 miles off the coast of Massachusetts. When policies and laws were searched, it was determined that the only leasing program in place was associated with oil and gas exploration and production. The prospective aquaculture facility was never pursued, but Congress did submit a request to the Office of Technology Assessment (OTA) to develop some options with respect to offshore aquaculture policy. That's where I came in.

I received a call from the OTA in 1993 and was asked to serve on a committee that was looking into what the role of the U.S. government might be with respect to aquaculture in general. I was also told that some additional committees were being formed to address specific topics such as closed systems, the economics of aquaculture, international aquaculture as it relates to U.S. interests, and offshore aquaculture (to mention a few). I was asked who the experts in offshore aquaculture are, and indicated that I didn't think there were any experts in our country. The response was:

> If there are no experts, then you know as much as anyone. Would you be willing to chair a committee to look into the offshore aquaculture situation and come up with some policy options for Congress?

I accepted the challenge and began forming my committee during the summer, 1993. It turned out that a few people had been involved in some preliminary investigations with respect to facilities that were either located in the EEZ or were in state waters but exposed to conditions not much different than what could be expected in the EEZ. I recruited John Forster (then a Washington state salmon producer), who had been working with a private company in the Strait of Juan de Fuca in testing an offshore net-pen design. I also tapped Russ Miget and Granvil Treece from the Texas Sea Grant Program. Miget was working with experimental offshore cages moored to drilling platforms—one of them was located about 80 miles offshore—and Treece was involved to the extent that he was a knowledgeable resource. Bob Blumberg of the Texas General Land Office was also involved with the drilling platform aquaculture research activity, so I added him to the committee. I recruited people involved with marine policy and law (Harlyn Halvorson and Robert Wilder) and with experience in the oil and gas arena (James Harding). To provide insight on governmental interests and the options that might be reasonable from the agency point of view, I asked Conrad Mahnken (National Marine Fisheries Service) and James McVey (Sea Grant) to join the committee. We met in Washington, D.C., in the fall, 1993, after which the report was drafted, reviewed, and submitted to the OTA.[910] With that background, let's look at more detail at the constraints associated with offshore aquaculture.

The bottom line, of course, is that the U.S. government needs to develop a policy by which the orderly development of offshore aquaculture can take place, should it actually develop. The committee was instructed to provide policy options, not recommendations—after all, it is up to Congress to decide what it wants to do after weighing the alternatives. We did point out that, unlike oil and gas leasing, offshore aquaculture would not add appreciably to the coffers of the U.S. government. There just isn't going to be that much profit involved from offshore aquaculture. A simple leasing program would seem to be more realistic. Also, Congress might wish to work collaboratively with the coastal states in developing its program since some facilities would certainly be sited in places where state and federal jurisdictions meet.

The committee indicated that an opportunity was available to utilize existing drilling platforms as aquaculture sites. Under current law, platforms have to be removed soon after production ceases, whereas the structures themselves may have a useful life of several more decades. There are presently something like 4,000 existing rigs in the Gulf of Mexico. Once they are removed, the opportunity to employ them for a second useful function will be lost.

The major constraint from what I could see was associated with the level of current technology. Various designs for offshore facilities have been conceived, and in some cases built.[913,914] Designs have taken the form of modified traditional net-pens, ship hulls with large openings through them from which net-pens are suspended, submersible net-pens, and semisubmersible net-pens. There have also been designs drawn up for large offshore algae farms, and the concept of hanging shellfish from drilling platforms and other offshore stationary mooring facilities has been the subject of some research.

My wife, Carolan, and I had the opportunity to visit a facility jointly operated by the University of Sterling and a private company in Scotland during 1992. The three semisubmerged net-pens were actually located in an embayment, but there was a fetch of something like 300 miles from which the wind blows during much of the year, so the exposure of the net-pens was similar to what would be experienced offshore. Twenty-foot waves are not uncommon, we were told. Fortunately, from our standpoint, the day we visited was dead calm. In any event, the cages are designed with automatic feeders and feed hoppers sufficiently large that the cages will operate without human intervention for a period of a few days during storms, when attempts to service them would be foolhardy and futile.

Thus, the technology is being developed, but the expense involved in designing structures that can withstand the constant and highly variable motion associated with the open sea are enormously higher than those associated with constructing facilities in protected waters. I am confident that the engineering problems can be solved, but whether the economic ones can be overcome is a significant question. It is clear that offshore aquaculturists will be required to grow products with high commercial value. Atlantic salmon would be an appropriate choice, but mullet would certainly not.

Finding people willing to work on offshore facilities may also pose a significant problem. The work will be arduous and extremely dangerous at times. It will probably be necessary to develop facilities that virtually maintain themselves most of the time. It is not reasonable to expect people to live on them constantly, unless they can be developed in conjunction with drilling platforms that have virtually all the comforts of home. (At least the larger ones can make that claim.) Just getting people to and from the sites will pose a significant problem. It will be necessary to maintain one or more fairly large vessels for servicing offshore sites, and that will be an extra cost not necessarily incurred by inshore culturists. Harvesting systems will have to be developed that can work efficiently in open water under at least moderately rough conditions. Certainly, there are many logistical problems that must be faced.

Interference, or possible interference with shipping and commercial fishing represent a relatively minor constraint. Proper marking with buoys and lights will warn ships of the presence of offshore aquaculture facilities. It would be necessary to issue leases only to sites that are outside of the shipping lanes, and it would also be necessary to take traditional fishing grounds into consideration when leases are granted. It can be expected that those with interests other than aquaculture will complain bitterly when applications for offshore fish farming sites are made, but there is plenty of area offshore to accommodate all present and future users, no matter what those in opposition might believe.

I am not particularly optimistic about the potential for offshore aquaculture development. I think it is technically feasible, but the economics do not currently appear favorable. It does not look like the price of fish will increase sufficiently in the next several years to change the situation. Imports of salmon and other high-priced species will keep the costs to consumers down and bode ill for those who want to compete by growing fish off the coast of the United States.

FINAL THOUGHTS

Our excursion through the history of U.S. aquaculture is now complete. Writing this book has been an enjoyable and interesting experience for me. In fact, the book essentially wrote itself. I was intrigued, amused, and thoroughly caught up in the early literature, in particular. I've even toyed with the idea of a novel based on the life and times of Livingston Stone, my hero.

Aquaculture has been an ongoing activity in the United States for well over 100 years now, and while it is not a subject on the tongues of every American citizen, its story is interesting, and its contribution to society has been significant. I think aquaculture will continue to play an important role for society in a number of ways, but I also think that the contributions of the commercial foodfish industry will soon peak because of the constraints that exist.

Until the United States becomes a hungry nation—and I honestly hope that time does not come anytime soon—the potential that exists for aquaculture production will not be fully realized. Imports will satisfy much of the nation's need for seafoods (which again, is not growing appreciably). There will be increasing pressure for fish to supply the fast-food markets. If our capture fisheries for cod, pollock, and similar species are not turned around, we may see the demise of the fish sandwich at fast-food restaurants. If tilapia were available, and the price was right, the demand for that fish by fast-food restaurants could be in the tens of thousands of tons annually. Whether U.S. aquaculture can meet that demand in a cost-competitive way remains to be seen, but certainly the attempt should be made.

We will probably not see much growth in salmon farming on the West Coast unless the fishermen in Alaska follow the lead of their counterparts in Norway and other nations and become aquaculturists during the greater part of the year when salmon are not available to them. If that unlikely change in attitude were to occur, perhaps commercial fishermen in Washington and Oregon would also become involved in aquaculture. Maine's industry could continue to grow appreciably, but it is unclear at present how much development will be allowed in that state.

The United States is unlikely to ever become much of a player in the shrimp farming industry, though small pockets of production will continue to occur. Catfish will continue to lead the way among the fish species cultured in the United States and the crawfish industry can be expected to see at least modest growth. We certainly haven't reached peak production in the overall aquaculture industry within the United States as yet, but I think that the rate of increase has already slowed and that it will slow further in the future. That statement is not meant to be pessimistic, only realistic. There will continue to be job opportunities for commercial aquaculturists in the United States, and we should not abandon our programs of training individuals to assume those positions. At the same time we cannot fool ourselves into thinking that U.S. aquaculture represents the wave of the future in food production for our nation. It will play its part, but that will only be a relatively small part in the world's most highly developed agricultural enterprise.

Literature

1. Shell, E. W. 1993. The development of aquaculture: an ecosystems perspective. Alabama Agricultural Experiment Station, Auburn, Ala. 265 p.

2. Stickney, R. R. 1994. Principles of aquaculture. John Wiley & Sons, NY. 502 p.

3. Ryther, J. H. 1969. Photosynthesis and fish production in the sea. Science, 166: 72–76.

4. Baird, S. F. 1873. Report of the Commissioner. pp. VII–XLI, In: Report on the condition of the sea fisheries of the south coast of New England in 1871 and 1872. Government Printing Office, Washington, D.C.

5. Bowen, J. T. 1970. A history of fish culture as related to the development of fishery programs. pp. 71–93, In: N. G. Benson (ed.). A century of fisheries in North America. American Fisheries Society, Washington, D.C.

6. Baird, S. F. 1874. Report of the Commissioner. pp. I–XCII, In: Report of the Commissioner for 1872 and 1873. U.S. Commission of Fish and Fisheries, Washington, D.C.

7. Herber, E. C. 1963. Correspondence between Spencer Fullerton Baird and Louis Agassiz—two pioneer American naturalists. Smithsonian Institution, Washington, D.C. 237 p.

8. Baird, S. F. 1880. Report of the Commissioner. pp. XV–LXIV, In: Report of the Commissioner for 1878. U.S. Commission of Fish and Fisheries, Washington, D.C.

9. Milner, J. W. 1874. The progress of fish-culture in the United States. pp. 523–566, In: Report of the Commissioner for 1872 and 1873. U.S. Commission of Fish and Fisheries, Washington, D.C.

10. Green, S. 1874. Experiences of a practical fish culturist. Proc. Am. Fish Culturists' Assoc. 3: 22–24.

11. Green, S. 1875. Stocking depleted waters. Proc. Am. Fish Culturists' Assoc. 4: 19–22.

12. Davis, H. S. 1967. Culture and diseases of game fishes. University of California Press, Berkeley. 332 p.

13. Roosevelt, R. B. 1877. Introductory address by R. B. Roosevelt, Esq., President. Trans. Am. Fish Culturists' Assoc. 6: 46–48.

14. Anonymous. 1873. p. 279, In: Report on the condition of the sea fisheries of the south coast of New England in 1871 and 1872. Government Printing Office, Washington, D.C.

15. Hedgpeth, J. W. 1941. Livingston Stone. Prog. Fish-Cult. 55: 11–14.

16. Atkins, C. G. 1874. The salmon of eastern North America, and its artificial culture. pp. 226–335, In: Report of the Commissioner for 1872 and 1873. U.S. Commission of Fish and Fisheries, Washington, D.C.

17. Kendall, W. C. 1936. Charles Grandison Atkins—a pioneer who blazed the trails. Prog. Fish-Cult. 25: 12–17.

18. Goode, G. B. 1881. Epochs in the history of fish culture. Trans. Am. Fish Cult. Assoc. 10: 34–57.

19. Baird, S. F. 1876. pp. VII–XLVI, In: Report of the Commissioner for 1873–4 and 1874–5. U.S. Commission of Fish and Fisheries, Washington, D.C.

20. Smiley, C. W. 1884. A statistical review of the production and distribution to public waters of young fish, by the United States Fish Commission, from its organization in 1871 to the close of 1880. pp. 825–1035, In: Propagation and inquiry of food fishes of the United States. U.S. Commission of Fish and Fisheries, Washington, D.C.

21. Mather, F. 1884. Account of eggs repacked and shipped to foreign countries, under the direction of the United States Fish Commission, during the winter of 1882–'83. pp. 809–812, In: Report to the Commissioner. U.S. Commission of Fish and Fisheries, Washington, D.C.

22. Stone, L. 1877. California salmon. Trans. Am. Fish Culturists' Assoc. 6: 73–77.

23. Mather, F. 1876. Voyage to Bremerhaven, Germany, with shad. pp. 328–330, In: Report of the Commissioner for 1873–4 and 1874–5. U.S. Commission of Fish and Fisheries, Washington, D.C.

24. Milner, J. W. 1876. Experiments with a view to transporting shad in sea-water. pp. 363–369, In: Report of the Commissioner for 1873–4 and 1874–5. U.S. Commission of Fish and Fisheries, Washington, D.C.

25. Tanner, Z. L. 1884. Report on the construction and work in 1880 of the Fish Commission steamer *Fish-Hawk*. pp. 1–53, In: Propagation and inquiry of food fishes of the United States. U.S. Commission of Fish and Fisheries, Washington, D.C.

26. Tanner, Z. L. 1884. Report on the work of the United States Fish Commission steamer *Fish-Hawk*, for the year ending December 31, 1881. pp. 1–53, In: Propagation and inquiry of food fishes of the United States. U.S. Commission of Fish and Fisheries, Washington, D.C.

27. Stone, L. 1876. Report of operations in California in 1873. pp. 377–423, In: Report of the Commissioner for 1873–4 and 1874–5. U.S. Commission of Fish and Fisheries, Washington, D.C.

28. Baird, S. F. 1884. Report of the Commissioner for 1881. pp. XIII–LXXI, In: Propagation and inquiry of food fishes of the United States. U.S. Commission of Fish and Fisheries, Washington, D.C.

29. Baird, S. F. 1884. Report of the Commissioner. pp. XVII–XCII, In: Report of the Commissioner for 1882. U.S. Commission of Fish and Fisheries, Washington, D.C.

30. Eastman, F. S. 1884. The description of the United States Commission car no. 2, designed for the distribution of young fish. pp. 39–41, In: Report to the Commissioner. U.S. Commission of Fish and Fisheries, Washington, D.C.

31. Atkins, C. G. 1876. Report on the collection and distribution of Penobscot salmon in 1873–'74 and 1874–'75. pp. 485–530, In: Report of the Commissioner for 1873–4 and 1874–5. U.S. Commission of Fish and Fisheries, Washington, D.C.

32. Atkins, C. G. 1884. Report on the propagation of Penobscot salmon in 1882–'83. pp. 857–862, In: Report to the Commissioner. U.S. Commission of Fish and Fisheries, Washington, D.C.

33. Atkins, C. G. 1882. Report on the propagation of Penobscot salmon in 1979–'80. pp. 721–731, In: Report of the Commissioner for 1879. U.S. Commission of Fish and Fisheries, Washington, D.C.

34. Atkins, C. G. 1882. Report on the propagation of Schoodic salmon in 1879–'80. pp. 733–766, In: Report of the Commissioner for 1879. U.S. Commission of Fish and Fisheries, Washington, D.C.

35. Atkins, C. G. 1884. Report on the propagation of Schoodic salmon in 1881–82. pp. 1091–1105, In: Propagation and inquiry of food fishes of the United States. U.S. Commission of Fish and Fisheries, Washington, D.C.

36. Atkins, C. G. 1884. Report on the propagation of Penobscot salmon in 1881–82. pp. 1085–1089, In: Propagation and inquiry of food fishes of the United States. U.S. Commission of Fish and Fisheries, Washington, D.C.

37. Davidson, F. A., and S. J. Hutchinson. 1937. The influence of natural conditions on the geographic distribution of the Pacific salmon. Prog. Fish Cult. 30: 24–34.

38. Stone, L. 1876. Report of operations during 1874 at the United States salmon-hatching establishment on the M'Cloud River, California. pp. 437–478, In: Report of the Commissioner for 1873–4 and 1874–5. U.S. Commission of Fish and Fisheries, Washington, D.C.

39. Stone, L. 1874. Salmon breeding. Proc. Fish Culturists' Assoc. 3: 9–22.

40. Stone, L. 1874. Report of operations during 1872 at the United States salmon-hatching establishment on the M'Cloud River, and on the California salmonidæ generally; with a list of specimens collected. pp. 168–215, In: Report of the Commissioner for 1872 and 1873. U.S. Commission of Fish and Fisheries, Washington, D.C.

41. Stone, L. 1876. A list of McCloud Indian words supplementary to a list contained in the report of 1872. pp. 428–429, In: Report of the Commissioner for 1873–4 and 1874–5. U.S. Commission of Fish and Fisheries, Washington, D.C.

42. Hubbard, W. F. 1880. Report of salmon-hatching operations in 1878, at the Clackamas hatchery. pp. 771–772, In: Report of the Commissioner for 1878. U.S. Commission of Fish and Fisheries, Washington, D.C.

43. Pratt, K. B. 1880. Report of salmon-hatching operations on Rogue River, Oregon, 1877–'78. pp. 773–774, In: Report of the Commissioner for 1878. U.S. Commission of Fish and Fisheries, Washington, D.C.

44. Stone, L. 1880. Report on operations at the United States salmon-hatching station on the M'Cloud River, California, in 1878, pp. 741–770, In: Report of the Commissioner for 1878. U.S. Commission of Fish and Fisheries, Washington, D.C.

45. Stone, L. 1882. Report of operations at the United States salmon-breeding station on the McCloud River, California, during the season of 1879. pp. 695–708, In: Report of the Commissioner for 1879. U.S. Commission of Fish and Fisheries, Washington, D.C.

46. Baird, S. F. 1882. Report of the Commissioner. pp. XI–LI, In: Report of the Commissioner for 1879. U.S. Commission of Fish and Fisheries, Washington, D.C.

47. Dunn, H. D. 1880. Do the spawning salmon ascending the Sacramento River all die without returning to sea? pp. 815–818, In: Report of the Commissioner for 1878. U.S. Commission of Fish and Fisheries, Washington, D.C. [With notes by Livingston Stone.]

48. Stone, L. 1884. The report of operations at the United States salmon-breeding station on the McCloud River, California, during the season of 1881. pp. 1063–1078, In: Propagation and inquiry of food fishes of the United States. U.S. Commission of Fish and Fisheries, Washington, D.C.

49. Stone, L. 1884. Report of operations at the salmon-breeding station of the United States Fish Commission on the M'Cloud River, California, during the season of 1882. pp. 839–850, In: Report to the Commissioner. U.S. Commission of Fish and Fisheries, Washington, D.C.

50. Panek, F. M. 1987. Biology and ecology of carp. pp. 1–15, In: Carp in North America. American Fisheries Society, Bethesda, Md.

51. Fritz, A. W. 1987. Commercial fishing for carp. pp. 17–30, In: Carp in North America. American Fisheries Society, Bethesda, Md.

52. Poppe, R. A. 1880. The introduction and culture of the carp in California. pp. 661–666, In: Report of the Commissioner for 1878. U.S. Commission of Fish and Fisheries, Washington, D.C.

53. Anonymous. 1948. Washington monument lot: fish ponds. Prog. Fish-Cult. 10: 198.

54. Finsch, O. 1882. Report on the transportation of a collection of living carp from Germany. pp. 667–670, In: Report of the Commissioner for 1879. U.S. Commission of Fish and Fisheries, Washington, D.C.

55. Smiley, C. W. 1884. Report on the distribution of carp to July 1, 1881, from young reared in 1879 and 1880. pp. 943–988, In: McDonald, M. 1884. Report on the distribution of carp during the season of 1882. Report to the Commissioner. U.S. Commission of Fish and Fisheries, Washington, D.C.

56. McDonald, M. 1884. Report of distribution of carp, during the season of 1881–'82, by the United States Fish Commission. pp. 1121–1126, In: Propagation and inquiry of food fishes of the United States. U.S. Commission of Fish and Fisheries, Washington, D.C.

57. McDonald, M. 1884. Report on the distribution of carp during the season of 1882. pp. 915–942, In: Report to the Commissioner. U.S. Commission of Fish and Fisheries, Washington, D.C.

58. Smiley, C. W. 1886. Some results of carp culture in the United States, pp. 657–881, In: Report of the Commissioner for 1884. U.S. Commission of Fish and Fisheries, Washington, D.C.

59. Clift, W. 1872. Shad culture. Proc. Am. Fish Culturists' Assoc. 1: 21–28.

60. Milner, J. W. 1874. Report on the propagation of the shad (*Alosa sapidissima*) and its introduction into new waters by the U.S. Fish Commissioner in 1873. pp. 419–451, In: Report of the Commissioner for 1872 and 1873. U.S. Commission of Fish and Fisheries, Washington, D.C.

61. Anderson, G. A. 1876. Report of shad-hatching in New Jersey. p. 327, In: Report of the Commissioner for 1873–4 and 1874–5. U.S. Commission of Fish and Fisheries, Washington, D.C.

62. Slack, J. H. 1872. Practical trout culture. (Reprinted in 1914 under the heading: On selecting a location for a fish farm. Prog. Fish-Cult. 11: 132.)

63. Slack, J. H. 1872. Practical trout culture. (Reprinted in 1951: Prog. Fish-Cult. 13: 152).

64. Griswold, C. D. 1876. Experiments with a view to transporting shad a few months old. pp. 370–372, In: Report of the Commissioner for 1873–4 and 1874–5. U.S. Commission of Fish and Fisheries, Washington, D.C.

65. Smiley, C. W. 1884. Statistics of the shad-hatching operations conducted by the United States Fish Commission in 1881. pp. 1107–1119, In: Propagation and inquiry of food fishes of the United States. U.S. Commission of Fish and Fisheries, Washington, D.C.

66. Rice, H. J. 1884. Experiments upon retarding the development of eggs of the shad made in 1879, at the United States shad-hatching station of Havre de Grace, Md. pp. 787–794, In: Propagation and inquiry of food fishes of the United States. U.S. Commission of Fish and Fisheries, Washington, D.C.

67. Clark, F. N. 1884. Report of experiments for determining the smallest amount of water in which young shad and eggs can be kept. pp. 783–786, In: Propagation and inquiry of food fishes of the United States. U.S. Commission of Fish and Fisheries, Washington, D.C.

68. Babcock, W. C. 1885. Report of shad operations conducted at Fort Washington, Maryland, in the spring of 1883. pp. 1035–1044, In: Report of the Commissioner for 1883. U.S. Fish and Fisheries Commission, Washington, D.C.

69. Green, S. 1876. Propagation of fish. Trans. Am. Fish Culturists' Assoc. 5: 8–13.

70. Milner, J. W. 1874. Report on the fisheries of the Great Lakes; the result of inquiries prosecuted in 1871 and 1872. pp. 1–75, In: Report of the Commissioner for 1872 and 1873. U.S. Commission of Fish and Fisheries, Washington, D.C.

71. Clark, F. N. 1885. Results of planting whitefish in Lake Erie. Trans. Am Fish. Soc. 14: 40–50.

72. Stone, L. 1872. Trout culture. Proc. Am. Fish Culturists' Assoc. 1: 46–56.

73. Stone, L. 1884. Report of operations at the United States trout ponds, McCloud River, California, for the season of 1881. pp. 1079–1089, In: Propagation and inquiry of food fishes of the United States. U.S. Commission of Fish and Fisheries, Washington, D.C.

74. Stone, L. 1885. Report of operations at the United States trout-breeding station on the McCloud River, California, during the year 1883. pp. 1001–1006, In: Report of the Commissioner for 1883. U.S. Fish and Fisheries Commission, Washington, D.C.

75. Earll, R. E. 1880. A report on the history and present condition of the shore cod-fisheries of Cape Ann, Mass., together with notes on the natural history and artificial propagation of the species. pp. 685–740, In: Report of the Commissioner for 1878. U.S. Commission of Fish and Fisheries, Washington, D.C.

76. Brice, J. J. 1898. A manual of fish-culture. pp. 1–340, In: Report of the Commissioner for the year ending June 30, 1897. U.S. Commission of Fish and Fisheries, Washington, D.C.

77. McDonald, M. 1884. Spanish mackerel—investigations at Cherrystone, Va., during the summer of 1881. pp. 1131–1133, In: Propagation and inquiry of food fishes of the United States. U.S. Commission of Fish and Fisheries, Washington, D.C.

78. Kite, J. A. 1885. Report upon apparatus and facialities needed for hatching Spanish mackerel. pp. 1095–1100, In: Report of the Commissioner for 1883. U.S. Fish and Fisheries Commission, Washington, D.C.

79. Wood, W. M. 1885. Report of operations in hatching eggs of Spanish mackerel in Chesapeake Bay by Steamer *Fish Hawk* during the summer of 1883. pp. 1089–1094, In: Report of the Commissioner for 1883. U.S. Fish and Fisheries Commission, Washington, D.C.

80. Green, S. 1874. Frog-culture. pp. 587–588, In: Report of the Commissioner for 1872 and 1873. U.S. Commission of Fish and Fisheries, Washington, D.C.

81. Chamberlain, F. M. 1898. Notes on the edible frogs of the United States and their artificial propagation. pp. 249–261, In: Report of the Commissioner for the year ending June 30, 1897. U.S. Commission of Fish and Fisheries, Washington, D.C.

82. Smith, H. M. 1895. Report of the division of statistics and methods of the fisheries. pp. 52–77, In: Report of the Commissioner for the year ending June 30, 1893. United States Commission of Fish and Fisheries, Washington, D.C.

83. Johnson, S. M. 1883. Lobster culture. Trans. Am. Fish-Cultural Assoc. 12: 18–21.

84. Perrin, M. L. 1876. Transportation of lobsters to California. pp. 259–265, In: Report of the Commissioner for 1873–4 and 1874–5. U.S. Commission of Fish and Fisheries, Washington, D.C.

85. Blackford, E. G. 1885. The oyster beds of New York. Trans. Am. Fish. Soc. 14: 85–89.

86. Ryder, J. A. 1887. An exposition of the principles of a rational system of oyster culture, together with an account of a new and practical method of obtaining oyster spat on a scale of commercial importance. pp. 381–421, In: Report of the Commissioner for 1885. U.S. Commission of Fish and Fisheries, Washington, D.C.

87. Mather, F. 1886. Oyster culture. Trans. Am. Fish. Soc. 15: 26–36.

88. Ryder, J. A. 1884. An account of experiments in oyster culture and observations relating thereto. pp. 763–778, In: Report to the Commissioner. U.S. Commission of Fish and Fisheries, Washington, D.C.

89. Winslow, F. 1884. Report of experiments in the artificial propagation of oysters, conducted at Beaufort, N.C., and Fair Haven, Conn., in 1882. pp. 741–761, In: Report to the Commissioner. U.S. Commission of Fish and Fisheries, Washington, D.C.

90. Ravenel, W. deC. 1887. Report of operations at Saint Jerome oyster-breeding station during 1885. pp. 165–183, In: Report of the Commissioner for 1885. U.S. Commission of Fish and Fisheries, Washington, D.C.

91. Moore, J. P. 1899. Report on mackerel investigations in 1897. pp. 3–22, In: Report of the Commissioner for the year ending June 30, 1898. U.S. Commission of Fish and Fisheries, Washington, D.C.

92. Wood, E. M. 1953. A century of American fish culture, 1853–1953. Prog. Fish-Cult. 15: 147–162.

93. Gay, J., and W. P. Seal. 1890. The past and present of fish culture, with an inquiry as to what may be done to further promote and develop the science. Trans. Am. Fish. Soc. 19: 66–79.

94. Anonymous. 1906. Robert Barnwell Roosevelt. Trans. Am. Fish. Soc. 35: 39.

95. Roosevelt, R. B. 1878. Reproductive power of eels. Trans. Am. Fish Culturists' Assoc. 7: 90–99.

96. Roosevelt, R. B. 1879. Reproductive habits of eels. Trans. Am. Fish Culturists' Assoc. 8: 32–44.

97. McDonald, M. 1891. Report of the Commissioner. pp. I–LXIII, In: Report of the Commissioner for 1887. U.S. Commissioner of Fish and Fisheries, Washington, D.C.

98. Kendall, W.C. 1939. Marshall McDonald, 1835–1895. Prog. Fish Cult. 43: 19–30.

99. McDonald, M. 1883. History of the experiments leading to the development of the automatic fish-hatching jar. Trans. Am. Fish-Cultural Assoc. 12: 34–46.

100. Patent applications by various fish culturists. pp. 1094–1097, In: Report of the Commissioner for 1885. U.S. Commission of Fish and Fisheries, Washington, D.C.

101. Anonymous. 1896. Report of the U.S. Commissioner of Fish and Fisheries for the fiscal year ending June 30, 1894. pp. 1–19, In: Report of the Commissioner of Fish and Fisheries for the year ending June 30, 1894. U.S. Commission on Fish and Fisheries, Washington, D.C.

102. Anonymous. 1896. Report of the United States Commissioner of Fish and Fisheries for the fiscal year ending June 30, 1895. pp. 1–5, In: Report of the Commissioner for the year ending June 30, 1895. U.S. Commission of Fish and Fisheries, Washington, D.C.

103. Brice, J. J. 1898. Report of the United States Commissioner of Fish and Fisheries for the fiscal year ending June 30, 1896. pp. 1–10, In: Report of the Commissioner for the year ending June 30, 1896. U.S. Commission of Fish and Fisheries, Washington, D.C.

104. Bowers, G. M. 1899. Report of the United States Commissioner of Fish and Fisheries for the fiscal year ending June 30, 1898. pp. VII–XXIX, In: Report of the Commissioner for the year ending June 30, 1898. U.S. Commission of Fish and Fisheries, Washington, D.C.

105. McDonald, M. 1892. Report of the Commissioner. pp. IX–XXXIX, In: Report of the Commissioner for 1888–89. U.S. Commissioner of Fish and Fisheries, Washington, D.C.

106. McDonald, M. 1894. Report of the United States Commission of Fish and Fisheries for the fiscal year ending June 30, 1892. pp. VII–LXXXVII, In: Report of the Commissioner of Fish and Fisheries for the year ending June 30, 1892. U.S. Commission on Fish and Fisheries, Washington, D.C.

107. McDonald, M. 1895. Report of the United States Commissioner of Fish and Fisheries for the fiscal year ending June 30, 1893. pp. 1–16, In: Report of the Commissioner for the year ending June 30, 1893. U.S. Commission of Fish and Fisheries, Washington, D.C.

108. Baird, S. F. 1885. Report of the Commissioner. pp. XVII–XCV, In: Report of the Commissioner for 1883. U.S. Fish and Fisheries Commission, Washington, D.C.

109. Baird, S. F. 1886. Report of the Commissioner. pp. XIII–LXXI, In: Report of the Commissioner for 1884. U.S. Commission of Fish and Fisheries, Washington, D.C.

110. Baird, S. F. 1887. Report of the Commissioner. pp. XIX–CXII, In: Report of the Commissioner for 1885. U.S. Commission of Fish and Fisheries, Washington, D.C.

111. Stone, L. 1885. Explorations on the Columbia River from the head of Clarke's Fork to the Pacific Ocean, made in the summer of 1883, with reference to the selection of a suitable place for establishing a salmon-breeding station. pp. 237–255, In: Report of the Commissioner for 1883. U.S. Commissioner of Fish and Fisheries, Washington, D.C.

112. Bean, T. H. 1896. Report on the propagation and distribution of food-fishes. pp. 20–80, In: Report of the Commissioner of Fish and Fisheries for the year ending June 30, 1894. U.S. Commission of Fish and Fisheries, Washington, D.C.

113. Ravenel, W. deC. 1896. Report on the propagation and distribution of food-fishes. pp. 6–72, In: Report of the Commissioner for the year ending June 30, 1895. U.S. Commission of Fish and Fisheries, Washington, D.C.

114. Brice, J. J. 1895. Establishment of stations for the propagation of salmon on the Pacific coast. pp. 387–392, In: Report of the Commissioner for the year ending June 30, 1893. U.S. Commission of Fish and Fisheries, Washington, D.C.

115. Smith, H. M. 1898. Report of the Division of Scientific Inquiry. pp. XCI–CXXIV, In: Report of the Commissioner for the year ending June 30, 1897. U.S. Commission of Fish and Fisheries, Washington, D.C.

116. Ravenel, W. deC. 1900. Report on the propagation and distribution of food-fishes. pp. XXXV–CXVII, In: Report of the Commissioner for the year ending June 30, 1899. U.S. Commission of Fish and Fisheries, Washington, D.C.

117. Ravenel, W. deC. 1901. Report on the propagation and distribution of food-fishes. pp. 25–118, In: Report of the Commissioner for the year ending June 30, 1900. U.S. Commission of Fish and Fisheries, Washington, D.C.

118. McDonald, M. 1889. Report of distribution of fish and eggs by the U.S. Fish Commission from January 1, 1886, to June 30, 1887. pp. 833–842, In: Report to the Commissioner for 1886. U.S. Commission of Fish and Fisheries, Washington, D.C.

119. Anonymous. 1891. Report of distribution of fish and eggs by the U.S. Fish Commission from July 1, 1887, to June 30, 1888. pp. 363–370, In: Report of the Commissioner for 1887. U.S. Commissioner of Fish and Fisheries, Washington, D.C.

120. Bowers, G. M. 1900. Report of the United States Commissioner of Fish and Fisheries for the fiscal year ending June 30, 1899. pp. VII–XXXIII, In: Report of the Commissioner for the year ending June 30, 1899. U.S. Commission of Fish and Fisheries, Washington, D.C.

121. Bowers, G. M. 1901. Report of the United States Commissioner of Fish and Fisheries for the fiscal year ending June 30, 1900. pp. 5–24, In: Report of the Commissioner for the year ending June 30, 1900. U.S. Commission of Fish and Fisheries, Washington, D.C.

122. McDonald, M. 1892. Report of the Commissioner for the fiscal year ending June 30, 1889. pp. IX–XXXIX, In: Report of the Commissioner for 1888. U.S. Commission on Fish and Fisheries, Washington, D.C.

123. Worth, S. G. 1895. Report on the propagation and distribution of food-fishes. pp. 78–138, In: Report of the Commissioner for the year ending June 30, 1893. U.S. Commission of Fish and Fisheries, Washington, D.C.

124. Clark, F. N. 1884. Report of operations at the Northville and Alpena (Mich.) stations, for the season of 1882–'83. pp. 813–835, In: Report to the Commissioner. U.S. Commission of Fish and Fisheries, Washington, D.C.

125. Ravenel, W. deC. 1898. Report on the propagation and distribution of food-fishes. pp. XVIII–XC, In: Report of the Commissioner for the year ending June 30, 1897. U.S. Commission of Fish and Fisheries, Washington, D.C.

126. Stone, L. 1882. Report on overland trip to California with living fishes, 1879. pp. 637–644, In: Report of the Commissioner for 1879. U.S. Commission of Fish and Fisheries, Washington, D.C.

127. Stone, L. 1885. Report of operations at the United States salmon-breeding station on the McCloud River, California, during the year 1883. pp. 989–1000, In: Report of the Commissioner for 1883. U.S. Fish and Fisheries Commission, Washington, D.C.

128. Stone, L. 1886. Report of operations at the U.S. salmon-breeding station, on the M'Cloud River, California, during the season of 1884. p. 169, In: Report of the Commissioner for 1894. U.S. Commission of Fish and Fisheries, Washington, D.C.

129. Stone, L. 1886. Report of operations at the U.S. trout-breeding station on the M'Cloud River, California, during the season of 1884. pp. 171–176, In: Report of the Commissioner for 1894. U.S. Commission of Fish and Fisheries, Washington, D.C.

130. Stone, L. 1889. Report of operations at the U.S. salmon and trout stations on the M'Cloud River, California, for the years 1885–87. pp. 737–745, In: Report to the Commissioner for 1886. U.S. Commission of Fish and Fisheries, Washington, D.C.

131. Atkins, C. G. 1880. Cheap fixtures for the hatching of salmon. pp. 945–965, In: Report of the Commissioner for 1878. U.S. Commission of Fish and Fisheries, Washington, D.C.

132. Atkins, C. G. 1885. The biennial spawning of salmon. Trans. Am. Fish. Soc. 14: 89–94.

133. Atkins, C. G. 1886. Report on the propagation of Penobscot salmon in 1884–'85. pp. 177–179, In: Report of the Commissioner for 1884. U.S. Commission of Fish and Fisheries, Washington, D.C.

134. Atkins, C. G. 1886. Report on the propagation of Schoodic salmon in 1884–85. pp. 181–187, In: Report of the Commissioner for 1884. U.S. Commission of Fish and Fisheries, Washington, D.C.

135. Atkins, C. G. 1887. Report on the propagation of Penobscot salmon in 1885–'86. pp. 141–144, In: Report of the Commissioner for 1885. U.S. Commission of Fish and Fisheries, Washington, D.C.

136. Atkins, C. G. 1887. Report on the propagation of Schoodic salmon in 1885–'86. pp. 145–156, In: Report of the Commissioner for 1885. U.S. Commission of Fish and Fisheries, Washington, D.C.

137. Atkins, C. G. 1889. Report on the propagation of Penobscot salmon in 1886–'87. pp. 747–749, In: Report to the Commissioner for 1886. U.S. Commission of Fish and Fisheries, Washington, D.C.

138. Atkins, C. G. 1889. Report on the propagation of Schoodic salmon at Grand Lake Steam, Maine, in 1886–'87. pp. 751–759, In: Report to the Commissioner for 1886. U.S. Commission of Fish and Fisheries, Washington, D.C.

139. Mather, F. 1887. Report of operations at Cold Spring Harbor, New York, during the season of 1885. pp. 109–115, In: Report of the Commissioner for 1885. U.S. Commission of Fish and Fisheries, Washington, D.C.

140. McDonald, M. 1887. Report of operations at the trout-breeding station at Wytheville, Va., from its occupation in January, 1882, to the close of 1884. pp. 103–108, In: Report of the Commissioner for 1885. U.S. Commission of Fish and Fisheries, Washington, D.C.

141. Clark, F. N. 1887. Report of operations at the Northville and Alpena (Mich.) stations for the season of 1885–'86. pp. 121–129, In: Report of the Commissioner for 1885. U.S. Commission of Fish and Fisheries, Washington, D.C.

142. Bower, S. 1910. Memorial of Frank Nelson Clark. 1910. Trans. Am. Fish Soc. 40: 85–92.

143. Clark, F. N. 1891. Rearing and distributing trout at the Northville station, U.S. Fish Commission. Trans. Am. Fish. Soc. 20: 30–33.

144. Clark, F. N. 1910. Personal fish-cultural reminiscences. Trans. Am. Fish. Soc. 40: 319–321.

145. McDonald, M. 1889. Report of shad propagation on the Potomac River during the season of 1886. pp. 815–817, In: Report to the Commissioner for 1886. U.S. Commission of Fish and Fisheries, Washington, D.C.

146. McDonald, M. 1889. Report of shad distribution for the season of 1886. pp. 801–806, In: Report to the Commissioner for 1886. U.S. Commission of Fish and Fisheries, Washington, D.C.

147. Grabill, L. R. 1889. Report of operations at the shad-hatching station on Battery Island, near Havre de Grace, Md., during the season of 1886. pp. 807–814, In: Report to the Commissioner for 1886. U.S. Commission of Fish and Fisheries, Washington, D.C.

148. Carswell, J. 1889. Report on the artificial propagation of the codfish at Wood's Holl, Mass., for the season of 1885–86. pp. 779–782, In: Report to the Commissioner for 1886. U.S. Commission of Fish and Fisheries, Washington, D.C.

149. Atkins, C. G. 1889. Report on the artificial propagation of codfish at Wood's Holl, Mass., for the season of 1886–'87. pp. 783–791, In: Report to the Commissioner for 1886. U.S. Commission of Fish and Fisheries, Washington, D.C.

150. Bower, W. T. 1921. Charles G. Atkins, 1841–1921. Trans. Am. Fish. Soc. 51: 31–32.

151. Moore, J. P. 1899. Report on mackerel investigations in 1897. pp. 3–22, In: Report of the Commissioner for the year ending June 30, 1898. U.S. Commission of Fish and Fisheries, Washington, D.C.

152. Rathbun, R. 1890. The transplanting of lobsters to the Pacific coast of the United States. Bull. U.S. Fish Comm. 8: 453–472.

153. Townsend, C. H. 1896. The transplanting of eastern oysters to Willapa Bay, Washington, with notes on the native oyster industry. pp. 193–202, In: Report of the Commissioner for the year ending June 30, 1895. U.S. Commission of Fish and Fisheries, Washington, D.C.

154. Evermann, B. W. 1899. Report on investigations by the U.S. Fish Commission in Mississippi, Louisiana, and Texas, in 1897. pp. 287–310, In: Report of the Commissioner for the year ending June 30, 1898. U.S. Commission of Fish and Fisheries, Washington, D.C.

155. Clark, F. N. 1900. Methods and results in connection with the propagation of commercial fishes for the Great Lakes. Trans. Am. Fish. Soc. 29: 88–98.

156. Bean, T. H. 1896. Report of the representative of the United States Fish Commission at the World's Columbian Exposition. pp. 177–196, In: Report of the Commissioner of Fish and Fisheries for the year ending June 30, 1894. U.S. Commission on Fish and Fisheries, Washington, D.C.

157. Ravenel, W. deC. 1898. Report of the representative of the United States Fish Commission at the Cotton States and International Exposition at Atlanta, Georgia, in 1895. pp. 147–167, In: Report of the Commissioner for the year ending June 30, 1896. U.S. Commission of Fish and Fisheries, Washington, D.C.

158. Ravenel, W. deC. 1902. The Pan-American Exposition. Report of the Commissioner for the year ending June 30, 1901. U.S. Commission of Fish and Fisheries, Washington, D.C.

159. Smith, H. M. 1897. A review of the history and results of the attempts to acclimatize fish and other water animals in the Pacific states. Bull. U.S. Fish Comm. 15: 379–472.

160. Bowers, G. M. 1905. Report of the Commissioner of Fisheries for the fiscal year ending June 30, 1904. pp. 1–23, In: Report of the Bureau of Fisheries 1904. Department of Commerce and Labor, Washington, D.C.

161. Bowers, G. M. 1902. Report of the United States Commissioner of Fish and Fisheries for the fiscal year ending June 30, 1901. pp. 1–20, In: Report of the Commissioner for the year ending June 30, 1901. U.S. Commission of Fish and Fisheries, Washington, D.C.

162. Bowers, G. M. 1904. Report of the United States Commissioner of Fish and Fisheries for the fiscal year ending June 30, 1902. pp. 1–21, In: Report of the Commissioner for the year ending June 30, 1902. U.S. Commission of Fish and Fisheries, Washington, D.C.

163. Bowers, G. M. 1905. Report of the United States Commissioner of Fish and Fisheries for the fiscal year ending June 30, 1903. pp. 1–28, In: Report of the Commissioner for the year ending June 30, 1903. U.S. Commission of Fish and Fisheries, Washington, D.C.

164. Titcomb, J. W. 1905. Report on the propagation and distribution of food fishes. pp. 29–74, In: Report of the Commissioner for the year ending June 30, 1903. U.S. Commission of Fish and Fisheries, Washington, D.C.

165. Titcomb, J. W. 1905. Report on the propagation and distribution of food fishes. pp. 25–80, In: Report of the Bureau of Fisheries 1904. Department of Commerce and Labor, Washington, D.C.

166. Bowers, G. M. 1907. Report of the Commissioner of Fisheries for the fiscal year ending June 30, 1905. pp. 1–46, In: Report of the Commissioner of Fisheries for the fiscal year 1905, and special papers. Bureau of Fisheries, Washington, D.C.

167. Bowers, G. M. 1909. Report of the Commissioner of Fisheries for the fiscal year ending June 30, 1907. pp. 1–20, In: Report of the Commissioner of Fisheries for the fiscal year 1907. pp. 1–20, In: Report of the Commissioner of Fisheries for the fiscal year 1907, and special papers. Bureau of Fisheries, Washington, D.C.

168. Bowers, G. M. 1911. Report of the Commissioner of Fisheries for the fiscal year ending June 30, 1909. pp. 1–38, In: Report of the Commissioner of Fisheries for the fiscal year 1909, and special papers. Bureau of Fisheries, Washington, D.C.

169. Bowers, G. M. 1911. Report of the Commissioner of Fisheries for the fiscal year ended June 30, 1910. pp. 1–40, In: Report of the Commissioner of Fisheries for the fiscal year 1910 and special papers. Bureau of Fisheries, Washington, D.C.

170. Clark, F. N. 1902. A successful year in the artificial propagation of the whitefish. Trans. Am. Fish. Soc. 31: 97–106.

171. Anonymous. 1970. The distribution of food fishes during the fiscal year 1906. pp. 9–78, In: Report of the Commissioner of Fisheries for the fiscal year 1906 and special papers. Bureau of Fisheries, Washington, D.C.

172. Johnson, R. S. 1910. The magnitude and scope of the fish-culture work of the U.S. Bureau of Fisheries, 1910. Trans. Am. Fish. Soc. 40: 169–171.

173. Fearnow, E. C. 1923. Fish distribution by the federal government. Trans. Am. Fish. Soc. 53: 160–170.

174. Evermann, B. W. 1905. Report of the Division of Statistics and Methods of the Fisheries. pp. 101–122, In: Report of the Commissioner for the year ending June 30, 1903. U.S. Commission of Fish and Fisheries, Washington, D.C.

175. Anonymous. 1907. The propagation and distribution of food fishes in 1905. Report of the Bureau of Fisheries 1905. Department of Commerce and Labor, Washington, D.C. 64 p.

176. Bower, W. T. 1910. History of the American Fisheries Society. Trans. Am. Fish. Soc. 40: 323–358.

177. Blackford, E. G. 1887. Treasurer's report. Trans. Am. Fish. Soc. 16: 65.

178. Willard, C. W. 1904. Treasurer's report. Trans. Am. Fish. Soc. 33: 18.

179. Bissell, J. H. 1886. Fish-culture—a practical art. Trans. Am. Fish. Soc. 15: 36–50.

180. Bissell, J. H. 1888. Co-operation in fish-culture. Trans. Am. Fish. Soc. 17: 89–99.

181. Stranahan, J. J. 1902. Fish culture on the farm. Trans. Am. Fish. Soc. 31: 130–137.

182. James, B. W. 1895. Impoverishment of the food-fish industries. Trans. Am. Fish. Soc. 24: 36–48.

183. Stearns, R. E. C. 1886. Intentional and unintentional distribution of species. Trans. Am. Fish. Soc. 15: 50–56.

184. Whitaker, H. 1895. Some observations on the moral phases of modern fishculture. Trans. Am. Fish. Soc. 24: 59–74.

185. Bower, S. 1897. Fish protection and fish production. Trans. Am. Fish. Soc. 26: 58–65.

186. Fullerton, S. F. 1906. Protection as an aid to propagation. Trans. Am. Fish. Soc. 35: 59–86.

187. Parker, J. C. 1901. Man as a controlling factor in aquatic life. Trans. Am. Fish. Soc. 30: 48–61.

188. Leary, J. L. 1907. Planting fish vs. fry. Trans. Am. Fish. Soc. 36: 140–153.

189. Stranahan, J. J. 1902. Fish culture on the farm. Trans. Am. Fish. Soc. 31: 130–137.

190. Allen, G. R. 1905. Note on the feeding of parent trout, with reference to virility of eggs produced. Trans. Am. Fish. Soc. 34: 122–123.

191. O'Brien, M. E. 1888. The propagation of natural food for fish, with special reference to fish-culture. Trans. Am. Fish. Soc. 17: 29–33.

192. Henshall, J. A. 1901. Practical hints on fish culture. Trans. Am. Fish. Soc. 30: 101–104.

193. Henshall, J. A. 1904. Experiments in feeding fry. Trans. Am. Fish. Soc. 33: 76–78.

194. Atkins, C. H. 1905. The early feeding of salmonoid fry. Trans. Am. Fish. Soc. 34: 75–89.

195. Meehan, W. E. 1904. A year's work of the fisheries interest in Pennsylvania. Trans. Am. Fish. Soc. 33: 82–85.

196. Meehan, W. E. 1906. Fish distributed by Pennsylvania from January 1, 1906, to July 1, 1906. Trans. Am. Fish. Soc. 35: 97.

197. Seal, W. P. 1891. Transportation of live fishes. Trans. Am. Fish. Soc. 20: 55–60.

198. Marsh, M. C. 1904. Danger in shipping cans. Trans. Am. Fish. Soc. 33: 53–54.

199. Atkins, C. G. 1901. The study of fish diseases. Trans Am. Fish. Soc. 30: 82–89.

200. Sykes. A. 1902. Inbreeding pond-rear trout. Trans. Am. Fish. Soc. 31: 116–124.

201. Bowers, G. M. 1913. Report of the Commissioner of Fisheries for the fiscal year ended in June 30, 1911. pp. 1–69, In: Report of the Commissioner of Fisheries for the fiscal year 1911 and special papers. Bureau of Fisheries, Washington, D.C.

202. Hildebrand, S. F. 1941. Hugh McCormick Smith. Prog. Fish-Cult. 55: 23–24.

203. O'Malley, H. 1923. Report of the United States Commissioner of Fisheries for the fiscal year ended June 30, 1922. In: Report of the United States Commissioner of Fisheries for the fiscal year 1922 with appendixes. Bureau of Fisheries, Washington, D.C. 50 p.

204. Bell, F. T. 1934. Bureau of Fisheries. pp. 71–95, In: Report of the United States Commissioner of Fisheries for the fiscal year 1933 with appendixes. Bureau of Fisheries, Washington, D.C.

205. Jackson, C. E. 1941. Bureau of Fisheries. pp. 135–163, In: Report of the United States Commissioner of Fisheries for the fiscal year 1939 with appendixes. Bureau of Fisheries, Washington, D.C.

206. Smith, H. M. 1914. Report of the Commissioner of Fisheries. pp. 5–78, In: Annual report of the Commissioner of Fisheries to the Secretary of Commerce for the fiscal year ended June 30, 1913. Bureau of Fisheries, Washington, D.C.

207. Johnson, R. S. 1915. The distribution of fish and fish eggs during the fiscal year 1914. Appendix 1 to the Report of the U.S. Commissioner of Fisheries for 1914. Bureau of Fisheries, Washington, D.C. 114 p.

208. Smith, H. M. 1917. Report of the United States Commissioner of Fisheries for the fiscal year ending June 30, 1915. pp. 1–81, In: Report of the United States Commissioner of Fisheries for the fiscal year 1914 with appendixes. U.S. Bureau of Fisheries, Washington, D.C.

209. Smith, H. M. 1916. Report of the Commissioner of Fisheries. pp. 5–114, In: Annual report of the Commissioner of Fisheries to the Secretary of Commerce for the fiscal year ended June 30, 1916. Bureau of Fisheries, Washington, D.C.

210. Smith, H. M. 1919. Report of the United States Commissioner of Fisheries for the fiscal year ended June 30, 1917. In: Report of the United States Commissioner of Fisheries for the fiscal year 1917 with appendixes. Bureau of Fisheries, Washington, D.C. 104 p.

211. Smith, H. M. 1920. Report of the United States Commissioner of Fisheries for the fiscal year ended June 30, 1918. In: Report of the United States Commissioner of Fisheries for the fiscal year 1918 with appendixes. Bureau of Fisheries, Washington, D.C. 94 p.

212. Smith, H. M. 1921. Annual report of the Commissioner of Fisheries to the Secretary of Commerce for the fiscal year ended June 30, 1919. In: Report of the United States Commissioner of Fisheries for the fiscal year 1919 with appendixes. Bureau of Fisheries, Washington, D.C. 57 p.

213. Smith, H. M. 1921. Annual report of the United States Commissioner of Fisheries to the Secretary of Commerce for the fiscal year ended June 30, 1920. In: Report of the United States Commissioner of Fisheries for the fiscal year 1920 with appendixes. Bureau of Fisheries, Washington, D.C. 66 p.

214. Smith, H. M. 1922. Report of the United States Commissioner of Fisheries for the fiscal year ended June 30, 1921. In: Report of the United States Commissioner of Fisheries for the fiscal year 1921 with appendixes. Bureau of Fisheries, Washington, D.C. 50 p.

215. Leach, G. C. 1923. Propagation and distribution of food fishes, 1922. In: Report of the United States Commissioner of Fisheries for the fiscal year 1922 with appendixes. Bureau of Fisheries, Washington, D.C. 116 p.

216. O'Malley, H. 1924. Annual report of the Commissioner of Fisheries to the Secretary of Commerce for the fiscal year ended June 30, 1923. In: Report of the Commissioner of Fisheries for the fiscal year 1923 with appendixes. Bureau of Fisheries, Washington, D.C. 47 p.

217. O'Malley, H. 1925. Report of the Commissioner of Fisheries. pp. I–XL, In: Report of the United States Commissioner of Fisheries for the fiscal year 1924 with appendixes. Bureau of Fisheries, Washington, D.C.

218. O'Malley, H. 1926. Report of the Commissioner of Fisheries. pp. I–XL, In: Report of the United States Commissioner of Fisheries for the fiscal year 1925 with appendixes. Bureau of Fisheries. Washington, D.C.

219. O'Malley, H. 1927. Report of the Commissioner of Fisheries. pp. I–XLVI, In: Report of the United States Commissioner of Fisheries for the fiscal year 1926 with appendixes. Bureau of Fisheries, Washington, D.C.

220. Leach, G. C. 1928. Propagation and distribution of food fishes, fiscal year 1927. pp. 685–736, In: Report of the United States Commissioner of Fisheries for the fiscal year 1927 with appendixes. Bureau of Fisheries, Washington, D.C.

221. Leach, G. C. 1929. Propagation and distribution of food fishes, fiscal year 1928. pp. 339–399, In: Report of the United States Commissioner of Fisheries for the fiscal year 1928 with appendixes. Part I. Bureau of Fisheries, Washington, D.C.

222. O'Malley, H. 1930. Report of the Commissioner of Fisheries. pp. I–XXX, In: Report of the United States Commissioner of Fisheries for the fiscal year 1929 with appendixes. Bureau of Fisheries, Washington, D.C.

223. Leach, G. C. 1931. Propagation and distribution of food fishes, fiscal year 1930. pp. 1123–1191, In: Report of the United States Commissioner of Fisheries for the fiscal year 1930 with appendixes. Bureau of Fisheries, Washington, D.C.

224. Leach, G. C. 1932. Propagation and distribution of food fishes, fiscal year 1931. pp. 629–689, In: Report of the United States Commissioner of Fisheries for the fiscal year 1931 with appendixes. Bureau of Fisheries, Washington, D.C.

225. Leach, G. C. 1933. Propagation and distribution of food fishes, fiscal year 1932. pp. 531–569, In: Report of the United States Commissioner of Fisheries for the fiscal year 1932 with appendixes. Bureau of Fisheries, Washington, D.C.

226. Leach, G. C. 1934. Propagation and distribution of food fishes, fiscal year 1933. pp. 451–484, In: Report of the United States Commissioner of Fisheries for the fiscal year 1933 with appendixes. Bureau of Fisheries, Washington, D.C.

227. Leach, G. C. 1936. Propagation and distribution of food fishes, fiscal year 1935. pp. 401–427, In: Report of the United States Commissioner of Fisheries for the fiscal year 1935 with appendixes. Bureau of Fisheries, Washington, D.C.

228. Leach, G. C. 1938. Propagation and distribution of food fishes, fiscal year 1936. pp. 349–379, In: Report of the United States Commissioner of Fisheries for the fiscal year 1936 with appendixes. Bureau of Fisheries, Washington, D.C.

229. Leach, G. C., and M. C. James. 1939. Propagation and distribution of food fishes, fiscal year 1937. pp. 461–492, In: Report of the United States Commissioner of Fisheries for the fiscal year 1937 with appendixes. Bureau of Fisheries, Washington, D.C.

230. Leach, G. C., M. C. James, and E. J. Douglass. 1940. Propagation and distribution of food fishes, fiscal year 1938. pp. 461–494, In: Report of the United States Commissioner of Fisheries for the fiscal year 1938 with appendixes. Bureau of Fisheries, Washington, D.C.

231. Leach, G. C., M. C. James, and E. J. Douglass. 1941. Propagation and distribution of food fishes for the fiscal year 1939. pp. 555–598, In: Report of the United States Commissioner of Fisheries for the fiscal year 1939 with appendixes. Bureau of Fisheries, Washington, D.C.

232. Smith, H. M. 1915. Report of the United States commissioner of Fisheries for the fiscal year ending June 30, 1914. pp. 1–81, In: Report of the United States Commissioner of Fisheries for the fiscal year 1914 with appendixes. U.S. Bureau of Fisheries, Washington, D.C.

233. O'Malley, H. 1928. Report of the Commissioner of Fisheries. pp. I–XXXIII, In: Report of the United States Commissioner of Fisheries for the fiscal year 1927 with appendixes. Bureau of Fisheries, Washington, D.C.

234. O'Malley, H. 1929. Report of the Commissioner of Fisheries. pp. I–XXXII, In: Report of the United States commissioner of Fisheries for the fiscal year 1928 with appendixes. Bureau of Fisheries, Washington, D.C.

235. O'Malley, H. 1932. Report of the Commissioner of Fisheries. pp. I–XXXIII, In: Report of the United States Commissioner of Fisheries for the fiscal year 1931 with appendixes. Bureau of Fisheries, Washington, D.C.

236. O'Malley, H. 1933. Report of the Commissioner of Fisheries. pp. I–XXVII, In: Report of the United States Commissioner of Fisheries for the fiscal year 1932 with appendixes. Bureau of Fisheries, Washington, D.C.

237. Bell, F. T. 1936. Bureau of Fisheries. pp. 87–111, In: Report of the Unites States Commissioner of Fisheries for the fiscal year 1935 with appendixes. Bureau of Fisheries, Washington, D.C.

238. Bell, F. T. 1938. Bureau of Fisheries. pp. 81–107, In: Report of the United States Commissioner of Fisheries for the fiscal year 1936 with appendixes. Bureau of Fisheries, Washington, D.C.

239. Bell, F. T. 1940. Bureau of Fisheries. pp. 95–122, In: Report of the United States Commissioner of Fisheries for the fiscal year 1938 with appendixes. Bureau of Fisheries, Washington, D.C.

240. Leach, G. C. 1930. Propagation and distribution of food fishes, fiscal year 1929. pp. 759–823, In: Report of the United States Commissioner of Fisheries for the fiscal year 1929 with appendixes. Bureau of Fisheries, Washington, D.C.

241. Leach, G. C. Propagation and distribution of food fishes, fiscal year 1925. pp. 439–500, In: Report of the United States Commissioner of Fisheries for the fiscal year 1925 with appendixes. Bureau of Fisheries, Washington, D.C.

242. Leach, G. C. 1922. Propagation and distribution of food fishes, 1921. In: Report of the United States Commissioner of Fisheries for the fiscal year 1921 with appendixes. Bureau of Fisheries, Washington, D.C. 94 p.

243. O'Malley, H. 1917. The distribution of fish and fish eggs during the fiscal year 1917. Appendix I to the Report of the United States Commissioner of Fisheries for the fiscal year 1917 with appendixes. Bureau of Fisheries, Washington, D.C. 99 p.

244. Henshall, J. A. 1883. On the distribution of the black bass. Trans. Am. Fish-Cult. Assoc. 12: 21–27.

245. Coker, R. E. 1916. The utilization [sic] and preservation of fresh-water mussels. Trans. Am. Fish. Soc. 46: 39–49.

246. Landau, M. 1990. Experimental culture of freshwater mussels during the early twentieth century. Aquacult. Mag. 16(3): 66–69.

247. Lydell, D. 1919. Fresh water mussels as a fish food. Trans. Am. Fish. Soc. 49: 24–28.

248. Culler, C. F. 1932. Progress in fish culture. Trans. Am. Fish. Soc. 62: 114–118.

249. Canfield, H. L. 1923. Production of the fresh water mussel. Trans. Am. Fish. Soc. 53: 171–175.

250. Howard, A. D. 1917. A second generation of artificially reared fresh-water mussels. Trans. Am. Fish. Soc. 46: 89–92.

251. Palmer, T. S. 1911. Licenses for hook and line fishing. Trans. Am. Fish. Soc. 41: 91–97.

252. Downing, S. W. 1911. Are the hatcheries on the Great Lakes of benefit to the commercial fishermen? Trans. Am. Fish. Soc. 41: 127–133.

253. Radcliffe, L. 1929. Status of the Great Lakes fisheries. Trans. Am. Fish Soc. 59: 45–52.

254. Smith. L. 1936. Observations on natural versus artificial propagation of commercial species of fish in the Great Lakes region. Trans. Am. Fish. Soc. 66: 56–62.

255. Breder, C. M., Jr. 1922. The problem of marine fish culture. Trans. Am. Fish. Soc. 51: 210–218.

256. Booth, D. C. 1925. Some observations on fish culture. Trans. Am. Fish. Soc. 55: 161–166.

257. Leach, G. C. 1925. Co-operative fish culture. Trans. Am. Fish. Soc. 55: 102–115.

258. Leach, G. C. 1926. Co-operative fish culture. Trans. Am. Fish. Soc. 56: 161–168.

259. Radcliffe, L. 1927. The status of aquiculture. Trans. Am. Fish. Soc. 57: 171–176.

260. Gordon, S. 1932. Scientific management—our future fisheries job. Trans. Am. Fish. Soc. 62: 73–79.

261. Earle, S. 1936. Fish increase responds to government and state cooperation. Trans. Am. Fish. Soc. 66: 392–394.

262. James, M. C. 1942. Glen C. Leach—1872–1942. Trans. Am. Fish. Soc. 72: 21–24.

263. Earle, S. 1937. Fish culture is big business in the United States. Prog. Fish Cult. 31: 1–29.

264. Leach, G. C. 1923. Artificial propagation of whitefish, grayling, and lake trout. In: Report of the Commissioner of Fisheries for the fiscal year 1923 with appendixes. Bureau of Fisheries, Washington, D.C. 32 p.

265. Leach, G. C. 1925. Artificial propagation of shad. pp. 459–486, In: Report of the United States Commissioner of Fisheries for the fiscal year 1924 with appendixes. Bureau of Fisheries, Washington, D.C.

266. Leach, G. C. 1928. Artificial propagation of pike perch, yellow perch, and pikes. pp. 1–27, In: Report of the United States Commissioner of Fisheries for the fiscal year 1927 with appendixes. Bureau of Fisheries, Washington, D.C.

267. Leach, G. C. 1923. Artificial propagation of brook trout and rainbow trout, with notes on three other species. In: Report of the Commissioner of Fisheries for the fiscal year 1923 with appendixes. Bureau of Fisheries, Washington, D.C. 74 p.

268. Cranston, C. K. 1912. The fish and game laws of Oregon. Trans. Am. Fish. Soc. 42: 75–88.

269. Mayhall, J. 1928. The green river hatchery. Trans. Am. Fish. Soc. 58: 87–88.

270. Dyche, L. L. 1911. A new and enlarged fish hatchery for the state of Kansas. Trans. Am. Fish. Soc. 41: 155–179.

271. Dyche, L. L. 1912. Report on progress in the construction of the new pond-fish hatchery in Kansas. Trans. Am. Fish. Soc. 42: 145–146.

272. Dyche, L. L. 1913. One year's work at the Kansas state fish hatchery. Trans. Am. Fish. Soc. 43: 77–85.

273. Buller, N. R. 1916. The work of the Pennsylvania fish commission. Trans. Am. Fish. Soc. 45: 145–153.

274. Buller, N. R. 1913. Work of the Pennsylvania fisheries department. Trans. Am. Fish. Soc. 43: 111–119.

275. Anonymous. 1939. Distribution highlights in Pennsylvania. Prog. Fish-Cult. 43: 34–35.

276. Surber, T. 1929. Fish culture in Minnesota, past, present and future. Trans. Am. Fish. Soc. 59: 224–233.

277. Westers, H., and R. R. Stickney. 1993. Northern pike and muskellunge. pp. 199–213, In: R. R. Stickney (ed.). Culture of nonsalmonid freshwater fishes. CRC Press, Boca Raton, Fla.

278. Surber, T. 1931. Fish cultural successes and failures in Minnesota. Trans. Am. Fish. Soc. 61: 240–246.

279. Earle, S. 1930. What Maryland is doing to stock her streams. Trans. Am. Fish. Soc. 60: 260–261.

280. Viosca, P., Jr. 1929. Fish culture in Louisiana. Trans. Am. Fish. Soc. 59: 207–216.

281. Viosca, P., Jr. 1928. Fish culture in Louisiana. Trans. Am. Fish. Soc. 58: 165–166.

282. Hayford, C. O. 1926. Is the general plan of stocking with fingerlings in thickly populated sections worth while? Trans. Am. Fish. Soc. 56: 181–189.

283. Lockwood, K. F. 1929. The relation between angler and fish culturist. Trans. Am. Fish. Soc. 59: 246–249.

284. Richan, F. J. 1936. What becomes of fish after they are planted? Prog. Fish-Cult. 23: 26–29.

285. Wall, R. 1940. A sports writer poses a question for fish-culturists. Prog. Fish-Cult. 49: 26.

286. Lord, R. F. 1934. Hatchery trout as foragers and game fish. Trans. Am. Fish. Soc. 64: 339–345.

287. Lord, R. F. 1935. The celebrated case of hatchery trout vs. average angler. Are hatchery trout inferior to wild trout? Prog. Fish Cult. 11: 1–9.

288. Needham, P. R. 1939. Natural propagation versus artificial propagation in relation to angling. Prog. Fish Cult. 44: 28–29.

289. Radcliffe, L. 1926. Aquaculture—or water farming. Trans. Am. Fish. Soc. 56: 83–90.

290. Agersborg, H. P. K. 1934. Aquiculture and agriculture. Trans. Am. Fish. Soc. 64: 266–269.

291. Culler, C. F. 1932. Progress in fish culture. Trans. Am. Fish. Soc. 62: 114–118.

292. Davis. H. S. 1938. Fish-cultural developments during recent years. Trans. Am. Fish. Soc. 68: 234–239.

293. Canfield, H. L. 1926. Open water fish culture. Trans. Am. Fish. Soc. 56: 190–194.

294. Anonymous. 1959. H. S. Davis 1875–1958. Trans. Am. Fish. Soc. 88: 74.

295. Bartlett, S. P. 1901. Discussion on carp. Trans. Am. Fish. Soc. 30: 114–132.

296. Townsend, C. H. 1904. Report of the Division of Statistics and Methods of the Fisheries. pp. 143–160, In: Report of the Commissioner for the year ending June 30, 1902. U.S. Commission of Fish and Fisheries, Washington, D.C.

297. Bartlett, S. P. 1905. Carp, as seen by a friend. Trans. Am. Fish. Soc. 34: 207–216.

298. Cole, L. J. 1905. The German carp in the United States. pp. 525–641, In: Report of the Bureau of Fisheries 1904. Department of Commerce and Labor, Washington, D.C.

299. Hunt, W. T. 1911. As to the carp. Trans. Am. Fish. Soc. 41: 189–193.

300. Henshall, J. A. 1919. Indiscriminate and inconsiderate planting of fish. Trans. Am. Fish. Soc. 48: 166–169.

301. Bartlett, S. P. 1910. The future of the carp. Trans. Am. Fish. Soc. 39: 151–154.

302. Leary, J. L. 1910. Propagation of crappie and catfish. Trans. Am. Fish. Soc. 39: 143–148.

303. Ryder, J. A. 1883. Preliminary notice of the development and breeding habits of the Potomac catfish, *Amiurus albidus* (Laseur) Gill. Bull. U.S. Fish Comm. 3: 225–230.

304. Kendall, W. C. 1902. Habits of some of the commercial cat-fishes. Bull. U.S. Bur. Fish. 22: 401–409.

305. Shira, A. F. 1917. Notes on the rearing, growth, and food of the channel catfish, *Ictalurus punctatus*. Trans. Am. Fish. Soc. 46: 77–88.

306. Shira, A. F. 1917. Additional notes on rearing the channel catfish, *Ictalurus punctatus*. Trans. Am. Fish. Soc. 47: 45–47.

307. Mobley, B. E. 1931. The culture of channel catfish (*Ictalurus punctatus*). Trans. Am. Fish. Soc. 61: 171–173.

308. Doze, J. B. 1925. The barbed trout of Kansas. Trans. Am. Fish. Soc. 55: 167–183.

309. Clapp, A. 1929. Some experiments in rearing channel catfish. Trans. Am. Fish. Soc. 59: 114–117.

310. Morris, A. G. 1939. Propagation of channel catfish. Prog. Fish. Cult. 44: 23–27.

311. Langlois, T. H. 1935. The production of small-mouth bass under controlled conditions. Prog. Fish Cult. 7: 1–7.

312. Surber, E. W. 1935. Production of bass fry. Prog. Fish Cult. 8: 1–7.

313. Nevin, J. 1910. Reminiscences of forty-one years' work in fish culture. Trans. Am. Fish. Soc. 40: 313–318.

314. Lydell, D. 1902. The habits and culture of the black bass. Trans. Am. Fish. Soc. 31: 45–73.

315. Riley, M. 1917. Bass rearing in Texas. Trans. Am. Fish. Soc. 46: 107–112.

316. Davis, H. S. 1930. Some principles of bass culture. Trans. Am. Fish. Soc. 60: 48–52.

317. Powell, A. M. 1931. Maryland's activities in raising of trout and bass. Trans. Am. Fish. Soc. 61: 80–82.

318. Meehean, O. L. 1939. A method for the production of largemouth bass on natural food in fertilized ponds. Prog. Fish-Cult. 47: 1–19.

319. Sprecher, G. E. 1938. Artificial spawning and propagation of black bass in Wisconsin. Prog. Fish Cult. 42: 21–24.

320. O'Malley, H. 1931. Report of the Commissioner of Fisheries. pp. I–XXXII, In: Report of the United States Commissioner of Fisheries for the fiscal year 1930 with appendixes. Bureau of Fisheries, Washington, D.C.

321. Denmead, T. 1931. The federal black bass law. Trans. Am. Fish. Soc. 61: 258–261.

323. Buller, N. R. 1905. Propagation and care of yellow perch. Trans. Am. Fish Soc. 34: 223–231.

323. Hile, R. 1936. The increase in the abundance of the yellow pike-perch, *Stizostedion vitreum* (Mitchill), in lakes Huron and Michigan, in relation to the artificial propagation of the species. Trans. Am. Fish. Soc. 66: 143–159.

324. Nevin, J. 1887. Hatching the wall-eyed pike. Trans. Am. Fish. Soc. 66: 143–159.

325. Nevin, J. 1897. Wall-eyed pike. Trans. Am. Fish. Soc. 26: 126–127.

326. Speaker, E. B. 1938. Pond rearing of wall-eyed pike. Prog. Fish Cult. 36: 1–6.

327. Thompson, W. T. 1911. Is irrigation detrimental to trout culture. Trans. Am. Fish. Soc. 41: 103–114.

328. Westerman, F. A. 1929. Progress in trout propagation in Michigan. Trans. Am. Fish. Soc. 59: 234–237.

329. Fish, F. F. 1939. An evaluation of trout culture. Trans. Am. Fish. Soc. 69: 85–89.

330. Babcock, J. P. 1910. Some experiments in the burial of salmon eggs—suggesting a new method of hatching salmon and trout. Trans. Am. Fish. Soc. 40: 393–395.

331. Radcliffe, L. 1932. Water-farming as exemplified by the oyster industry. Trans. Am. Fish. Soc. 62: 154–157.

332. Rowe, H. C. 1919. The great decline of the oyster industry in Connecticut and Rhode Island. Trans. Am. Fish. Soc. 49: 33–37.

333. Gutsell, J. S. 1924. Oyster-cultural problems on Connecticut. In: Report of the Commissioner of Fisheries for the fiscal year 1923 with appendixes. Bureau of Fisheries, Washington, D.C. 10 p.

334. Wells, W. F. 1920. Artificial propagation of oysters. Trans. Am. Fish. Soc. 50: 301–306.

335. Prytherch, H. F. 1924. Experiments in the artificial propagation of oysters. In: Report of the Commissioner of Fisheries for the fiscal year 1923 with appendixes. Bureau of Fisheries, Washington, D.C. 14 p.

336. Kincaid, T. E. 1928. Development of oyster industry of Pacific. Trans. Am. Fish. Soc. 58: 117–122.

337. Galtsoff, P. S. 1930. Oyster industry of the Pacific coast of the United States. pp. 367–400, In: Report of the United States Commissioner of Fisheries for the fiscal year 1929 with appendixes. Bureau of Fisheries, Washington, D.C.

338. Stickney, R. R. 1989. Flagship: a history of fisheries at the University of Washington. Kendall-Hunt, Dubuque, Iowa. 153 p.

339. Smith, H. M. 1913. The need for a national institution for the technical instruction of fisherfolk. Trans. Am. Fish. Soc. 43: 41–45.

340. Kincaid, T. 1914. A proposed school of fisheries. Trans. Pac. Fish. Soc. 1: 29–38.

341. Cobb, J. N. 1920. Development of the College of Fisheries. Trans. Am. Fish. Soc. 50: 56–68.

342. Needham, P. R. 1939. George C. Embody. Prog. Fish Cult. 44: 38–39.

343. Cobb, J. N. 1928. Fisheries education. Trans. Am. Fish. Soc. 58: 146–153.

344. Hines, N. O. 1962. Proving ground. University of Washington Press, Seattle. 336 p.

345. Hines, N. O. 1976. Fish of rare breeding: salmon and trout of the Donaldson strains. Smithsonian Institution Press, Washington, D.C. 167 p.

346. Shell, E. W. 1984. Introduction. pp. 1–4, In: Proc. Auburn University Symposium on Fisheries and Aquaculture. Auburn University, Auburn, Ala.

347. Anonymous. 1973. Tribute to Homer Scout Swingle 1902–1973. Proc. World Maricult. Soc. 4: 13–14.

348. Swingle, H. S., and E. V. Smith. 1939. Increasing fish production in pounds [sic]. Prog. Fish Cult. 44: 32–33.

349. Anonymous. 1986. Aquaculture at Auburn University. Aquacult, Mag. 12(4): 30ff.

350. Bost, J. E. 1936. Why is there no textbook on fish culture? Trans. Am. Fish. Soc. 66: 298–300.

351. Deason, H. J. 1940. A survey of academic qualifications for fishery biologists and of institutional facilities for training fishery biologists. Trans. Am. Fish. Soc. 70: 128–142.

352. Anonymous. 1941. A school for fish-culturists. Prog. Fish-Cult. 54: 36–37.

353. Higgins, E. 1934. Prospectus. Prog. Fish Cult. 1:1.

354. Anonymous. 1939. Is our face red? Prog. Fish Cult. 47: 56.

355. Anonymous. 1941. War and the P.F.C. Prog. Fish-Cult. 56: 1.

356. Anonymous. 1947. Peace and the P.F.C. Prog. Fish-Cult. 9: 1.

357. Mitchell, C. R. 1941. The potato ricer in the fish hatchery. Prog. Fish-Cult. 55: 28–30.

358. Burrows, R. E., and J. E. Manning. 1947. Enlarged ricer for pond feeding. Prog. Fish-Cult. 9: 42–45.

359. Bryant, M., Jr. 1949. That potato ricer again. Prog. Fish-Cult. 11: 136–137.

360. Anonymous. 1952. Trout feed dispenser. Prog. Fish-Cult. 14: 124.

361. Fish. F. F. 1940. A report of the prevention and control of Costia necatrix [sic]. Prog. Fish-Cult. 48: 1–10.

362. Foster, F. J. 1936. The use of malachite green as a fish fungicide and antiseptic. Prog. Fish. Cult. 18: 7–9.

363. Fish, F. F. 1935. The bacterial diseases of fish. Prog. Fish-Cult. 5: 1–9.

364. Fish. F. F. 1935. The protozoan diseases of hatchery fish. Prog. Fish Cult. 6: 1–4.

365. Fish, F. F. 1940. Additional notes on the control of ecto-parasitic protozoans. Prog. Fish-Cult. 49: 31–32.

366. Fish, F. F. 1935. The Bureau of Fisheries' disease service. Prog. Fish-Cult. 3: 1–7.

367. Fish, F. F. 1935. The Bureau of Fisheries' disease service. Prog. Fish-Cult. 8: 9–12.

368. Higgins, E. 1935. Hatchery disease service established. Prog. Fish-Cult. 8: 8.

369. James, M. C. 1935. Hatchery records and accounting. Prog. Fish-Cult. 2: 1–4.

370. James, C. C. 1936. Floating fish nests. Prog. Fish-Cult. 21: 19–21.

371. Moy, P. B., and R. R. Stickney. 1987. Suspended spawning cans for channel catfish in a surface-mine lake. Prog. Fish-Cult. 49: 76–77.

372. Cottam, C., and F. M. Uhler. 1936. The role of fish-eating birds. Prog. Fish Cult. 14: 1–14.

373. Carman, L. G. 1936. Fish-eating birds. Prog. Fish-Cult. 19: 16–18.

374. Lucas, C. R. 1936. "The role of fish-eating birds"—a reply. Prog. Fish-Cult. 19: 7–10.

375. Hoover, E. E. 1936. Fish-eating birds. Prog Fish-Cult. 21: 21–22.

376. Lord, R. F. 1936. Heron provides novel instrument. Prog. Fish-Cult. 24: 28.

377. James, M. C. 1940. Requirements of federal waters to receive first consideration. Prog. Fish-Cult. 52: 8–11.

378. Surber, E. W. 1938. A comparison of the growth of fingerling rainbow trout in concrete and dirt-bottomed circular pools of the same size. Prog. Fish Cult. 42: 29–31.

379. Davis, H. S. 1927. Some results of feeding experiments with trout fingerlings. Trans. Am. Fish. Soc. 57: 281–287.

380. Einarsen, A. S., and L. Royal. 1929. What shall we feed? Trans. Am. Fish. Soc. 59: 268–271.

381. James, M. C. 1928. More about fish food. Trans. Am. Fish. Soc. 58: 170–174.

382. Fiedler, R. H., and V. J. Samson. 1935. Survey of fish hatchery foods and feeding practices. Trans. Am. Fish. Soc. 65: 376–402.

383. Connell, F. H., and E. D. Palmer. 1939. Starch digestion in trout. Prog. Fish Cult. 44: 18–19.

384. Bing, F. C. 1927. A progress report upon feeding experiments with brook trout fingerlings at the Connecticut state fish hatchery, Burlington, Conn. Trans. Am. Fish. Soc. 57: 266–280.

385. Thompson, W. T. 1929. A rival to liver. Trans. Am. Fish. Soc. 59: 168–173.

386. Davis, H. S., and R. F. Lord. 1929. The use of substitutes for fresh meat in the diet of trout. Trans. Am. Fish. Soc. 59: 160–167.

387. McCay, C. M., J. W. Titcomb, E. W. Cobb, M. F. Crowell, and A. Tunison. 1930. The nutritional requirements of trout. Trans. Am. Fish. Soc. 60: 127–139.

388. Davis, H. S. 1932. The use of dry foods in the diet of trout. Trans. Am. Fish. Soc. 62: 189–196.

389. Lord, R. F. 1935. Practical methods of feeding various classes of hatchery foods with a minimum loss. Prog. Fish Cult. 4: 1–7.

390. Donaldson, L. R. 1935. The use of salmon by-products as food for young king salmon. Trans. Am. Fish. Soc. 65: 165–171.

391. Thomas, W. H. 1935. Comments from our readers. Prog. Fish Cult. 6: 6.

392. Brass, J. L. 1937. Teaching baby small mouth and baby large mouth bass to take artificial food. Prog. Fish Cult. 32: 17–20.

393. Donaldson, L. R., and R. F. Foster. 1939. Some experiments on the use of salmon viscera, seal meal, beef pancreas, and apple flour in the diet of young chinook salmon (*Oncorhynchus tschawytscha*). Prog. Fish Cult. 44: 10–17.

394. Donaldson, L. R., and F. J. Foster. 1939. Some experiments on the use of beef liver, seal meal salmon meal and apple flour in the diet of young chinook salmon. Prog. Fish Cult. 46: 12–15.

395. Davis, H. S. 1935. Cheaper trout foods. Prog. Fish Cult. 9: 7–10.

396. Donaldson, L. R., and F. J. Foster. 1938. A summary table of some experimental tests in feeding young salmon and trout. Trans. Am. Fish. Soc. 67: 262–270.

397. McCay, C. M., F. C. Bing, and W. S. Dilley. 1927. The effect of variations in vitamins, protein, fat and mineral matter in the diet upon the growth and mortality of eastern brook trout. Trans. Am. Fish. Soc. 57: 240–249.

398. McCay, C. M., and W. E. Dilley. 1927. Factor H in the nutrition of trout. Trans. Am. Fish. Soc. 57: 250–260.

399. Ellis, J. N. 1993. In search of history: Idaho's thousand springs area. Aquacult. Mag. 19(6): 39–44.

400. Tunison, A. V., and C. M. McCay. 1935. The nutritional requirements of trout. Trans. Am. Fish. Soc. 65: 359–375.

401. Castell, J. D., R. O. Sinhuber, J. H. Wales, and D. J. Lee. 1972. Essential fatty acids in the diet of rainbow trout (*Salmo gairdneri*): growth, feed conversion and some gross deficiency symptoms. J. Nutr. 102: 77–86.

402. Castell, J. D., D. J. Lee, and R. O. Sinhuber. 1972. Essential fatty acids in the diet of rainbow trout (*Salmo gairdneri*): lipid metabolism and fatty acid composition. J. Nutr. 102: 93–99.

403. Reisner, M. 1986. Cadillac desert. Penguin Books, New York. 582 p.

404. Ward, H. B. 1920. Atlantic and Pacific salmon. Trans. Am. Fish. Soc. 50: 92–95.

405. Anonymous. 1976. Columbia basin salmon & steelhead analysis. Pacific Northwest Regional Commission, Portland, Oregon.

406. Hasler, A. D., and W. J. Wisby. 1950. Use of fish for the olfactory assay of pollutants (phenols) in water. Trans. Am. Fish. Soc. 70: 64–70.

407. Hassler, A. D. 1954. Odour perception and orientation in fishes. J. Fish. Res. Board Can. 11: 107–129.

408. Mayhall, L. E. 1927. Propagation work and problems in perpetuating the salmon fisheries of Washington. Trans. Am. Fish. Soc. 57: 165–170.

409. Synnestvedt, H. 1928. The work of the Western Food and Game Fish Protective Association. Trans. Am. Fish. Soc. 58: 78–79.

410. Anonymous. 1939. Salvaging the salmon runs obstructed by the Grand Coulee dam. Prog. Fish-cult. 47: 25–30.

411. Anonymous. 1941. Salvaging the Columbia River salmon. Prog. Fish-Cult. 54: 13.

412. Needham, P. R., O. R. Smith, and H. A. Hanson. 1940. Salmon salvage problems in relation to Shasta dam, California, and notes on the biology of the Sacramento River salmon. Trans. Am. Fish. Soc. 70: 55–69.

413. Smith, R. T. 1939. The reactions of silver salmon, Oncorhynchus kisutch [sic], to a 190 foot fall. Prog. Fish-Cult. 47: 51–52.

414. Cobb, E. W. 1910. Commercial trout hatcheries and their influence on public hatcheries. Trans. Am. Fish. Soc. 40: 173–178.

415. Townsend, C. H. 1913. The private fish pond—a neglected resource. Trans. Am. Fish. Soc. 43: 87–92.

416. Gordon, M. 1926. Tropical aquarium fish-culture in industry and scientific research. Trans. Am. Fish. Soc. 56: 227–237.

417. Gutermuth, C. R. 1938. Club-operated fish hatcheries—a part of Indiana's state-wide plan. Trans. am. Fish. Soc. 68: 118–125.

418. Tunison, A. V., S. M. Mullin, and O. L. Meehean. 1949. Survey of fish culture in the United States. Prog. Fish-Cult. 11: 31–69.

419. Gottschalk, J. S. 1971. Abram Vorhis Tunison, 1909–1971. J. Wildl. Manage. 35: 589–591.

420. Tunison, A. V., S. M. Mullin, and O. L. Meehean. 1949. Extended survey of fish culture in the United States. Prog. Fish-Cult. 11: 253–262.

421. Anonymous. 1950. Prog. Fish-Cult. 12: 25.

422. Kinney, E. C. 1969. Summary of the national survey of needs for hatchery fish. Proc. Southeast. Assoc. Game Fish Comm. 22: 364–367.

423. Osborne, L. 1941. Is wild trout fishing doomed? Prog. Fish-Cult. 54: 24–29.

424. Williamson, R. F. 1947. From egg to trout brook. Pa. Angler 16(2): 2.

425. Baeder, H. A., P. I. Tack, and A. S. Hazzard. 1948. A comparison of the palatability of hatchery-reared and wild brook trout. Trans. Am. Fish. Soc. 75: 181–185.

426. Schuck, H. A. 1948. Survival of hatchery trout in streams and possible methods of improving the quality of hatchery trout. Prog. Fish-Cult. 10: 3–14.

427. Stickney, R. R. 1994. Use of hatchery fish in enhancement programs. Fisheries, 19(5): 6–13.

428. Farley, J. L. 1954. Policies and procedures of the U.S. Fish and Wildlife Service. Trans. Am. Fish. Soc. 83: 13–19.

429. Anonymous. 1941. War and the P.F.C. Prog. Fish-Cult. 56: 1.

430. Lagler, K. F., and C. K. Fisher, Jr. 1947. Selected bibliography of North American fresh-water fishery biology, 1941–1946. Prog. Fish-Cult. 9: 213–230.

431. Anonymous. 1947. Retread on rubber boots. Prog. Fish-Cult. 9: 170.

432. Anonymous. 1948. Paint mixer. Prog. Fish-Cult. 10: 116.

433. Portland Cement Association. 1948. Cement: its preparation and use in fish-management projects. Prog. Fish-Cult. 10: 163–175.

434. Anonymous. 1950. Lanolin to the rescue. Prog. Fish-Cult. 12: 32.

435. Anonymous. 1950. Prog. Fish-Cult. 12: 28.

436. Anonymous. 1950. Prog. Fish-Cult. 12: 11.

437. Anonymous. 1948. Mallards and bream. Prog. Fish-Cult. 10: 61.

438. Anonymous. 1952. Prog. Fish–Cult. 14: 55.

439. Duncan, L. M. 1947. Use of bentonite for stopping seepage in ponds. Prog. Fish-Cult. 9: 37–39.

440. Balch, A. C. 1947. Aluminum hatching troughs. Prog. Fish-Cult. 9: 167–168.

441. Sharp, R. W. 1951. Chlorine-removal unit at full-scale operating hatchery. Prog. Fish-Cult. 13: 146–148.

442. Anonymous. 1953. Plastic hatching jars. Prog. Fish-Cult. 15: 169.

443. Burrows, R. E., and D. D. Palmer. 1955. A vertical egg and fry incubator. Prog. Fish-Cult. 17: 147–155.

444. Saila, S. B. 1955. T-O$_2$, an instrument for the estimation of temperature and dissolved oxygen in natural waters. Prog. Fish-Cult. 17: 162–165.

445. Sneed, K. E., and H. P. Clemens. 1956. Survival of fish sperm after freezing and storage at low temperatures. Prog. Fish-Cult. 18: 99–102.

446. Ihering, R. von 1937. A method for inducing fish to spawn. Prog. Fish Cult. 34: 15–16.

447. Sneed, K. E., and H. P. Clemens. 1960. Hormone spawning of warm-water fishes: its practical and biological significance. Prog. Fish-Cult. 22: 109–113.

448. Sneed, K. E., and H. P. Clemens. 1959. The use of human chorionic gonadotrophin to spawn warm-water fishes. Prog. Fish-Cult. 21: 117–120.

449. Anonymous. 1987. Kermit E. Sneed. Aquacult. Mag. 13(5): 63ff.

450. Anonymous. 1947. More 2,4-D. Prog. Fish-Cult. 9: 184.

451. Reeve, J. H. 1948. Comments. Prog. Fish-Cult. 10: 37.

452. Lawrence, J. M. 1950. Toxicity of some new insecticides to several species of pondfish. Prog. Fish-Cult. 12: 141–146.

453. Shell, E. W. 1966. Comparative evaluation of plastic and concrete pools and earthen ponds in fish-cultural research. Prog. Fish-Cult. 28: 201–205.

454. Madsen, D. H. 1926. Modern methods of fish planting. Trans. Am. Fish. Soc. 56: 195–202.

455. Haskell, D. C. 1940. An investigation of the use of oxygen in transporting trout. Trans. Am. Fish. Soc. 70: 149–160.

456. Copeland, T. H. 1947. Fish distribution units. Prog. Fish-Cult. 9: 193–209.

457. Clark, C. F. 1959. Experiments in the transportation of live fish in polyethylene bags. Prog. Fish-Cult. 21: 177–182.

458. Hulsey, A. H. 1965. Transportation of channel catfish fry in plastic bags. Proc. Southeast. Assoc. Game Fish Comm. 16: 354–356.

459. Feast, C. N., and C. E. Hagie. 1948. Colorado's glass fish tank. Prog. Fish-Cult. 10: 29–30.

460. Griffiths, F. P. 1939. Considerations of the introduction and distribution of exotic fishes in Oregon. Trans. Am. Fish. Soc. 69: 240–243.

461. Miller, R. R. 1945. The introduced fishes of Nevada, with a history of their introduction. Trans. Am. Fish. Soc. 72: 173–193.

462. Anonymous. 1940. King salmon in Maine. Prog. Fish-Cult. 52: 7.

463. Nielson, R. S. 1953. Should we stock brown trout? Prog. Fish-Cult. 15: 125–126.

464. Swingle, H. 1957. Control of pond weeds by use of herbivorous fishes. Proc. South. Weed Conf. 10: 11–17.

465. Sills, J. 1970. A review of herbivorous fish for weed control. Prog. Fish-Cult. 32: 158–161.

466. Guillory, V., and R. D. Gasaway. 1978. Zoogeography of the grass carp in the United States. Trans. Am. Fish. Soc. 107: 105–112.

467. Pentelow, F. T. K., and B. Stott. 1965. Grass carp for weed control. Prog. Fish-Cult. 27: 210.

468. Avault, J. W., Jr. 1965. Preliminary studies with grass carp for aquatic weed control. Prog. Fish-Cult. 27: 207–209.

469. Stevenson, J. H. 1965. Observations on grass carp in Arkansas. Prog. Fish-Cult. 27: 203–206.

470. Stickney, R. R. 1993. Tilapia. pp. 81–115, In: R. R. Stickney (ed.). Culture of nonsalmonid freshwater fishes. CRC Press, Boca Raton, Fla.

471. Anonymous. 1955. *Tilapia.* Prog. Fish-Cult. 17:86.

472. Kelley, H. D. 1957. Preliminary studies on *Tilapia mossambica* Peters relative to experimental pond culture. Proc. Southeast Assoc. Game Fish Comm. 10: 139–149.

473. Swingle, H. S. 1958. Further experiments with *Tilapia mossambica* as a pondfish. Proc. Southeast. Assoc. Game Fish Comm. 11: 152–154.

474. Swingle, H. S. 1960. comparative evaluation of two tilapias as pondfishes in Alabama. Trans. Am. Fish. Soc. 89: 142–148.

475. Lahser, C. W., Jr. 1967. *Tilapia mossambica* as a fish for aquatic weed control. Prog. Fish-Cult. 29: 48–50.

476. Hastings, W. H. 1993. In search of history. Aquacult. Mag. 19(1): 43–46.

477. Brockway, D. R. 1953. Fish food pellets show promise. Prog. Fish-Cult. 15: 92–93.

478. Willoughby, H. 1953. Use of pellets as trout food. Prog. Fish-Cult. 15: 127–128.

479. Wolf, H. 1953. Colored fish-food pellets. Prog. Fish-Cult. 15: 182.

480. Hagen, W., Jr. 1957. Fish foods purchased for federal hatcheries in 1955, compared with 1945 and 1949. Prog. Fish-Cult. 19: 32–39.

481. Maxwell, J. M. 1958. Advance in use of pelleted trout food. Prog. Fish-Cult. 20: 182.

482. Grassl. E. F. 1957. Pelleted dry rations for most propagation in Michigan hatcheries. Trans. Am. Fish. Soc. 86: 307–322.

483. Grassl, E. F. 1957. Possible value of continuous feeding of medicated dry diets to prevent and control pathogens in hatchery-reared trout. Prog. Fish-Cult. 19: 85–88.

484. Tiemeier, O. W. 1962. Supplemental feeding of fingerling channel catfish. Prog. Fish-Cult. 24: 88–90.

485. Carnes, W. C. 1967. Preliminary observations on supplementary feeding of pond fishes. Proc. Southeast. Assoc. Game Fish Comm. 20: 292–296.

486. Graff, D. R., and L. Sorenson. 1970. The successful feeding of a dry diet to esocids. Prog. Fish-Cult. 32: 31–35.

487. Phillips, A. M., Jr., G. L. Hammer, and E. A. Pyle. 1964. Dry concentrates as complete trout foods. Prog. Fish-Cult. 26: 21–24.

488. Phillips, A. M., Jr., G. L. Hammer, J. P. Edwards, and H. F. Hostking. 1964. Dry concentrates as complete trout foods for growth and egg production. Prog. Fish-Cult, 26: 155–159.

489. Fowler, L. G., and R. E. Burrows. 1971. The Abernathy salmon diet. Prog. Fish-Cult. 33: 67–75.

490. Hublou, W. F. 1963. Oregon pellets. Prog. Fish-Cult. 25: 175–180.

491. Snow, J. R., and J. I. Waxwell. 1970. Oregon moist pellet as a production ration for largemouth bass. Prog. Fish-Cult. 32: 101–102.

492. Hastings, W., and H. K. Dupree. 1969. Formula feeds for channel catfish. Prog. Fish-Cult. 31: 187–196.

493. Hoffman, G. L. 1984. Dr. S. F. Snieszko 1902–1984. Aquacult. Mag. 10(3): 12ff.

494. Burrows, R. E. 1949. Prophylactic treatment for control of fungus (*Saprolegnia parasitica*) on salmon eggs. Prog. Fish-Cult. 11: 97–103.

495. Sharp, R. W., L. H. Bennett, and E. C. Saegling. 1952. A preliminary report on the control of fungus in the eggs of the pike (*Esox lucius*) with malachite green. Prog. Fish-Cult. 14: 30.

496. Allison, L. N. 1954. Advancement in prevention and treatment of parasitic diseases of fish. Trans. Am. Fish. Soc. 83: 221–228.

497. Wolf, L. E. 1954. Development of disease-resistant strains of fish. Trans. Am. Fish. Soc. 83: 342–349.

498. Post, G. 1965. A review of advances in the study of diseases of fish: 1954–64. Prog. Fish-Cult. 27: 3–12.

499. Snieszko, S. F. 1959. Antibiotics in fish disease and fish nutrition. Antibiotics and Chemotherapy, 9: 541–545.

500. Swingle, H. S. 1957. Preliminary results on the commercial production of channel catfish in ponds. Proc. Southeast. Assoc. Game Fish Comm. 10: 160–162.

501. Johnson, M. C. 1959. Food-fish farming in the Mississippi delta. Prog. Fish-Cult. 21: 154–160.

502. Hulsey, A. H. 1967. Trends in commercial fish farming practices in Arkansas. Proc. Southeast. Assoc. Game Fish Comm. 18: 313–324.

503. Swingle, H. S. 1957. Revised procedures for commercial production of bigmouth buffalo fish in ponds in the southeast. Proc. Southeast Assoc. Game Fish Comm. 10: 162–165.

504. Swingle, H. S. 1954. Experiments on commercial fish production in ponds. Proc. Southeast. Assoc. Game Fish Comm. 7: 69–74.

505. Walker, M. C., and P. T. Frank. 1952. The propagation of buffalo. Prog. Fish-Cult. 14: 129–130.

506. Brady, L., and A. Hulsey. 1959. Propagation of buffalo fishes. Proc. Southeast. Assoc. Game Fish Comm. 13: 80–90.

507. Green, B. L., and T. Mullins. 1959. Use of reservoirs for production of fish in the rice areas of Arkansas. Agricultural Experiment Station, University of Arkansas, Fayetteville. Special Report No. 9. 13 p.

508. Meyer, F. P., D. L. Gray, W. P. Mathis, J. M. Martin, and B. R. Wells. 1968. Production and returns from the commercial production of fish in Arkansas during 1966. Proc. Southeast. Assoc. Game Fish Comm. 21: 525–531.

509. Anonymous. 1970. Major catfish farming areas in 5 states. Am. Fish Farmer, 1(6): 23.

510. White, J. T. 1969. Louisiana catfish harvest. Am. Fish Farmer, 1(8): 10–12.

511. Anonymous. 1969. Fish farming today. A rapidly expanding multimillion-dollar business. Am. Fish Farmer, 1(1): 11.

512. Hubbs, C. 1970. Scientist cites dangers in aquacultural control methods. Am. Fish Farmer, 1(6): 18–19.

513. Prather, E. E. 1961. A comparison of production of albino and normal channel catfish. Proc. Southeast. Assoc. Game Fish Comm. 15: 302–303.

514. Snow, J. R. 1959. Notes on the propagation of the flathead catfish, *Pilodictis* [*sic*] *olivaris* (Rafinesque). Prog. Fish-Cult. 21: 75–80.

515. Sneed, K. E., H. K. Dupree and O. L. Green. 1961. Observations on the culture of flathead catfish (*Pylodictis olivaris*) fry and fingerlings in troughs. Proc. Southeast. Assoc. Game Fish Comm. 15: 298–302.

516. Henderson, H. 1965. Observations on the propagation of flathead catfish in the San Marcos state fish hatchery, Texas. Proc. Southeast. Assoc. Game Fish Comm. 17: 173–177.

517. Swingle, H. S. 1957. Commercial production of red cats (speckled bullheads) in ponds. Proc. Southeast. Assoc. Game Fish Comm. 10: 156–160.

518. Lenz, G. 1947. Propagation of catfish. Outdoor Nebraska. 25(1): 4–6. (Reprinted in the Prog. Fish-Cult. 9: 231–233.)

519. Anonymous. 1947. Feeding fingerling catfish. Prog. Fish-Cult. 9: 93.

520. Canfield, H. L. 1947. Artificial propagation of those channel cats. Prog. Fish-Cult. 9: 27–30.

521. Toole, M. 1951. Channel catfish culture in Texas. Prog. Fish-Cult. 13:3–10.

522. Nelson, B. 1957. Propagation of channel catfish in Arkansas. Proc. Southeast. Assoc. Game Fish Comm. 10: 165–168.

523. Crawford, B. 1958. Propagation of channel catfish (*Ictalurus lacustris*) at state fish hatchery. Proc. Southeast. Assoc. Game Fish Comm. 11: 132–141.

524. Swingle, H. S. 1959. Experiments on growing fingerling channel catfish to marketable size in ponds. Proc. Southeast. Assoc. Game Fish Comm. 12: 63–72.

525. Grizzell, R. A., Jr. 1968. Pond construction and economic considerations in catfish farming. Proc. Southeast. Assoc. Game Fish Comm. 21: 459–472.

526. Mullins, T. 1970. Capital requirements for initiating a catfish production enterprise. Am. Fish Farmer, 1(3): 12–14.

527. Andrews, J. W. 1970. Scientists study intensive fish culture. Am. Fish Farmer, 1(6): 14–16.

528. Collins, R. A. 1970. Culturing catfish in cages. Am. Fish Farmer, 1(3): 5–8.

529. Schmittou, H. R. 1970. The culture of channel catfish, Ictalurus punctatus (Rafinesque), in cages suspended in ponds. Proc. Southeast. Assoc. Game Fish Comm. 23: 226–244.

530. Anonymous. 1969. New farming techniques make Roosevelt's dream come true. Am. Fish Farmer, 1(1): 17–18.

531. Novotny, A. J. 1975. Net-pen culture of Pacific salmon in marine waters. Mar. Fish. Rev. 37: 36–47.

532. Anonymous. 1970. Salmon aquaculture deemed feasible. Am. Fish Farmer, 1(7): 13.

533. DeWitt, J. W., Jr., 1954. A survey of private trout enterprises in the west. Prog. Fish-Cult. 16: 147–152.

534. Belcher, D. D. 1958. Problems in commercial trout production. Trans. Am. Fish. Soc. 87: 368–373.

535. Rucker, R. R. 1958. Some problems of private trout hatchery operators. Trans. Am. Fish. Soc. 87: 374–379.

536. Greene, A. F. C. 1958. Some state problems associated with the commercial trout industry. Trans. Am. Fish. Soc. 87: 365–367.

537. Swingle, H. S. 1960. Report of Joint Committee with the U.S. Trout Farmers Association. Trans. Am. Fish. Soc. 89: 109–111.

538. Buss, K. 1959. Trout and trout hatcheries of the future. Trans. Am. Fish. Soc. 88: 75–80.

539. Anonymous. 1969. World's largest trout farm. Am. Fish Farmer, 1(1:): 6–9.

540. Anonymous. 1969. Rearing 20,000 trout in a silo. Am. Fish Farmer, 1(1): 19.

541. Buss, K., D. R. Graff, and E. R. Miller. 1970. Trout culture in vertical units. Prog. Fish-Cult. 32: 187–192.

542. Seaman, W. R. 1951. The feeding of cooked carp to hatchery trout in Colorado. Prog. Fish-Cult. 13: 227–228.

543. Sneed, K. E. 1971. The white amur: a controversial biological control. Am. Fish Farmer World Aquacult. News, 2(6): 6–9.

544. Bailey, W. M., and R. L. Boyd. 1971. A preliminary report on spawning and rearing of grass carp (Ctenopharyngodon idella) in Arkansas. Proc. Southeast. Assoc. Game Fish Comm. 24: 560–569.

545. Lynch, T. 1979. White amur experience leads to development of grass carp hybrid. Aquacult. Mag. 6(1): 33–36.

546. Worth, S. G. 1904. The recent hatching of striped bass and possibilities with other commercial species. Trans. Am. Fish. Soc. 33: 223–230.

547. Worth, S. G. 1910. Progress in hatching striped bass. Trans. Am. Fish Soc. 39: 155–159.

548. Anderson, J. C. 1966. Production of striped bass fingerlings. Prog. Fish-Cult. 28: 162–164.

549. Stevens, R. E., O. D. May, Jr., and J. L. Herschell. 1965. An interim report on the use of hormones to ovulate striped bass (Roccus saxatilis). Proc. Southeast. Assoc. Game Fish Comm. 17: 226–237.

550. Stevens, R. E. 1966. Hormone-induced spawning of striped bass for reservoir stocking. Prog. Fish-Cult. 28: 19–28.

551. Stevens, R. E. 1967. A final report on the use of hormones to ovulate striped bass, Roccus saxatilis (Walbaum). Proc. Southeast Assoc. Game Fish Comm. 18: 523–538.

552. Sandoz, O., and K. H. Johnston. 1966. Culture of striped bass Roccus saxatilis (Walbaum). Proc. Southeast Assoc. Game Fish Comm. 19: 390–394.

553. Kelley, J. R., Jr. 1967. Preliminary report on methods for rearing striped bass, Roccus saxatilis (Walbaum), fingerlings. Proc. Southeast. Assoc. Game Fish Comm. 20: 341–346.

554. Logan, H. J. 1968. Comparison of growth and survival rates of striped bass and striped bass × white bass hybrids under controlled environments. Proc. Southeast. Assoc. Game Fish Comm. 21: 260–263.

555. Bishop, R. D. 1968. Evaluation of the striped bass (*Roccus saxatilis*) and white bass (*R. chrysops* hybrids after two years. Proc. Southeast Assoc. Game Fish Comm. 21: 245–254.

556. Markus, H. C. 1934. The fate of our forage fish. Trans. Am. Fish. Soc. 64: 93–96.

557. Radcliffe, L. 1931. Propagation of minnows. Trans. Am. Fish. Soc. 61: 131–137.

558. Hubbs, C. L. 1932. Some experiences and suggestions on forage fish culture. Trans. Am. Fish. Soc. 62: 53–63.

559. Surber, E. W., and G. E. Klak. 1939. Experiments with forage minnows in bass ponds. Prog. Fish-Cult. 47: 31–37.

560. Markus, H. C. 1939. Propagation of bait and forage fish. U.S. Bur. Fish. Fishery Circ. 28. 19 p. Washington, D.C.

561. Surber, T. 1940. Propagation of minnows. Minn. Dep. Conserv. Div. Game Fish. 22 p. Minneapolis.

562. Dobie, J. R., O. L. Meehan, and G. N. Washburn. 1948. Propagation of minnows and other bait species. U.S. Fish Wildl. Serv. Circ. 12. 113 p. Washington, D.C.

563. Dobie, J. R., O. L. Meehan, and G. N. Washburn. 1956. Raising bait fishes. U.S. Fish Wildl. Serv. Circ. 35. 123 p.

564. Prather, E. E. 1957. Experiments on the commercial production of golden shiners. Proc. Southeast. Assoc. Game Fish Comm. 10: 150–155.

565. Prather, E. E. 1959. Further experiments of feeds for fathead minnows. Proc. Southeast. Assoc. Game Fish Comm. 12: 176–178.

566. Davis, J. T. 1993. Baitfish. pp. 307–321, In: R. R. Stickney (ed.). Culture of nonsalmonid freshwater fishes. CRC Press, Boca Raton, Fla.

567. White, J. T. 1970. Minnows . . . by the million.. Am. Fish Farmer. 1(9): 8ff.

568. Avault, J. W., Jr., and E. W. Shell. 1968. Preliminary studies with the hybrid Tilapia nilotica × Tilapia mossambica. FAO Fish. Rep. 44: 237–242.

569. Meyer, F. P., and J. H. Stevenson. 1962. Studies on the artificial propagation of the paddlefish. Prog. Fish-Cult. 24: 65–67.

570. Purkett, C. A., Jr. 1963. Artificial propagation of paddlefish. Prog. Fish-Cult. 25: 31–33.

571. Purkett, C. A., Jr. 1961. Reproduction and early development of paddlefish. Trans. Am. Fish. Soc. 90: 125–129.

572. Finucane, J. H. 1970. Pompano mariculture in Florida. Am. Fish Farmer, 1(4): 5–10.

573. Moore, H. F. 1904. Progress of experiments with sponge culture. Trans. Am. Fish. Soc. 33: 231–243.

574. Smith, H. M. 1905. Remarks on sponge cultivation. Trans. Am. Fish Soc. 34: 256.

575. Moore, H. F. 1910. The commercial sponges and the sponge fisheries. Bull. U.S. Bur. Fish. 28: 399–511.

576. Moore, H. F. 1910. A practical method of sponge culture. Bull. U.S. Bur. Fish. 28: 545–585.

577. Radcliffe, L. 1938. Sponge farming: a necessity. Trans. Am. Fish. Soc. 67: 46–51.

578. MacKenzie, C. L., Jr. 1970. Oyster culture modernization in Long Island Sound. Am. Fish Farmer, 1(6): 7–10.

579. Loosanoff, V. L. 1959. You, too, can now hatch clams. Prog. Fish-Cult. 21: 35.

580. Cook, H. L., and M. A. Murphy. 1966. Rearing penaeid shrimp from eggs to postlarvae. Proc. Southeast. Assoc. Game Fish Comm. 19: 283–286.

581. de la Bretonne, L. W., Jr., and J. W. Avault, Jr. 1970. Shrimp mariculture methods tested. Am. Fish Farmer, 1(12): 8ff.

582. Wheeler, R. S. 1969. Culture of penaeid shrimp in brackish-water ponds, 1966–67. Proc. Southeast Assoc. Game Fish Comm. 22: 387–391.

583. Anonymous. 1970. First cultured shrimp harvested at Florida farm. Am. Fish Farmer & World Aquacult. News, 2(1): 7.

584. Smith, H. M. 1905. General account of the lobster and clam investigations. pp. 141–147, In: Report of the Commissioner for the year ending June 30, 1903. U.S. Commission of Fish and Fisheries, Washington, D.C.

585. Mead, A. D. 1901. Experiments in lobster culture. Trans. Am. Fish. Soc. 30: 94–100.

586. Mead, A. D. 1905. The problem of lobster culture. Trans Am. Fish. Soc. 34: 156–173.

587. Smith, H. M. 1904. Report on the inquiry respecting food-fishes and the fishing-grounds. pp. 111–142, In: Report of the Commissioner for the year ending June 30, 1902. U.S. Commission of Fish and Fisheries, Washington, D.C.

588. Sherwood, G. H. Experiments in lobster rearing. pp. 149–174, In: Report of the Commissioner for the year ending June 30, 1903. U.S. Commission of Fish and Fisheries, Washington, D.C.

589. Kensler, C. B. 1970. The potential of lobster culture. Am. Fish Farmer, 1(11): 8ff.

590. Jahnig, C. E. 1973. Growing the American lobster. Proc. World Maricult. Soc. 4: 171–181.

591. Botsford, L. W. 1977. Current economic status of lobster culture research. Proc. World Maricult. Soc., 723–739.

592. Crawford, D. R. 1920. Spawning habits of the spiny lobster (*Panulirus argus*), with notes on artificial hatching. Trans. Am. Fish. Soc. 50: 312–319.

593. Ting, R. Y. 1973. Culture potential of spiny lobster. Proc. World Maricult. Soc. 4: 165–170.

594. Avault, J. W., Jr., L. de la Bretonne, Jr., and E. J. Jaspers. 1970. Louisiana crustacean king. Am. Fish Farmer, 1(10): 8ff.

595. Thomas C. H. 1965. A preliminary report in the agricultural production of the red-swamp crawfish (*Procambarus clarki*) (Girard) in Louisiana rice fields. Proc. Southeast Assoc. Game Fish Comm. 17: 180–186.

596. Anonymous. 1970. Crawfish: a Louisiana aquaculture crop. Am. Fish Farmer, 1(9): 12–15.

597. Viosca, P., Jr. 1931. Principles of bullfrog (*Rana catesbiana*) culture. Trans. Am. Fish. Soc. 61: 262–269.

598. Safford, W. H. 1910. Some observations in frog culture. Trans. Am. Fish. Soc. 40: 289–292.

599. Meehan, W. E. 1905. Frog culture. Trans. Am. Fish. Soc. 34: 257–264.

600. Culley, D. D., and C. T. Gravois. 1970. Frog culture. Am. Fish Farmer, 1(5): 5–10.

601. Webster, D. A. 1950. Fishery research program at Cornell University. Prog. Fish-Cult. 12: 77–80.

602. Anonymous. 1957. The Cortland, New York, fishery station: a summary of twenty-five years of work. Prog. Fish-Cult. 19: 172–178.

603. Van Cleve, R. 1952. The school of fisheries. Prog. Fish-Cult. 14: 159–164.

604. Carlander, K. D. 1959. A survey of technical fishery personnel. Trans. Am. Fish. Soc. 88: 18–22.

605. Swingle, H. S. 1959. Careers in fisheries. Trans. Am. Fish. Soc. 88: 16–17.

606. Moss, D. D., J. H. Gorver, H. R. Schmittou, E. W. Shell, and F. L. Lichtkoppler. 1979. Auburn University's philosophy and strategy for international aquacultural development and technology transfer. Proc. World Maricult. Soc. 10: 68–78.

607. Avault, J. W., Jr., W. Guthrie Perry, Jr., and J. Massey. 1994. History of the World Aquaculture Society. World Aquaculture Society, Baton Rouge, La. 6 p.

608. Leary, T. R. 1970. History of the World Mariculture Society. Proc. World Maricult. Soc. 1: 13–14.

609. Bente, P. F., Jr. 1970. Mariculture on the move. Proc. World Maricult Soc. 1: 18–26.

610. Hendry, R. R. 1970. The Florida mariculture law. Proc. World Maricult. Soc. 1: 47–49.

611. Bente, P. F., Jr. 1970. Application of the first mariculture law, operation under the law. Proc. World Maricult. Soc. 1: 50–54.

612. Davies, C. B., and H. W. Shields. 1970. Mariculture and the law. Proc. World Maricult. Soc. 1: 55–59.

613. Pausina, B. V. 1970. Louisiana oyster culture. Proc. World Maricult. Soc. 1: 29–34.

614. Shaw, W. N. 1970. Oyster farming in North America. Proc. World Maricult. Soc. 1: 39–44.

615. Finucane, J. H. 1970. Progress in pompano mariculture in the United States. Proc. World Maricult. Soc. 1: 69–72.

616. Klontz, G. W. 1970. Mariculture medicine. Proc. World Maricult. Soc. 1: 129–131.

617. Mock, C. R., and M. A. Murphy. 1970. Techniques for raising penaeid shrimp from the egg to postlarvae. Proc. World Maricult. Soc. 1: 143–156.

618. Taub, F. B. 1970. Algal culture as a source of feed. Proc. World Maricult. Soc. 1: 101–117.

619. Stevens, R. E. 1970. Striped bass culture. Proc. World Maricult. Soc. 1: 159–162.

620. Nash, C. E. 1991. Turn of the millennium aquaculture: navigating troubled waters or riding the crest of the wave? World Aquacult. 22(3): 28–49.

621. Abel, R. B. 1971. The future of aquaculture—A manic-depressive view. Proc. World Maricult. Soc. 2: 13–19.

622. Provenzano, A. J., Jr. 1971. Research priorities for non-penaeid crustaceans. Proc. World Maricult. Soc. 2: 123–124.

623. Landers, W. S. 1971. Research priorities for molluskan mariculture. Proc. World Maricult. Soc. 2: 125–126.

624. Sparks, A. K. 1971. Research priorities for penaeid shrimp mariculture. Proc. World Maricult. Soc. 2: 121–122.

635. Cobb, B. F., III. 1971. The use of mariculture to produce a quality food product, a challenge for the future. Proc. World Maricult. Soc. 2: 87–92.

636. Mulvihill, P. 1971. the academic-industrial relationship in mariculture. Proc. World Maricult. Soc. 2: 93–95.

627. Anonymous. 1985. Aquaculture at Louisiana State University. Aquacult. Mag. 11(3): 42–45.

628. Conrad, J. 1987. Mississippi State's aquaculture program. Aquacult. Mag. 13(5): 32–35.

629. Anonymous. 1988. Auburn University's role in catfish development. Aquacult. Mag. 14(4): 71–74.

630. Anonymous. 1993. Aquaculture a the University of Wisconsin. Aquacult. Mag. 19(3): 98–100.

631. Anonymous. 1993. Study in fisheries and aquaculture at Louisiana State University. Aquacult. Mag. 19(5): 96–97.

632. Glude, J. B. (ed.). 1977. NOAA aquaculture plan. U.S. Department of Commerce, Washington, D.C. 41 p.

633. Glude, J. B. 1976. The role of federal agencies in aquaculture. Proc. World Maricult. Soc. 7: 33–47.

634. National Academy of Sciences. 1978. Aquaculture in the United States. Report to the Senate Committee on Agriculture, Nutrition, and Forestry. Washington, D.C. 93 p.

635. FAO. 1991. FAO Fisheries Circular 815, revision 2. Food and Agriculture Organization of the United Nations, Rome. 136 p.

636. JSA. 1983. National aquaculture development plan, vol. 1. Joint Subcommittee on Aquaculture, Washington, D.C. 67 p.

637. JSA. 1983. National aquaculture development plan, vol. 2. Joint Subcommittee on Aquaculture, Washington, D.C. 196 p.

638. Kane, T. E. 1972. Mariculture: raising legal problems. Proc. World Maricult. Soc. 3: 39–44.

639. Conrad, J. 1984. Washington state aquaculture faces regulatory quagmire. Aquacult. Mag. 10(6A): 22–23.

640. Culley, D. D., Jr. 1973. Raceways; exotic species most affected by proposed E.P.A. discharge permits. Am. Fish Farmer World Aquacult. News, 4(8): 9–12.

641. Rubino, M. C., and C. A. Wilson. 1993. Issues in aquaculture regulation. Bluewaters, Inc., Bethesda, Md. 72 p.

642. Davis, J. T. 1993. Survey of aquaculture effluent permitting and 1993 standards in the south. Southern Regional Aquaculture Center, Stoneville, Mississippi. Publ. 465. 4 p.

643. Stickney, R. R. 1988. Aquaculture on Trial. World Aquacult. 19: 16–18.

644. Stickney, R. R. 1990. Controversies in salmon aquaculture and projections for the future of the aquaculture industry. pp. 455–461, In: Proc. Fourth Pacific Congress on Marine Science and Technology, Tokyo, Japan, July 16–20. PACON International, Tokyo.

645. Madewell, C. E., and R. J. Ballew. 1972. Historical development of catfish farming. Am. Fish Farmer & World Aquacult. News, 3(3): 8–11.

646. Anonymous. 1971. Catfish production: some regional comparisons. Am. Fish Farmer & World Aquacult. News, 2(9): 11–13.

647. Roberts, K., and L. Bauer. 1978. United States mariculture development and the economic evolution of catfish systems as paradigms of relevance. Proc. World Maricult. Soc. 9: 375–382.

648. Avault, J. W., Jr. 1980. Six tons of catfish with aeration in one acre of water. Aquacult. Mag. 7(1): 44.

649. Anonymous. 1979. Production-inventory-sales data (national processors). Aquacult. Mag. 6(1): 52.

650. Avault, J. W., Jr. 1986. New advances in catfish farming research. Aquacult. Mag. 12(3): 41–43.

651. Collins, C. 1989. Some of the old and new in channel catfish farming. Aquacult. Mag. 15(4): 47–54.

652. Hollerman, W. D., and C. E. Boyd. 1985. Effects of annual draining on water quality and production of channel catfish in ponds. Aquaculture, 46: 45–54.

653. Anonymous. 1981. Process production report on farm-raised catfish. Aquacult. Mag. 7(3): 57.

654. Anonymous. 1982. Production-inventory-sales data (national processors). Aquacult. Mag. 8(3): 65.

655. Anonymous. 1983. Production-inventory-sales data (national processors). Aquacult. Mag. 9(3): 49.

656. Anonymous. 1984. Production-inventory-sales data (national processors). Aquacult. Mag. 10(6A): 49.

657. Anonymous. 1985. Production-inventory-sales data (national processors). Aquacult. Mag. 11(4): 65.

658. Anonymous. 1986. Production-inventory-sales data (national processors). Aquacult. Mag. 12(3): 57.

659. Anonymous. 1987. Catfish processing up 64 percent. Aquacult Mag. 13(4): 69.

660. Anonymous. 1988. Catfish processing up 30 percent. Aquacult. Mag. 14(4): 73.

661. Anonymous. 1989. Catfish processing up four percent. Aquacult. Mag. 15(3): 84.

662. Anonymous. 1990. Catfish processing up 13 percent. Aquacult. Mag. 16(4): 95.

663. Anonymous. 1991. Catfish processing up 1 percent. Aquacult. Mag. 17(4): 96.

664. Anonymous. 1992. Catfish processing up 15 percent. Aquacult. Mag. 18(6): 96.

665. Anonymous. 1994. Catfish processing up 8 percent. Aquacult. Mag. 20(2): 99–100.

666. Anonymous. 1985. Profile of catfish industry. Aquacult Mag. 11(1): 4.

667. Hastings, W. H., B. Hinson, D. Tackett, and B. Simco. 1972. Monitoring channel catfish use of a demand feeder. Prog. Fish-Cult. 34: 204–206.

668. Collins, R. A. 1971. Cage culture of catfish in reservoir lakes. Proc. Southeast Assoc. Game Fish Comm. 24: 489–496.

669. Kelley, J. R., Jr. 1973. An improved cage design for use in culturing channel catfish. Prog. Fish-Cult. 35: 167–169.

670. Newton, S. H. 1980. Review of cage culture activity indicates continuing interest. Aquacult. Mag. 7(1): 32–36.

671. Yingst, W. L. III, and R. R. Stickney. 1980. Growth and survival of caged channel catfish (*Ictalurus punctatus*) fingerlings on diets containing various lipids. Prog. Fish-Cult. 42: 24–26.

672. Stickney, R. R., and P. B. Moy. 1985. Production of caged channel catfish in an Illinois surface mine lake. J. World Maricult Soc. 16: 32–39.

673. Buttner, J. K. 1992. Cage culture of black bullhead. Aquacult. Mag. 18(2): 32ff.

674. Anonymous. 1985. Status of fish farming in Mississippi. Aquacult. Mag. 11(3): 4.

675. Anonymous. 1986. Fish farming in Mississippi. Aquacult. Mag. 12(2): 4.

676. Anonymous. 1987. Status of fish farming in Mississippi. Aquacult. Mag. 13(2): 8.

677. Anonymous. 1991. Status of fish farming in Mississippi, May 1991. Aquacult. Mag. 17(6): 14ff.

678. Anonymous. 1991. Catfish industry's growth continues in 1990. Aquacult. Mag. 17(2): 6ff.

679. Collins, C. 1991. Fish farming in Arkansas. Aquacult. Mag. 17(4): 69–71.

680. Bush, M. J., and J. L. Anderson. 1993. Northeast region aquaculture industry situation and outlook report. Northeastern Aquaculture Center, North Dartmouth, Mass. 60 p.

681. Keenum, M. E., and J. E. Waldrop. 1988. Economic analysis of farm-raised catfish production in Mississippi. Mississippi Agricultural & Forestry Experiment Station, Starkville, Mississippi. Tech. Bull. 155. 27 p.

682. Erickson, J. D. 1981. American trout farming marks 100 years plus—still growing. Aquacult. Mag. 7(3): 14–17.

683. Anonymous. 1970 Washington trout hatchery. Am. Fish Farmer, 1(2): 13–16.

684. Hoffman, T. E. 1972. Idaho trout industry threatened. Am. Fish Farmer & World Aquacult. News, 3(12): 8–9.

685. Conrad, J. 1985. Trout farming in Idaho. Aquacult. Mag. 11(1): 32ff.

686. Chew, K. K., and D. R. Toba. 1993. Western regional aquaculture industry situation and outlook report, volume 2 (1986 through 1991). Western Regional Aquaculture Center, Seattle. 34 p.

687. U.S. Department of Agriculture. 1993. Aquaculture situation and outlook report. U.S. Department of Agriculture, Washington, D.C. 46 p.

688. Stolte, L. W. 1974. Introduction of coho salmon into the coastal waters of New Hampshire. Prog. Fish-Cult. 36: 29–32.

689. Mahnken, C. V. W., A. J. Novotny, and T. Joiner. 1970. Salmon mariculture potential assessed. Am. Fish Farmer & World Aquacult. News, 2(1): 12ff.

690. Anonymous. 1971. Washington salmon culture project well under way. Am. Fish Farmer & World Aquacult. News, 2(8): 17.

691. Lindsay, C. E. 1980. Salmon farming in Washington moves closer to industry status. Aquacult. Mag. 6(3): 20–27.

692. McNeil, W. J. 1975. Perspectives on ocean ranching of Pacific salmon. Proc. World Maricult. Soc. 6: 299–307.

693. Himsworth, D. 1981. Impact on wild salmon runs by Oregon's ranching industry. Aquacult. Mag. 7(2): 12–16.

694. Anonymous. 1981. Oregon halts new salmon permits. Aquacult. Mag. 7(3): 13.

695. Stickney, R. R. 1988. Commercial fishing and net-pen salmon aquaculture: turning conceptual antagonism toward a common purpose. Fisheries, 13(4): 9–13.

696. Guerrero, R. D. 1973. Tilapia cultured at Auburn. Am. Fish Farmer & World Aquacult. News, 4(6): 12.

697. Fassler, R. C. 1983. Hawaii marketing program pays off as tilapia enjoys popularity surge. Aquacult. Mag. 9(2): 41.

698. Kirk, R. G. 1972. A review of recent developments in *Tilapia* culture, with special reference to fish farming in the heated effluents of power stations. Aquaculture, 1: 45–60.

699. Habel, M. L. 1975. Overwintering of the cichlid, *Tilapia aurea,* produces fourteen tons of harvestable size fish in a South Alabama bass-bluegill public fishing lake. Prog. Fish-Cult. 37: 31–32.

700. Chervinski, J., and R. R. Stickney. 1981. Overwintering facilities for tilapia in Texas. Prog. Fish-Cult. 43; 20–21.

701. Stickney, R. R., H. B. Simmons, and L. O. Rowland. 1977. Growth responses of *Tilapia aurea* to feed supplemented with dried poultry waste. Tex. J. Sci. 28: 93–99.

702. Stickney, R. R., L. O. Rowland, and J. H. Hesby. 1977. Water quality—*Tilapia aurea* interactions in ponds receiving swine and poultry wastes. Proc. World Maricult. Soc. 8: 55–71.

703. Stickney, R. R., and J. R. Hesby. 1978. Tilapia culture in ponds receiving swine waste. pp. 90–101, In: R. O. Smitherman, W. L. Shelton, and J. H. Grover (eds.). Culture of Exotic Fishes Symposium Proceedings. Fish Culture Section, American Fisheries Society, Auburn, Alabama.

704. Davis, A. T., and R. R. Stickney. 1978. Growth of responses of *Tilapia aurea* to dietary protein quality and quantity. Trans. Am. Fish. Soc. 107: 469–483.

705. Stickney, R. R., J. H. Hesby, R. B. McGeachin, and W. A. Isbell. 1979. Growth of *Tilapia nilotica* in ponds with differing histories of organic fertilization. Aquaculture 17: 189–194.

706. Redner, B. D., and R. R. Stickney. 1979. Acclimation to ammonia by *Tilapia aurea*. Trans. Am. Fish. Soc. 108: 383–388.

707. Henderson-Arzapalo, A., R. R. Stickney, and D. H. Lewis. 1980. Immune hypersensitivity in intensively cultured *Tilapia* species. Trans. Am. Fish Soc. 109: 244–247.

708. Burns, R. P., and R. R. Stickney. 1980. Growth of *Tilapia aurea* in ponds receiving poultry wastes. Aquaculture. 20: 117–121.

709. Winfree, R. A., and R. R. Stickney. 1981. Effects of dietary protein and energy on growth, feed conversion efficiency and body composition of *Tilapia aurea*. J. Nutr. 111: 1001–1012.

710. McGeachin, R. B., and R. R. Stickney. 1982. Manuring rates of production of blue tilapia in simulated sewage lagoons receiving laying hen waste. Prog. Fish-Cult. 44: 25–28.

711. Rouse, D. B., and R. R. Stickney. 1982. Evaluation of the production potential of *Macrobrachium rosenbergii* in monoculture and in polyculture with *Tilapia aurea*. Proc. World Maricult. Soc. 13: 73–85.

712. Stickney, R. R., R. B. McGeachin, E. H. Robinson, G. Arnold, and L. Suter. 1983. Growth of *Tilapia aurea* as a function of degree of dietary lipid saturation. Proc. Southeast. Assoc. Fish Wildl. Agencies, 36: 172–181.

713. Stickney, R. R., and R. B. McGeachin. 1984. Effects of dietary lipid quality on growth and food conversion of *Tilapia aurea*. Proc. Southeast. Assoc. Fish Wildl. Agencies, 37: 352–357.

714. Stickney, R. R., and R. B. McGeachin. 1984. Responses of *Tilapia aurea* to semi-purified diets of differing fatty acid composition. pp. 346–355, In: L. Fishelson and Z. Yaron (comps.), Proc. International Symposium on Tilapia in Aquaculture, Nazareth, Israel, May 1983. Tel Aviv University Press, Tel Aviv.

715. Stickney, R. R., and R. B. McGeachin. 1984. Growth, food conversion and survival of fingerling *Tilapia aurea* fed differing levels of dietary beef tallow. Prog. Fish-Cult. 46: 102–105.

716. Robinson, E. H., S. D. Rawles, P. W. Oldenburg, and R. R. Stickney. 1984. Effects of feeding glandless or glanded cottonseed products to *Tilapia aurea*. Aquaculture, 38: 145–154.

717. Stickney, R. R., and W. A. Wurts. 1986. Growth response of blue tilapia to selected levels of dietary menhaden and catfish oils. Prog. Fish-Cult. 48: 107–109.

718. Suffern, J. S. 1980. The potential of tilapia in United States aquaculture. Aquacult. Mag. 6(6): 14–18.

719. Sample, W. D. 1992. Tilapia culture in the United States: what are the prospects? Aquacult. Mag. 18(5): 75–76.

720. Chamberlain, G. W. and G. McCarty. 1985. Why choose redfish? Aquacult. Mag. 11(2): 35ff.

721. Arnold, C. R., W. H. Bailey, T. D. Williams, A. Johnson, and J. L. Lasswell. 1977. Laboratory spawning and larval rearing of red drum and southern flounder. Proc. Southeast. Assoc. Fish Wildl. Agencies, 31: 437–440.

722. Henderson-Arzapalo, A. 1992. Red drum aquaculture. Rev. Aquat. Sci. 6: 479–491.

723. Crocker, P. A., C. R. Arnold, J. A. DeBoer, and J. D. Holt. 1981. Preliminary evaluation of survival and growth of juvenile red drum (*Sciaenops ocellata*) in fresh and salt water. Proc. World Maricult. Soc. 12: 122–134.

724. Smith, T. I. J. 1988. Aquaculture of striped bass and its hybrids in North America. Aquacult. Mag. 14(1): 40–49.

725. Liao, D. S. 1985. The economic and market potential for hybrid bass aquaculture in estuarine waters: a preliminary evaluation. J. World Maricult. Soc. 16: 151–157.

726. Van Olst, J. C., and J. M. Carlberg. 1990. Commercial culture of hybrid striped bass: states and potential. Aquacult. Mag. 16(1): 49–59.

727. Smith, T. I. J. 1988. The culture potential of striped bass and its hybrids. World Aquacult. 20(1): 32–38.

728. Willis, D. W., and S. A. Flickinger. 1980. Survey of private and government hatchery success in raising largemouth bass. Prog. Fish-Cult. 42: 232–233.

729. Williamson, J. H., G. J. Carmichael, K. G. Graves, B. A. Simco, and J. R. Tomasso, Jr. 1993. Centrachids. pp. 145–197, In: R. R. Stickney (ed.). Culture of nonsalmonid freshwater fishes. CRC Press, Boca Raton, Fla.

730. Stone, N. 1994. Bighead carp. Aquacult. Mag. 20(4): 12ff.

731. Hassler, W. W., and W. T. Hogarth. 1977. The growth and culture of dolphin, *Coryphaena hippurus*, in North Carolina. Aquaculture, 12: 115–122.

732. Anonymous. 1981. Dolphin successfully spawned in Hawaii. Aquacult. Mag. 7(4): 6.

733. Hagood, R. W., G. N. Rothwell, M. Swafford, and M. Tosaki. 1981. Preliminary report on the aquacultural development of the dolphin fish, *Coryphaena hippurus* (Linnaeus). Proc. World Maricult. Soc. 12: 135–139.

734. Anonymous. 1992. Mahimahi research lead to diet formulation. Aquacult. Mag. 18(6): 87–88.

735. Petrocci, C. 1994. Tails from the road: paddlefish. Aquacult. Mag. 20(3): 63–67.

736. Bumpas, G. S. 1973. Paddlefish cultivation possible. Am. Fish Farmer World Aquacult. News, 4(6): 4–6.

737. Arnold, C. R., J. M. Wakeman, T. D. Williams, and G. D. Treece. 1978. Spawning of red snapper (*Lutjanus campechanus*) in captivity. Aquaculture, 15: 301–302.

738. Minton, R. V., J. P. Hawke, and W. M. Tatum. 1983. Hormone induced spawning of red snapper, *Lutjanus campechanus*. Aquaculture, 30: 363–368.

739. Miller, R. R. 1972. Threatened freshwater fishes of the United States. Trans, Am. Fish. Soc. 101: 239–252.

740. Green, S. 1879. Fish hatching and fish catching. Rochester, N.Y. (cited in ref 7.)

741. Carter, E. N. 1904. Notes on sturgeon culture in Vermont. Trans. Am. Fish. Soc. 33: 60–75.

742. Dean, B. 1894. Recent experiments in sturgeon hatching on the Delaware River. Fish. Bull. U.S. Fish Comm. 13: 335–339.

743. Post, H. 1890. Sturgeon experiments in hatching. Trans. Am. Fish. Soc. 19: 36–40.

744. Meehan, W. E. 1910. Experiments in sturgeon culture. Trans. Am. Fish. Soc. 39: 85–91.

745. Harkness, W. J. K., and J. R. Dymond. 1961. The lake sturgeon, the history of its fishery and problems of conservation. Ontario Department of Lands, Forests. 121 p.

746. Smith, T. I. J., E. K. Dingley, and D. E. Marchette. 1980. Induced spawning and culture of Atlantic sturgeon. Prog. Fish-Cult. 42: 147–151.

747. Smith, T. I. J., E. K. Dingley, and D. E. Marchette. 1981. Culture trials with Atlantic sturgeon, *Acipenser oxyrhynchus*, in the USA. J. World Maricult. Soc. 12: 78–87.

748. McGuire, A. B. 1980. First U.S. sturgeon hatchery started at University of California Davis campus. Aquacult. Mag. 6(6): 4–5.

749. Lindley, D. W. 1983. The white sturgeon comes to California. Aquacult. Mag. 10(1): 22–24.

750. Anonymous. 1982. Enthusiastic operation on west coast begins commercial sturgeon hatchery. Aquacult. Mag. 8(2): 4–5.

751. Anonymous. 1991. California farms using innovation to develop sturgeon. Aquacult. Mag. 17(4): 90–92.

752. Conte, F. S. 1990. California aquaculture. World Aquacult. 21(3): 33ff.

753. Smith, T. I. J., and E. K. Dingley. 1984. Review of biology and culture of Atlantic (*Acipenser oxyrynchus*) and shortnose sturgeon (*A. brevirostrum*). J. World Maricult. Soc. 15: 210–218.

754. Bailey, W. M., F. P. Meyer, J. M. Martin, and D. L. Gray. 1974. Farm fish production in Arkansas during 1972. Proc. Southeast. Assoc. Game Fish Comm. 27: 750–755.

755. Freeze, M., and S. Henderson. 1980. The aquaculture industry of Arkansas in 1979. Proc. Annu. Conf. Southeast Assoc. Fish Wildl. Agencies, 34: 115–126.

756. Collins, C. 1986. Present status of fish farmers in Arkansas. Aquacult. Mag. 12(3): 50ff.

757. Conrad, J. 1984. Visiting Anderson's minnow farms. Aquacult. Mag. 10(2): 32–33.

758. Helfrich, L. A., D. L. Weigmann, and D. L. Garling. 1979. Commercial aquaculture in Virginia in 1978. Proc. Annu. Conf. Southeast. Assoc. Fish Wildl. Agencies, 33: 318–323.

759. Anonymous. 1972. Ozark's goldfish breed in artificial nests. Am. Fish Farmer & World Aquacult. News, 3(3): 5ff.

760. Neils, K. E. 1979. The technical and commercial aspects of the goldfish industry. Aquacult. Mag. 6(1): 22–26.

761. McLarney, B. 1984. Native aquarium fish a new aquaculture crop, Part I. Aquacult. Mag. 10(2): 26–28.

762. Hennessy, T. K. 1986. Tropical fish production in Florida. Aquacult. Mag. 12(4): 12ff.

763. Boozer, D. 1973. Tropical fish farming. Am. Fish Farmer & World Aquacult. News, 4(8): 4–5.

764. Belleville, B. 1981. Large effort with small fish builds valuable Florida tropicals industry. Aquacult. Mag. 7(4): 12–17.

765. Knox, M. 1984. Freshwater aquaculture in Florida: highlights of a 1983 survey. Proc. Southeast. Assoc. Fish Wildl. Agencies, 38: 407–412.

766. Anonymous. 1992. Florida aquaculture sales total $54 million in 1991. Aquacult. Mag. 18(5): 55–61.

767. Winfree, R. A. 1989. Tropical fish. World Aquacult. 20(3): 24–30.

768. McLarney, B. 1985. Pioneers in saltwater aquarium fish. Aquacult. Mag. 11(6): 38ff.

769. McLarney, B. 1986. Pioneers in saltwater aquarium fish part II. Aquacult. Mag. 12(1): 31–33.

770. Chew, K. K. 1992. Cultivation of the geoduck, largest burrowing clam in the U.S. Aquacult. Mag. 18(6): 66–70.

771. Shaw, W. N. 1991. Abalone farm in northern California. Aquacult. Mag. 17(5): 92–97.

772. Chew, K. K. 1986. Review of recent molluscan culture. pp. 173–195, In: M. Bilio, H. Rosenthal, and C. J. Sindermann (eds.). Realism in aquaculture: achievements, constraints, perspectives. European Aquaculture Society, Bredene, Belgium.

773. Klopfenstein, D., and J. Klopfenstein. 1976. U.S. Abalone hatcheries adapt a Japanese technique. Fish Farm. Int. 3(3): 46–48.

774. Rutherford, D. 1976. California farm rears prized red abalone. Fish Farm. Int. 3(4): 9.

775. Menzel, W. 1971. The mariculture potential of clam farming. Am. Fish Farmer & World Aquacult. News, 2(8): 8–14.

776. Menzel, R. W. 1971. Quahog clams and their possible mariculture. Proc. World Maricult. Soc. 2: 23–36.

777. Stickney, R. R. 1990. In Memorium, Robert Winston Menzel, Sr., 1920–1989. J. Shellfish Res. 8(2).

778. Menzel, R. W. 1943. Albino catfish in Virginia. Copeia, 2: 124.

779. Menzel, R. w. 1958. Further notes on the albino catfish. J. Hered. 49: 1 p.

780. Aiken, D. 1993. The perfect clam. World Aquacult. 24(2): 6–15.

781. Skidmore, D., and K. K. Chew. 1985. Mussel aquaculture in Puget Sound. Washington Sea Grant Program, Seattle. Publication WSG 85-4. 57 p.

782. Chew, K. K., and D. Toba. 1991. Western region aquaculture industry situation and outlook report. Western Regional Aquaculture Center, Seattle, 23 p.

783. Wilson, J., and D. Fleming. 1989. Economics of the Maine mussel industry. World Aquacult. 20(4): 49–55.

784. Glude, J. B. 1983. Marketing and economics in relation to U.S. bivalve aquaculture. J. World Maricult. Soc. 14: 576–586.

785. Gunter, G., and W. J. Demoran, 1971. Mississippi oyster culture. Am. Fish Farmer & World Aquacult. News, 2(5): 8–12.

786. Anonymous. 1980. Aquaculture progressing in Maine with help of Coastal Enterprises, Inc. Aquacult. Mag. 6(5): 10–11.

787. Westley, R. E. 1975. Past, present, and future trends in cultural techniques and oyster production in the state of Washington. Proc. World Maricult. Soc. 6: 213–219.

788. Shaw, W. N. 1971. Oyster culture research—off-bottom growing techniques. Am. Fish Farmer & World Aquacult. News, 2(9): 16–21.

789. Matthiessen, G. C. 1972. Commercial culture of oysters in New England. Proc. World Maricult. Soc. 3: 319–323.

790. Aprill, G., and D. Maurer. 1976. The feasibility of oyster raft culture in east coast estuaries. Aquaculture, 7: 147–160.

791. Ogle, J., S. M. Ray, and W. J. Wardle. 1977. A summary of oyster mariculture utilizing an offshore petroleum platform in the Gulf of Mexico. Proc. World Maricult. Soc. 8: 447–455.

792. Bardach, J. E., J. H. Ryther, and W. O. McLarney. 1972. Aquaculture. Wiley-Interscience, New York. 868 p.

793. Dunstan, W. M., and K. R. Tenore. 1972. Intensive outdoor culture of marine phytoplankton with treated sewage effluent. Aquaculture, 2: 181–192.

794. Ryther, J. H., J. C. Goldman, C. E. Gifford, J. E. Huguenin, A. S. Wing, J. P. Clarner, L. D. Williams, and B. E. LaPointe. 1975. Physical models of integrated waste recycling—marine polyculture systems. Aquaculture, 5: 163–177.

795. Huguenin, J. E. 1975. Development of a marine aquaculture research complex. Aquaculture, 5: 135–150.

796. Mann, R., and J. H. Ryther. 1977. Growth of six species of bivalve molluscs in a waste recycling—aquaculture system. Aquaculture, 11: 231–245.

797. Lipschultz, F., and G. E. Krantz. 1980. Production optimization and economic analysis of an oyster (*Crassostrea virginica*) hatchery on the Chesapeake Bay, Maryland, U.S.A. Proc. World Maricult. Soc. 11: 580–591.

798. Lipovsky, V. P. 1980. An industry review of the operations and economics of molluscan shellfish hatcheries in Washington State. Proc. World Maricult. Soc. 11: 577–579.

799. Supan, J., and C. Wilson. 1993. Oyster seed alternatives for Louisiana. World Aquacult. 24(4): 79–82.

800. Chew, K. K. 1984. Recent advances in the cultivation of molluscs in the Pacific United States and Canada. Aquaculture, 39: 69–81.

801. Krantz, G. E., and S. V. Otto. 1981. More on oyster diseases. pp. 68–84, In: D. Webster and J. Greer (eds.). Oyster culture in Maryland 1980. University of Maryland. College Park, Md. Sea Grant Publication UM-SG-MAP-81-01.

802. Perret, W. S., R. J. Dugas, and M. F. Chatry. 1991. Louisiana oyster enhancing the resource through shell planting. World Aquacult. 22(4): 42–45.

803. Stanley, J. G., S. K. Allen, Jr., and H. Hidu. 1981. Polyploidy induced in the American oyster, *Crassostrea virginica*, with cytochalasin B. Aquaculture, 23: 1–10.

804. Allen, S. K., Jr., S. L. Downing, and K. K. Chew. 1989. Hatchery manual for producing triploid oysters. Washington Sea Grant Program, University of Washington Seattle. 27 p.

805. Waddy, S. L. 1988. Farming the homarid lobsters: state of the art. World Aquacult. 19(4): 63–71.

806. Lambert, S. S. 1980. Crawfish farming provides rewards if diligence, ingenuity are applied. Aquacult. Mag. 6(5): 14–18.

807. Lawson, T. B., and F. W. Wheaton. 1983. Crawfish culture systems and their management. J. World Maricult. Soc. 14: 325–335.

808. Huner, J. V. 1985. An update on crawfish aquaculture. Aquacult. Mag. 11(4): 33ff.

809. Massey, P. 1989. Louisiana. World Aquacult. 20(1): 24–29.

810. Johnson, S. K. 1990. Aquaculture of Texas. World Aquacult. 21(4): 52–56.

811. Liao, D. S. 1984. Market analysis for crawfish aquaculture in South Carolina. J. World Aquacult. Soc. 15: 106–107.

812. Pomeroy, R. S., D. B. Luke, and J. M. Whetstone. 1986. The economics of crawfish production in South Carolina. Aquacult. Mag. 12(2): 36–39.

813. Avault, J. W., Jr. 1992. A review of world crustacean aquaculture part two. Aquacult. Mag. 18(4): 83–92.

814. Heyman, J. J. 1971. Production of soft crayfish for bait. Am. Fish Farmer & World Aquacult. News, 3(1): 5–7.

815. Culley, D. D., Jr., M. Z. Said, and P. T. Culley. 1985. Procedures affecting the production and processing of soft-shelled crawfish. J. World Maricult. Soc. 16: 183–192.

816. Huner, J. V. 1990. A major Louisiana soft-shelled crawfish aquaculture endeavor. Aquacult. Mag. 16(2): 51–55.

817. Swank, R. 1991. A new aquacultrue operation. Aquacult. Mag. 17(3): 57–61.

818. Posadas, B. C., and J. Homziak. 1992. Economics of soft shell crawfish production in Mississippi. Aquacult. Mag. 18(4): 59–64.

819. Huner, J. V. 1990. New horizons for the crawfish industry. Aquacult. Mag. 16(5): 65–70.

820. Avault, J. W., Jr. 1990. Some recent advances in crawfish farming research. Aquacult. Mag. 16(5): 77–81.

821. Ham, B. G. 1971. Crawfish culture techniques. Am. Fish Farmer & World Aquacult. News, 2(5): 5ff.

822. Shireman, J. V. 1973. Experimental introduction of the Australian crayfish (*Cherax tenuimanaus*) into Louisiana. Prog. Fish-Cult. 35: 107–109.

823. Rouse, D. B., C. M. Austin, and P. B. Medley. 1991. Progress toward profits? Information on the Australian crayfish. Aquacult. Mag. 17(3): 46–56.

824. Rubino, M., N. Alon, C. Wilson, D. Rouse, and J. Armstrong. 1990. Marron aquaculture in the United States and the Caribbean. Aquacult. Mag. 16(3): 27ff.

825. Hanson, J. A., and H. L. Goodwin. 1977. Shrimp and prawn farming in the western hemisphere. Dowden, Hutchinson & Ross, Stroudsburg, Pa. 439 p.

826. Costello, T. J. 1971. Freshwater prawn culture techniques developed. Am. Fish Farmer & World Aquacult. News, 2(2): 8ff.

827. Anonymous. 1971. First Hawaiian fish farms begins commercial venture. Am. Fish Farmer & World Aquacult. News, 2(2): 11.

828. Hopkins, J. S. 1991. Status and history of marine and freshwater shrimp farming in South Carolina and Florida. pp. 17–35, In: P. A. Sandifer (ed.). Shrimp culture in North America and the Caribbean. World Aquaculture Society, Baton Rouge, La.

829. Smith, T. I. J., P. A. Sandifer, and W. C. Trimble. 1976. Pond culture of the Malaysian prawn, *Macrobrachium rosenbergii* (de Man), in South Carolina, 1974–1975. Proc. World Maricult. Soc. 7: 625–645.

830. Shang, Y. C., and T. Fujimura. 1977. The production economics of freshwater prawn (*Macrobrachium rosenbergii*) framing in Hawaii. Aquaculture, 11: 99–110.

831. Liao, D. S., and T. I. J. Smith. 1981. Test marketing of freshwater shrimp, *Macrobrachium rosenbergii*, in South Carolina. Aquaculture, 23: 373–379.

832. Liao, D. S., and T. I. J. Smith. 1983. Economic analysis of small-scale prawn farming in South Carolina. J. World Maricult. Soc. 14: 441–450.

833. Greenland, J. 1982. South Texas *Macrobrachium* ranch finds a home on the range. Aquacult. Mag. 9(1): 14–17.

834. Anonymous. 1984. Freshwater shrimp produced in Texas. Aquacult. Mag. 10(5): 30–33.

835. Neal, R. A. 1973. Shrimp culture research brings commercial farming closer. Am. Fish Farmer & World Aquacult. News, 4(11): 13–14.

836. Riley, J. R., K. T. Dorsey, and V. C. Supplee. 1974. The shrimp aquaculture program at the Environmental Research Laboratory, University of Arizona. Proc. World Maricult. Soc. 5: 421–430.

837. Cordover, R. 1972. Penaeid prawn production. Am. Fish Farmer World Aquacult. News, 3(11): 4–7.

838. New, M. B. 1976. A review of dietary studies with shrimp and prawns. Aquaculture, 9: 101–144.

839. New, M. B. 1980. A bibliography of shrimp and prawn nutrition. Aquaculture, 21: 101–128.

840. Neal, R. A. 1971. Shrimp culture research at the Galveston Biological Lab. Am. Fish Farmer & World Aquacult. News, 2(6): 10ff.

841. Anonymous. 1972. Shrimp harvest at Marifarms. Am. Fish. Farmer & World Aquacult. News, 3(2): 5–7.

842. Anonymous. 1973. Shrimp culture making progress at Texas A&M. Am. Fish Farmer & World Aquacult. News, 4(2): 18.

843. Parker, J. C., and H. W. Holcomb, Jr. 1973. Growth and production of brown and white shrimp (*Penaeus aztecus* and *P. setiferus*) from experimental ponds in Brazoria and Orange counties, Texas. Proc. World Maricult. Soc. 4: 215–234.

844. Caillouet, C. W. 1972. Ovarian maturation induced by eyestalk ablation in pink shrimp, *Penaeus durorarum* Burkenroad. Proc. World Maricult. Soc. 3: 205–225.

845. Wurts, W. A., and R. R. Stickney. 1984. An hypothesis on the light requirements for spawning penaeid shrimp, with emphasis on *Penaeus setiferus*. Aquaculture, 41: 93–98.

846. McVey, J. P. 1980. Current developments in the penaeid shrimp culture industry. Aquacult. Mag. 6(5): 20–25.

847. Chamberlain, G. W., and A. L. Lawrence. 1981. Effect of light intensitty and male and female eyestalk ablation on reproduction of *Penaeus stylirostris* and *P. vannamei*. Proc. World Maricult. Soc. 12: 357–372.

848. Griffin, W. L., J. S. Hanson, R. W. Brick, and M. A. Johns. 1981. Bioeconomic modeling with stochastic elements in shrimp culture. J. World Maricult. Soc. 12: 94–103.

849. Johns, M., W. Griffin, A. Lawrence, and D. Hutchins. 1981. Budget analysis of shrimp maturation facility. J. World Maricult. Soc. 12: 104–109.

850. Johns, M., W. Griffin, A. Lawrence, and J. Fox. 1981. Budget analysis of penaeid shrimp hatchery facilities. J. World Maricult. Soc. 12: 305–321.

851. Huang, H.-J., W. L. Griffin, and D. V. Aldrich. 1984. A preliminary economic feasibility analysis of a proposed commercial penaeid shrimp culture operation. J. World Maricult. Soc. 15: 95–105.

852. Hanson, J. S., W. L. Griffin, J. W. Richardson, and C. J. Nixon. 1985. Economic feasibility of shrimp farming in Texas: an investment analysis for semi-intensive pond grow-out. J. World Maricult. Soc. 16: 129–150.

853. Chamberlain, G. W. 1991. Status of shrimp farming in Texas. pp. 36–57, In: P. A. Sandifer (ed.). Shrimp culture in North American and the Caribbean. World Aquaculture Society, Baton Rouge, La.

854. Anonymous. 1984. Shrimp steal the show at Vancouver WMS meeting. Aquacult. Mag. 10(4): 4–7.

855. Pruder, G. D. 1991. Shrimp culture in North America and the Caribbean: Hawaii 1988. pp. 58–69, In: P. A. Sandifer (ed.). Shrimp culture in North America and the Caribbean. World Aquaculture Society, Baton Rouge, La.

856. Nieto, R. R. 1991. Aquaculure in Puerto Rico. World Aquacult. 22(1): 74ff.

857. Priddy, J. M., and D. D. Culley, Jr. 1972. Frog culture industry. Am. Fish Farmer & World Aquacult. News, 3(9): 4–7.

858. Priddy, J. M., and D. D. Culley, J. R. 1972. The frog culture industry, past and present. Proc. Southeast. Assoc. Game Fish Comm. 25: 597–602.

859. Culley, D. D., Jr., and C. Gravios, Jr. 1972. Recent developments in frog culture. Proc. Southeast Assoc. Game Fish Comm. 25: 583–597.

860. Culley, D. D., Jr. 1981. Have we turned the corner on bullfrog culture? Aquacult. Mag. 7(3): 20–24.

861. Culley, D. D. 1986. Bullfrog culture still a high risk venture. Aquacult. Mag. 12(5): 28ff.

862. Lester, D. 1988. Raising bullfrogs on nonliving food. Aquacult. Mag. 14(2): 20ff.

863. Stickney, R. R., D. B. White, and D. Perlmutter. 1973. Growth of sea turtles on natural and artificial diets. bull. Ga. Acad. Sci. 31: 37–44.

864. Hendrickson, J. R. 1974. Marine turtle culture—an overview. Proc. World Maricult. Soc. 5: 167–181.

865. Walker, T. A. 1992. The Cayman turtle farm. Aquacult. Mag. 18(2): 47–55.

866. Wood, J. R., and F. E. Wood. 1979. Artificial incubation of green sea turtle eggs (*Chelonia mydas*). Proc. World Maricult. Soc. 10: 215–221.

867. Binger, A. 1981. Cayman turtle farms suffers from ban on giant green sea turtle imports. Aquacult. Mag. 7(4): 8–10.

868. Avault, J. W., Jr. 1985. The alligator story. Aquacult. Mag. 11(4): 41–44.

869. Belleville, B. 1982. Alligator farming expands slowly but steadily as culture techniques develop. Aquacult. Mag. 8(6): 4–5.

870. Zajicek. P. 1993. Limited markets trap the alligator industry. Aquacult. Mag. 19(6): 50–59.

871. Anonymous. 1991. U.S. alligator production grows. Aquacult. Mag. 17(5): 6ff.

872. Anonymous. 1982. Nori culture possible for Washington state. Aquacult. Mag. 8(2): 15.

873. Anonymous. 1984. Puget Sound tests seaweed farming. Aquacult. Mag. 10(6): 16.

874. Harger, W. W., and M. Neushul. Test-farming of the giant kelp, Macrocystis, as a marine biomass producer. J. World Maricult. Soc. 14: 392–403.

875. Brinkhuis, B. H., V. A. Breda, S. Tobin, and B. A. Macler. 1983. New York marine biomass program—culture. J. World Maricult. Soc. 14: 360–379.

876. Webber, H. H., and P. F. Riordan. 1976. Criteria for candidate species for aquaculture. Aquaculture, 7: 107–123.

877. State of Hawaii. 1978. Aquaculture development for Hawaii. State Department of Planning and Economical Development, Honolulu. 222 p.

878. Corbin, J. S., and R. T. Gibson. 1979. Planning aquaculture development: the first time is always the hardest. Proc. World Maricult. Soc. 10: 21–27.

879. Stickney, R. R., and J. T. Davis (comps.). 1981. Aquaculture in Texas: a status report and development plan. Texas A & M University, College Station. Marine Information Service Publication TAMU-SG-81-119. 103 p.

880. Anonymous. 1983. Missouri approves aquaculture plan while Florida begins initial efforts. Aquacult. Mag. 9(2): 10.

881. Lindbergh, J., and K. Pryor. 1984. Six ways to lose money in aquaculture. Aquacult. Mag. 10(4): 24–25.

882. Parametrix, Inc. 1990. Final programmatic environmenal impact statement: fish culture in floating net pens. Washington Department of Fisheries, Olympia. 161 p.

883. Pullin, R. S. V., H. Rosenthal, and J. L. Maclean (eds.). 1993. Environment and aquaculture in developing countries. International Center for Living Aquatic Resources Management, Manila. 359 p.

884. Primmer, K. W., and J. P., and J. P. Clugston. 1975. Effects of effluents from trout hatcheries on the benthos and fish in receiving streams. Proc. Southeast. Assoc. Game Fish Comm. 28: 332–342.

885. Stechey, D. 1991. Build your own settling pond. North. Aquacult. 7(5): 22ff.

886. Avault, J. W., Jr. 1993. Aquaculture: environmental and conservation considerations. Aquacult. Mag. 19(1): 64–67.

887. Harrell, L. W., R. A. Elston, T. M., Scott and M. T. Wilkinson. 1986. A significant new systemic disease of net-pen reared chinook salmon (Oncorhynchus tshawytscha) brood stock. Aquaculture, 55: 249–262.

888. Kent, M. L. 1990. Netpen liver disease (NLD) of salmonid fisheries reared in sea water: species susceptibility, recovery, and probable cause. Dis. Aquat. Org. 8: 21–28.

889. Stickney, R. R. 1991. Growout of Pacific salmon in net-pens. pp. 71–83, In: R. R. Stickney (ed.), Culture of salmonid fishes. CRC Press, Boca Raton, Fla.

890. Hanebrink, E., and W. Byrd. 1989. Predatory birds in relation to aquaculture farming. Aquacult. Mag. 15(3): 47–51.

891. Hoy, M. D. 1994. Depredations by herons and egrets at bait fish farms in Arkansas. Aquacult. Mag. 20(1): 52–56.

892. Anonymous. 1991. Cormorant feeding rates on commercially grown catfish. Aquacult. Mag. 17(2): 89–90.

893. Meyer, F. P., R. A. Schnick, and K. B. Cumming. 1976. Registration status of fishery chemicals, February 1976. Prog. Fish-Cult. 38: 3–7.

894. Anonymous. 1979. Announcement of compounds registered for fishery uses. Prog. Fish-Cult. 41: 36–37.

895. Stefan, G. E. 1992. FDA regulation of animal drugs used in aquaculture. Aquacult. Mag. 18(5): 62–67.

896. Meyer, F. P., and R. A. Schnick. 1989. A review of chemicals used for the control of fish diseases. Rev. Aquat. Sci. 1: 693–710.

897. Anonymous. 1983. Aquaculture/WDC/83 attracts international audience. Aquacult. Mag. 9(3): 20–21.

898. Nosho, T., and K. Freeman (eds.). 1994. Proc. Conference on Marine Fish Culture and Enhancement. University of Washington Sea Grant Program, Seattle. 64 p.

899. McEachron, L. W., C. E. McCarty, and R. R. Vega. 1994. Texas red drum enhancement works. p. 29, In: T. Nosho and K. Freeman (eds.), Proc. Conference on Marine Fish Culture and Enhancement. University of Washington Sea Grant Program. Seattle.

900. Bromage, N. R. Shields, M. Gillespie, and R. Johnstone. 1994. U. K. mariculture: experiences and prospects. pp. 30–32, In: T. Nosho and K. Freeman (eds.), Proc. Conference on Marine Fish Culture and Enhancement. University of Washington Sea Grant Program, Seattle.

901. Stickney, R. R. 1994. A review of the research efforts on Pacific halibut, *Hippoglossus stenolepis,* with emphasis on research and development needs. pp. 39–40, In: T. Nosho and K. Freeman (eds.), Proc. Conference on Marine Fish Culture and Enhancement. University Washington Sea Grant Program, Seattle.

902. Holm, J. C. 1994. Production of juveniles with emphasis on Atlantic halibut. pp. 24–25, In: T. Nosho and K. Freeman (eds.). Proc. Conference on Marine Fish Culture and Enhancement. University of Washington Sea Grant Program, Seattle.

903. Avault, J. W., J. R. 1994. What is happening to our ocean fisheries/what is the implication for aquaculture? Aquacult. Mag. 20(4): 80–84.

904. Webber, H. H. 1973. Risks to the aquaculture enterprise. Aquaculture 2: 157–172.

905. Avault, J. W., Jr. 1986. Aquaculture potential in the United States. Aquacult. Mag. 12(5): 43–45.

906. Avault, J. W., Jr. 1990. Aquaculture in the United States the signs for the future. Aquacult. Mag. 16(2): 26ff.

907. Harvey, D. J. 1994. Outlook for U.S. aquaculture. Aquacult. Mag. 20(1): 40ff.

908. Anonymous. 1994. U.S. per capita seafood consumption growing again. Water Farming J. 8(7): 11.

909. Anonymous. 1993. Status of world aquaculture 1993. Aquacult. Mag. Buyer's Guide '94 6ff.

910. Stickney, R. R. 1995. Offshore aquaculture: Current technology and the need for policy development in the United States. pp. 339–348, In: Proceedings, Sustainable Agriculture 95. PACON 95, Honolulu, Hawaii, PACON International, Honolulu.

911. Sandifer, P. A. 1994. U.S. coastal aquaculture: flirting with opportunity. Water Farming J. 8(4): 3ff.

912. National Research Council. 1992. Marine aquaculture: opportunities for growth. Marine Board, Commission on Engineering and Technical Systems, National Research Council, Washington, D.C. 290 p.

913. McCoy, H. D., II. 1993. Open ocean fish farming: part one. Aquacult. Mag. 19(5): 66–74.

914. McCoy, H. D., II. 1993. Open ocean fish farming: part two. Aquacult. Mag. 19(6): 60–67.

Index

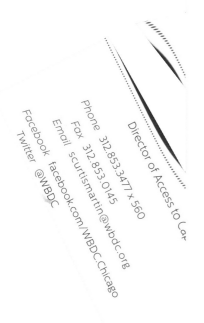